Microbial Fuel Cells

Emerging trends in electrochemical applications

Online at: https://doi.org/10.1088/978-0-7503-4791-4

Microbial Fuel Cells

Emerging trends in electrochemical applications

Edited by

Mohamad Nasir Mohamad Ibrahim

Materials Technology Research Group (MaTRec), School of Chemical Sciences, Universiti Sains Malaysia, Penang, Malaysia

Asim Ali Yaqoob

School of Chemical Sciences, Universiti Sains Malaysia, Penang, Malaysia

Akil Ahmad

Chemistry Department, College of Sciences and Humanities, Prince Sattam bin Abdulaziz University, Al-Kharj, Saudi Arabia

IOP Publishing, Bristol, UK

ISBN 978-0-7503-4791-4 (ebook)
ISBN 978-0-7503-4789-1 (print)
ISBN 978-0-7503-4792-1 (myPrint)
ISBN 978-0-7503-4790-7 (mobi)

DOI 10.1088/978-0-7503-4791-4

Version: 20221201

IOP ebooks

British Library Cataloguing-in-Publication Data: A catalogue record for this book is available from the British Library.

Published by IOP Publishing, wholly owned by The Institute of Physics, London

IOP Publishing, No.2 The Distillery, Glassfields, Avon Street, Bristol, BS2 0GR, UK

US Office: IOP Publishing, Inc., 190 North Independence Mall West, Suite 601, Philadelphia, PA 19106, USA

The editors are honored to dedicate this book to Mr Mohamad Yaqoob
(Father of Dr Asim Ali Yaqoob)

and

Dr Akil's grandparents, the late Abdul Aziz Khan, the late Haji Yaqoob Khan, the late Haji Daud Khan, and the late Haji Ayub Khan.

Allah SWT forgive their small and big mistakes and give them a higher place in Jannatul Firdous. Ameen.

Contents

Preface

Microbial fuel cells (MFCs) are central to an emerging and significant approach that converts chemical energy into electrical energy using bacteria that serve as catalysts. Significant achievements in power density have been reported. However, MFCs have not yet been implemented at a commercial scale due to their low energy production. There are several reasons for this low energy production efficiency. Owing to their sustainability, MFCs have been recognized as a viable technology for producing energy and removing toxic pollutants from wastewater resources. In addition to a proton exchange membrane (PEM), MFCs consist of two electrodes—an anode and a cathode—the former of which is responsible for providing sufficient space for bacterial growth and respiration in order to oxidize organic matter in wastewater and thereby generate protons and electrons. The approach has attracted attention for its capacity to generate renewable energy and treat wastewater simultaneously.

Research on energy production through MFC technology has received great attention. It is emerging as a promising approach which converts chemical energy into electrical energy in the presence of a biocatalyst and simultaneously removes toxic pollutants from water. The electroactive bacteria species degrade the organic waste and the released electrons are passed from the organic waste to an electrode for the generation of energy. Furthermore, the transportation of electrons from bacteria to the electrode surface is an essential step for energy generation. Hence, the poor transportation of electrons may greatly hinder the performance of MFCs. In the last decade, several efforts have been made to improve electron transportation between bacteria and electrodes which ultimately affects the efficiency of the MFCs. To achieve this objective, a large variety of materials was used to fabricate different improved extracellular electron transferable electrodes for high energy generation and achieve excellent durability performance of MFCs.

This book covers the electrochemical performance of MFCs, the basic mechanisms, affecting factors, required materials for high electrochemical performance, and a comparison of MFCs with other bioelectrochemical approaches in terms of energy generation. There is also significant coverage of the commercial status, trends, and performance of MFCs.

This book covers all aspects related to energy generation via MFCs, including the roles of the organic substrate, electrode, and biocathode, to provide the reader with clear mechanisms for advancement in MFCs.

We are grateful to all the qualified researchers, scholars, and leading experts who have contributed their valuable work. The chapters of this book provide cutting-edge and up-to-date research findings on MFCs. We have collected together information on MFCs and related research provided by eminent authors from the USA, Mexico, Spain, UK, Australia, India, Malaysia, Saudi Arabia, Nigeria, Sudan, Taiwan, South Africa, Japan, Pakistan, and Bangladesh, compiling this project in a fruitful way.

Mohamad Nasir Mohamad Ibrahim
Asim Ali Yaqoob
Akil Ahmad

Acknowledgement

The editors wish to recognise the support of Universiti Sains Malaysia and Prince Sattam bin Abdulaziz University, Saudi Arabia.

Editor biographies

Dr Mohamad Nasir Mohamad Ibrahim

Mohamad Nasir Mohamad Ibrahim obtained his BSc (1994), MSc (1997), and PhD (1999) from Missouri S&T (formerly known as University of Missouri-Rolla, USA). He is currently serving as an Associate Professor in the School of Chemical Sciences, Universiti Sains Malaysia (USM). He has published more than 120 journal articles, 17 book chapters, and 5 academic books throughout his 20 year career at USM, including *Graphene: A Versatile Advanced Material*. Currently his Scopus h-index is 35 with 4026 citations and he has been granted ten patents for his R&D products/processes, five of them at an international level. More than 30 research grants (one international grant and five industry grants) have been secured and utilized to support his team's research activities. He has received fourteen international awards for his research output and currently serves as a Guest Editor for *Frontiers in Chemistry*. He has supervised more than 20 graduate students. His main research areas are lignin and lignocellulosic materials, nanoparticles, graphene, biomaterials, and petroleum engineering. Recently, he has been busy working in the field of microbial fuel cells, in particular developing a novel electrode and has published several papers on this topic in well reputed journals such as *Chemical Engineering Journal*, *Journal of Cleaner Production*, etc. At the moment he is a pioneer in the field of MFC research in the School of Chemical Sciences, USM. He enjoys sharing his industrial experience, spending two and a half years worked as a R&D Manager at NSE Resources Corporation, in his classes.

Dr Asim Ali Yaqoob

Asim Ali Yaqoob obtained his BSc (2010), MSc (2014), and MPhil (2018) from Mirpur University of Science and Technology Mirpur, Mirpur, 10250, AJK Pakistan. He completed his PhD (2021) at the School of Chemical Sciences, Universiti Sains Malaysia (USM) under the supervision of Dr Mohamad Nasir Mohamad Ibrahim. His area of interest is energy generation coupled with wastewater treatment and his current project is on microbial fuel cells. He has published 22 articles in highly reputed journals and several book chapters. He is also the co-author of an academic book (*Graphene: A Versatile Advanced Material*) published by USM Press, Malaysia. He completed his MPhil in Material Chemistry (2018) with the topic 'Designing and characterizing of Ag@Polycarbazole nanocables' at the Mirpur University of Science and Technology (MUST), AJK Pakistan. His Google scholar H-index and number of citations are 14 and 695, respectively, and he has filed one patent for an industrial process. He is also an invited reviewer for many reputed international journals, such as *Nanotechnology Reviews*, *International Journal of Energy Research*, *Journal of Chemistry*, *PLOS ONE*, and many others. As a part of his academic work, he has different collaborations with well-known research laboratories in the world. He enjoys sharing his industrial experience, having spent a year working as a research assistant at Ekahala Resources, Malaysia.

Dr Akil Ahmad

Akil Ahmad currently works at Prince Sattam bin Abdulaziz University, Saudi Arabia, as an Assistant Professor in Chemistry and has seven years of experience as a Research Fellow, Teaching Fellow, Postdoc, and Visiting Researcher at the Universiti Teknologi Malaysia, Universiti Sains Malaysia, University of KwaZulu-Natal, South Africa, and Universiti Kebangsaan Malaysia. He completed his PhD in Analytical Chemistry (2011) with the topic 'Modification of resin for their use in the separation, preconcentration and determination of metal ions' from Aligarh Muslim University (AMU), India. His research interests lie in the areas of environmental pollutants and their safe removal, the synthesis of nanoparticles and nano-sorbents (GO, CNT), photo-degradation and antimicrobial effects, water and wastewater treatment, and adsorption and ion-exchange. He has published more than 100 research articles and chapters in journals of publishers of international repute, such as *Scientific Reports*, *Talanta*, *Chemical Engineering Journal*, *Journal of Industrial and Engineering Chemistry*, *Journal of Molecular Liquids* etc. He has also edited five books published by Springer and Elsevier. His H-index and citation number in Scopus are 22 and 1827, and in Google Scholar, his H-index and citation number are 25 and 2404. He is a Guest Editor for many reputed journals, namely *Adsorption Science and Technology* (Hindawi), *Polymers* (MDPI), *Frontiers in Environmental Chemistry*, and *Journal of Chemistry* (Hindawi).

List of contributors

Akil Ahmad
Chemistry Department, College of Sciences and Humanities, Prince Sattam bin Abdulaziz University, Al-Kharj, Saudi Arabia

Tanzina Akter
Microbial Biotechnology Division, National Institute of Biotechnology, Ganakbari, Ashulia, Savar, Dhaka, Bangladesh

Mohammad Al Mamun
Department of Chemistry, Jagannath University, Dhaka 1100, Bangladesh
and
Nanotechnology and Catalysis Research Centre, Institute for Advanced Studies, University of Malaya, Kuala Lumpur, Malaysia

Mohammad Ali
Department of Mechanical Engineering, Bangladesh University of Engineering and Technology (BUET), Dhaka, Bangladesh

Mohammed B Alshammari
Chemistry Department, College of Sciences and Humanities, Prince Sattam bin Abdulaziz University, Al-Kharj, Saudi Arabia

Wilgince Apollon
Facultad de Agronomía, Departamento de Ingeniería Agrícola y de los Alimentos, Universidad, Nuevo León, Mexico

María del Carmen Monterrubio Badillo
Centro Mexicano para la Producción más Limpia, Instituto Politécnico Nacional, Mexico City, Mexico

Claudio Guerrero Barajas
Laboratorio de Biotecnología Ambiental, Departamento de Bioprocesos, Unidad Profesional Interdisciplinaria de Biotecnología, Instituto Politécnico Nacional, Mexico City, Mexico

Showkat Ahmad Bhawani
Faculty of Resource Science and Technology, Universiti Malaysia Sarawak (UNIMAS), Kota Samarahan, Malaysia

Najwa Najihah Mohamad Daud
Materials Technology Research Group (MaTRec), School of Chemical Sciences, Universiti Sains Malaysia, Penang, Malaysia

Elsa M Arce Estrada
Departamento de Metalurgia y Materiales, Escuela Superior de Ingeniería Química e Industrias Extractivas, Instituto Politécnico Nacional, Mexico City, Mexico

Oluwakemi S Fasinu
Department of Mechanical Engineering, Olabisi Onabanjo University, Ago-Iwoye, Nigeria

Solomon O Giwa
Department of Mechanical Engineering, Olabisi Onabanjo University, Ago-Iwoye, Nigeria

Juan Florencio Gómez-Leyva
TecNM-Instituto Tecnológico de Tlajomulco (ITTJ), Laboratorio de Biología Molecular, Jalisco, Mexico

Abu Hashem
Nanotechnology and Catalysis Research Centre, Institute for Advanced Studies, University of Malaya, Kuala Lumpur, Malaysia

Al-Mustasin Abir Hossain
Chemistry Department, College of Sciences and Humanities, Prince Sattam bin Abdulaziz University, Al-Kharj, Saudi Arabia

Mohamad Nasir Mohamad Ibrahim
Materials Technology Research Group (MaTRec), School of Chemical Sciences, Universiti Sains Malaysia, Penang, Malaysia

Mustapha Omenesa Idris
Department of Pure and Industrial Chemistry, Kogi State University, Anyigba, Nigeria

Nor Asniza Isjak
School of Educational Studies, Universiti Sains Malaysia, 11800 Penang, Malaysia

Nur Syamimi Ab Jalil
School of Industrial Technology, Bioprocess Technology Division, Universiti Sains Malaysia, Penang, Malaysia

Jayapriya Jayaprakash
Department of Applied Science and Technology, AC Tech, Anna University, Chennai, India

Sathish-Kumar Kamaraj
TecNM-Instituto Tecnológico El Llano Aguascalientes (ITEL), Laboratorio de Medio Ambiente Sostenible, El Llano Ags., Mexico

Ansa Kanwal
Department of Chemistry and Molecular Biology, University of Gothenburg, Gothenburg, Sweden

Tayyaba Kanwal
Department of Chemistry, Mirpur University of Science and Technology (MUST), Mirpur, Azad Jammu and Kashmir, Pakistan

Rizwana Kausar
School of Biological Sciences, Universiti Sains Malaysia, Gelugor, Penang, Malaysia

Md Abu Sayem Khan
Department of Microbiology, University of Dhaka, Bangladesh

Abayomi T Layeni
Department of Mechanical Engineering, Olabisi Onabanjo University, Ago-Iwoye, Nigeria

Muaz Mohd Zaini Makhtar
School of Industrial Technology, Universiti Sains Malaysia, Penang, Malaysia
and
Fellow of Center for Global Sustainability Studies, Universiti Sains Malaysia, Penang, Malaysia

Alejandro Isabel Luna-Maldonado
Universidad Autónoma de Nuevo León, Facultad de Agronomía, Departamento de Ingeniería Agrícola y de los Alimentos, Nuevo León, Mexico

Nahian Masud
Department of Mechanical Engineering, Bangladesh University of Engineering and Technology (BUET), Dhaka, Bangladesh

Modaser Momin
Department of Mechanical Engineering, University of Kordofan, Sudan

Mary L Monisha
Department of Applied Science and Technology, AC Tech, Anna University, Chennai, India

Azizul Moqsud
Department of Civil and Environmental Engineering, Yamaguchi University Japan

Collins N Nwaokocha
Department of Mechanical Engineering, Olabisi Onabanjo University, Ago-Iwoye, Nigeria

Sunday Oyedepo
Department of Mechanical Engineering, Covenant University, Ota, Nigeria

Tabassum Parveen
Department of Pure and Industrial Chemistry, Kogi State University, PMB 1008 Anyigba, Kogi State, Nigeria

Faiz Ur Rahman
Unit of Microbiology, Department of Basic Health Sciences, IISPV, Universitat Rovira I Virgili, 43201, Reus, Spain

Merlyn Sujatha Rajakumar
Department of Applied Science and Technology, AC Tech, Anna University, Chennai, India

Idzni Nashuha Mohd Rashidy
School of Industrial Technology, Bioprocess Technology Division, Universiti Sains Malaysia, Penang, Malaysia

Humerto Rodriguez-Fuentes
Universidad Autónoma de Nuevo León, Facultad de Agronomía, Departamento de Ingeniería Agrícola y de los Alimentos, Nuevo León, Mexico

Mohammad Tajuddin Sabri
School of Industrial Technology, Bioprocess Technology Division, Universiti Sains Malaysia, Penang, Malaysia

Olusegun D Samuel
Department of Mechanical Engineering, Federal University of Petroleum Resources, Effurun, Nigeria

Modhusudon Shaha
Microbial Biotechnology Division, National Institute of Biotechnology, Ganakbari, Ashulia, Savar, Dhaka, Bangladesh
and
Faculty of Medicine, Nursing and Health Sciences, Monash University, Clayton-3800, Victoria, Australia

Mohsen Sharifpur
Department of Mechanical and Aeronautical Engineering, University of Pretoria, Pretoria, South Africa
and
Department of Medical Research, China Medical University Hospital, China Medical University, Taichung, Taiwan

Affia Siddique
School of Chemistry and Biosciences, Faculty of Life Sciences, University of Bradford, Bradford, West Yorkshire, UK

Khalid Umar
Materials Technology Research Group (MaTRec), School of Chemical Sciences, Universiti Sains Malaysia, Penang, Malaysia

Juan Antonio Vidales-Contreras
Universidad Autónoma de Nuevo León, Facultad de Agronomía, Departamento de Ingeniería Agrícola y de los Alimentos, Nuevo León, Mexico

Asim Ali Yaqoob
School of Chemical Sciences, Universiti Sains Malaysia, Penang, Malaysia

Sundas Bahar Yaqoob
Department of Zoology, Mirpur University of Science and Technology (MUST), 10250 Mirpur, Azad Jammu and Kashmir, Pakistan

IOP Publishing

Microbial Fuel Cells
Emerging trends in electrochemical applications
Mohamad Nasir Mohamad Ibrahim, Asim Ali Yaqoob and Akil Ahmad

Chapter 1

Introduction to microbial fuel cell technology

M Azizul Moqsud

Due to many variables, such as ecological changes, increased human populations, increasing demand for energy at both the commercial and home level, and high material costs, energy and water have become global concerns. To address these issues, microbial fuel cells (MFCs) are being promoted as an emerging innovative technology that can produce power while also improving the removal efficiency of various contaminants from wastewater, thus MFCs can create energy from natural waste sources and are also effective in wastewater treatment. This chapter addresses the fundamentals of MFC technology, concentrating on the energy production mechanism as well as wastewater treatment. This chapter also discusses some types of MFC approaches as well as potential future views, with a focus on the idea of electromicrobiology.

1.1 What is a microbial fuel cell?

A microbial fuel cell (MFC) is a type of bio-electrochemical fuel cell system that generates bioelectricity through the metabolic activities of micro-organisms [1–5]. The electricity generated by the decomposing organic substance travels from the anode to cathode through an external circuit [6–9]. The MFC is a promising technology for renewable energy production in specific applications such as the remediation of pollution or cleaning up wastewater. There are many other applications of MFCs in the fields of energy and the environment [10–15]. In microbial fuel cells, organic substances are degraded by micro-organisms and hence produce electrons. The external circuit, with an anode and cathode, is placed to collect this electron and continue the current. In previous research, it was observed that this MFC method can be used to clean wastewater, bioremediate sulfide contaminated sediment, and consequently generate bioelectricity. The benefit of this method is that it can generate bioelectricity while cleaning the environment [16–19]. In 2013 Moqsud *et al* showed that MFCs can also generate electricity from organic waste in a compost-type MFC.

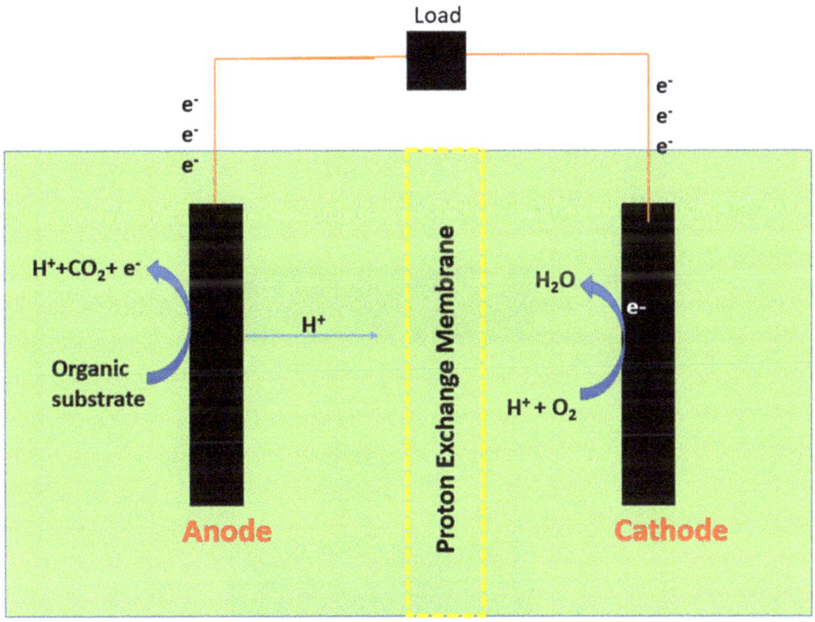

Figure 1.1. Basic schematic diagram of a microbial fuel cell.

Since then, other researchers have been trying to use this novel technology to generate bioelectricity and recycle organic waste [4, 5, 20–23].

Figure 1.1 shows a schematic diagram of the basic microbial fuel cell which was used in the laboratory. The anode and cathode are connected with the external circuit. The resistance was also used to complete the external circuit for the power output.

1.2 Variation of microbial fuel cells based on their construction and use

Currently, many researchers are using the basic MFC technology to serve their purposes or the objectives of their research. There are different types of MFCs which are used commonly by researchers all over the world. Some MFCs are used for wastewater treatment and some MFCs are using for bioremediation of contaminated soil or sediments. The main aspects of an MFC are the objective of its use and the medium of its use. For example, if it is used for the treatment of wastewater then it is used under a water environment. However, if it is used for soil/sediment then soil is the environmental condition. The name of an MFC refers to its use, for example if it is used in sediment then it is called a sediment MFC (SMFC). One of the most common types of MFCs is the SMFC. After the development of this type of MFC, the range of MFCs has increased significantly.

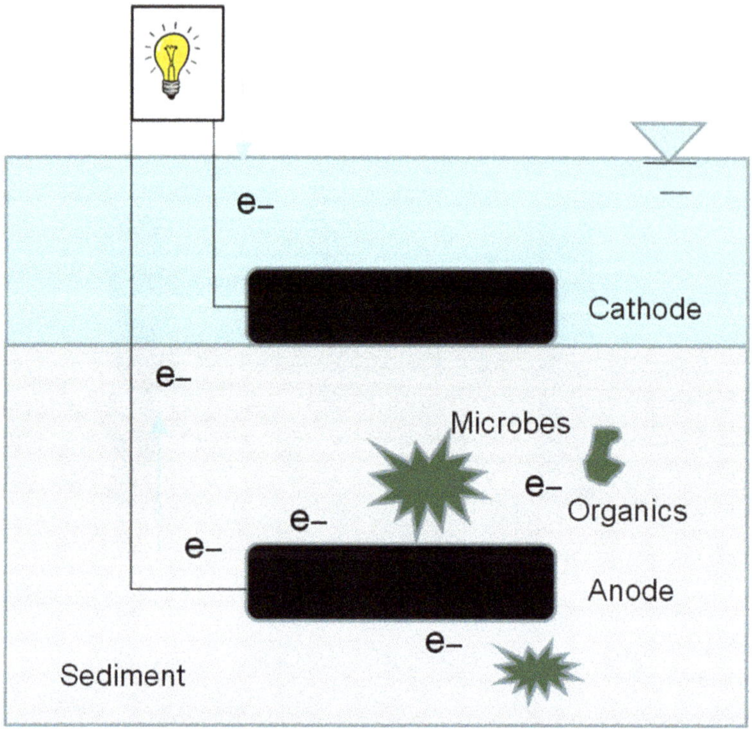

Figure 1.2. Schematic diagram of a sediment microbial fuel cell (SMFC).

1.3 Sediment microbial fuel cells

In an SMFC the anode is set inside the sediment and the cathode is set at the surface of the sediment. The organic matter is broken down by the geo-bacteria and the sulfate reducing bacteria (SRB) in the sediment and consequently electrons are released. The electron transfers from the anode to the cathode and thus electricity is generated. The cathode is placed on the surface or near the surface of the sediment. The availability of oxygen is required at the cathode areas. The benefits of this type of MFC, namely the ability to purify contamination while generating electricity, were found by Moqsud and Khong in 2021. Figure 1.2 shows a schematic diagram of the sediment microbial fuel cell.

1.4 Plant microbial fuel cells

Plant MFCs (PMFCs) are a type of SMFC in which a plant is used to supply the bioelectricity [4]. The mechanism of a PMFC is very interesting. The green leaves absorb sunlight from the Sun and produce carbohydrates through photosynthesis. The generated carbohydrates go to the root zone. More than 60% of the total generated carbohydrates are released at the rhizosphere in nature. For this reason, the number of geo-bacteria is greater in this area. The geo-bacteria use these carbohydrates as food and grow in number. Electrons are released while breaking

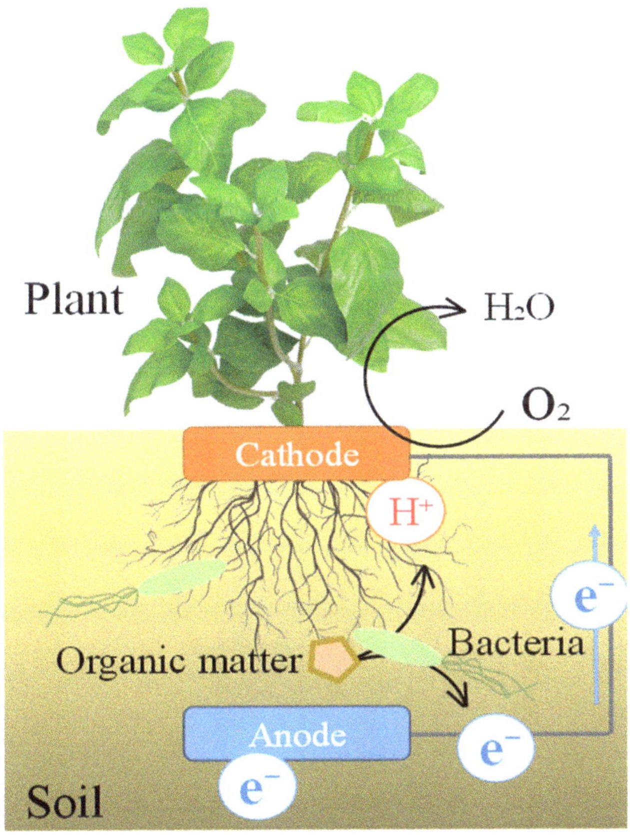

Figure 1.3. Schematic diagram of a plant microbial fuel cell (PMFC).

the carbohydrates. The anode catches these electrons and transports it to the cathode. This generates bioelectricity in the PMFC.

Figure 1.3 shows a schematic diagram of a PMFC. It is shown that due to photosynthesis, carbohydrates are produced in the green leaves. An excess of carbohydrates is released in the root areas and the geo-bacteria break down these carbohydrates as part of their regular activities in the rhizosphere.

1.5 Solid waste microbial fuel cells

An MFC in which organic waste is used as the biomass is a solid waste MFC (figure 1.4). Various types of solid organic waste can be used inside an MFC. Bioelectricity generation can be possible by using almost all types of organic waste MFCs [4]. Bioelectricity can be also generated by using bamboo waste and kitchen garbage, which has been confirmed in experiments. This could reduce the solid waste management problems in the world. Organic waste makes up a major portion when considering the total amount of solid waste, however, much of this organic waste is not recycled properly, both in developing countries as well as industrialized parts of the world. Thus,

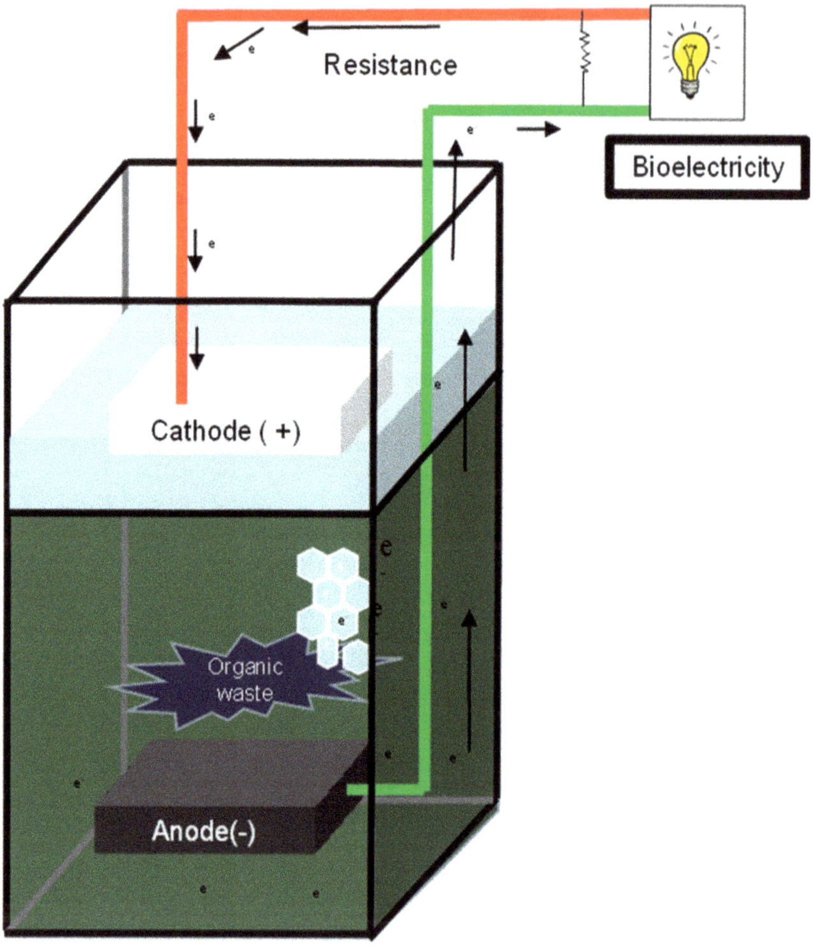

Figure 1.4. Schematic diagram of a solid waste MFC.

if it is possible to generate bioelectricity by using organic waste then this will help in managing solid waste problems all over the world.

Electricity from landfill leachate using MFCs has been also studied [30, 50–53].

1.6 Important components of microbial fuel cells

There are different types of MFCs in terms of their construction. The single chamber and the double chamber microbial fuel cells are the most common types of MFCs [24–29]. Normally, a single chamber MFC has two parts, namely the anode part and cathode part [1, 4, 30–33]. The anode part is responsible for the greater production of electricity as the electrons are released in this part due to the biodegradation of the organic matter. The cathode part is the part in where the electron receives oxygen and produces a water molecule. Normally, it is designed in such a manner that it can receive oxygen properly. Usually, the cathode is placed on the surface to receive enough oxygen during the process of bioelectricity generation [34–36]. Another important part of an MFC is the electrode materials and their connection with the copper wire. The external circuit, made of copper wire, a resistor, and a data logger are other major components for electricity generation from an MFC [37–41]. The MFC cell is normally an enclosed cell where the biomass and other organic substances can be set up. The biomass, sediment, and soil are placed inside the cell. Sometimes, there are covers on the surface of the cell to protect it from external disturbances [42–44].

1.7 Electrode materials used in a microbial fuel cell

Generally, carbon materials are used for microbial fuel cells. The reason for using carbon materials is that they are conductive materials and can be durable. Carbon materials do not react much with other materials even in different media [45, 46]. Commonly used electrode materials are carbon fiber, carbon felt, graphite felt, carbon brush, and carbon cloth. Different types of carbon materials are used for different purposes. Choosing the best carbon material is one of the challenges for the construction of an efficient MFC in different environmental conditions. Some researchers use bamboo charcoal as the electrode material in their MFCs [47–50].

1.8 Global energy, environment, and MFC contributions

The global energy demand is increasing and the reserves of fossil fuels are decreasing day by day. We need to find alternative sources of energy for future generations. In obtaining energy from fossil fuels, greenhouse gases are released all the time, which causes environmental pollution from these raw materials which are non-renewable sources. The consequence of this phenomenon is global warming and climate change. The trend of climate change has been more evident in recent years. Natural disasters have been increasing and the loss of life and property is increasing each year. To stop this global warming and climate change phenomenon, the use of environmentally friendly sources of energy generation must be implemented as soon as possible. To meet the ever-increasing energy demands and to cope with the problems of global warming are the biggest challenges for scientists and engineers. Due to the potential dangers of accidents at nuclear power plants, people are seeking green sources of energy that are also safe for the future generations.

Mother Earth is ready to embrace the fourth industrial revolution. The Earth stands on the brink of a digitalization and technological revolution that will fundamentally change the way we live, work, and relate to one another. The scale, scope, and complexity of this revolution will be so enormous that mankind has not experienced it before. The demand for energy is increasing faster than ever. Due to the use of the Internet, artificial intelligence, and wireless sensor networks for smart cities and the automated automobiles, energy demands will be increasing further soon [54]. The global population is increasing and many countries are developing at a faster pace. This trend of increasing energy demands is alarming and is a threat to Mother Nature.

Renewable energy is a good source of green energy which is good for humans and the environment. Solar energy, wind energy, thermal energy, and biofuels are some of the future substitutes for fossil fuels. However, the current sources of renewable energy are unable to meet the demands of the future. This additional demand may cause further trouble if the technologies cannot be expanded. Renewable energy can solve the problem of global warming and climate change related disasters in the future, however, there are many disadvantages to renewable energy. The most prominent disadvantage of renewable energy is that it is affected significantly by weather conditions. For example, solar power cannot work on the rainy days and wind power cannot work when there is no wind blowing. Thus renewable energy will need a lot of development before it can be considered as a major source of green energy [55, 56]. The first-generation biofuels are fuels which are generally produced from food grains such as soybean, corn, sugar cane, and other food products. However, the use of this huge amount of food is not a sustainable solution for biofuels as there are millions of people in the world who are hungry. It is completely unnecessary to destroy food products to fuel cars instead of feeding millions of hungry children.

Second-generation biofuel, which is mainly produced from different kinds of organic waste, is also not suitable or sustainable solution. This is because the amount of organic waste is not sufficient, and it is not possible to collect this huge amount of organic waste in a short period of time. The third-generation biofuels, produced my algae, are also not yet popular due to the process of collecting the raw sources and the amount of energy collected from a unit amount of the sources of algae.

The future of MFCs is promising. An MFC is a green and safe source of energy. The other major benefits of MFCs are that they can be used for environmental pollution removal, wastewater treatment, solid waste management, and desalination along with the bioelectricity generation. The MFCs need to be improved for the future and for use in the large-scale production of bioelectricity. Another noteworthy benefit of MFCs is that they can be used 24 h a day without being affected by the weather conditions or other external factors, in contrast to other types of renewable energy. The various types of MFCs can be used as the power sources for the monitoring sensors of smart cities to achieve the sustainable development goals and the fourth industrial revolutions [7, 57].

1.9 Conclusions

MFC technology has been developing day by day. Many researchers have contributed to developing this novel technology for practical uses. However, the large-scale construction of MFCs is still a challenging issue. The availability of low-cost electrode materials is another basic problem for the implementation of MFCs in the field. Nevertheless, MFC technology is one of the major renewable energy sources for future smart city sensors and a new source of green energy for a future sustainable world. The future challenge of increased organic waste management can be achieved using this novel technology [58–61].

References

[1] Yaqoob A A, Ibrahim M N M, Umar K, Parveen T, Ahmad A, Lokhat D and Setapar S H 2021 A glimpse into the microbial fuel cells for wastewater treatment with energy generation *Desalin. Water Treat.* **214** 379–89

[2] Moqsud M A, Omine K, Yasufuku N, Hyodo M and Nakata Y 2013 MFC for bioelectricity generation from organic waste *Waste Manag.* **33** 2465–71

[3] Moqsud M A, Omine K, Yasufuku N, Hyodo M and Nakata Y 2014 Bioelectricity from bamboo and kitchen garbage *Waste Manage. Res.* **32** 124–30

[4] Moqsud M A, Yoshitake J, Bushra Q, Hyodo M, Omine K and Srik D 2015 Compost in plant microbial fuel cell for bioelectricity generation *Waste Manage.* **36** 63–9

[5] Zhang Y J *et al* 2013 Electricity production from molasses wastewater in two-chamber microbial fuel cell *Water Sci. Technol.* **68** 494–8

[6] Zhuang L, Zhou S, Wang Y, Liu C and Geng S 2009 Membrane-less cloth cathode assembly (CCA) for scalable microbial fuel cells *Biosens. Bioelectron.* **249** 3652–6

[7] Moqsud M A 2020 Bioelectricity generation and remediation of sulfide contaminated tidal flat sediment *Int. J. Sediment Res.* **35** 91–6

[8] Logan B E and Regan J M 2006 Electricity-producing bacterial communities in microbial fuel cells *Trends Microbiol.* **14** 512–8

[9] Yaqoob A A, Ibrahim M N, Yaakop A S and Ahmad A 2021 Application of microbial fuel cells energized by oil palm trunk sap (OPTS) to remove the toxic metal from synthetic wastewater with generation of electricity *Appl. Nanosci.* **11** 1949–61

[10] Aelterman P, Rabaey K, Pham H T, Boon N and Verstraete W 2006 Continuous electricity generation at high voltages and currents using stacked microbial fuel cells *Env. Sci. Technol.* **40** 3388–94

[11] Moqsud M A 2021 *Strategies of Sustainable Solid Waste Management* (Rijeka: IntechOpen) p 129

[12] Daud N N, Ahmad A, Yaqoob A A and Ibrahim M N 2021 Application of rotten rice as a substrate for bacterial species to generate energy and the removal of toxic metals from wastewater through microbial fuel cells *Env. Sci. Pollut. Res.* **28** 62816–27

[13] Allen R M and Bennetto H P 1993 Microbial fuel-cells: electricity production from carbohydrates *Appl. Biochem. Biotechnol.* **39** 27–40

[14] Yaqoob A A, Serrà A, Ibrahim M N and Yaakop A S 2021 Self-assembled oil palm biomass-derived modified graphene oxide anode: an efficient medium for energy transportation and bioremediating Cd (II) via microbial fuel cells *Arabian J. Chem.* **14** 103121

[15] Behera M and Ghangrekar M M 2011 Electricity generation in low cost microbial fuel cell made up of earthenware of different thickness *Water Sci. Technol.* **64** 2468

[16] Moqsud M A, Gazali T A, Omine K and Nakata Y 2017 Green electricity by water plants in organic soil and marine sediment through microbial fuel cell *Energy Sources* A **39** 160–5

[17] Yaqoob A A, Ibrahim M N and Umar K 2021 Biomass-derived composite anode electrode: synthesis, characterizations, and application in microbial fuel cells (MFCs) *J. Environ. Chem. Eng.* **9** 106111

[18] Borole A P, Aaron D, Hamilton, Choo Y and Costas T 2010 Understanding long-term changes in microbial fuel cell performance using electrochemical impedance spectroscopy *Environ. Sci. Technol.* **44** 2740–5

[19] Capodaglio A G 2013 Microbial fuel cells for direct electrical energy recovery from urban wastewaters *Sci. World J.* **8** 63473

[20] Chen S, Patil S A and Schroder U 2018 A high-performance rotating graphite fiber brush air cathode for microbial fuel cells *Appl. Energy* **211** 1089–94
[21] Choi J and Ahn Y 2013 Continuous electricity generation in stacked air cathode microbial fuel cell treating domestic wastewater *J. Env. Manag.* **130** 146–52
[22] Corbella C, Puigagut J and Garf M 2017 Life cycle assessment of constructed wetland systems for wastewater treatment coupled with microbial fuel cells *Sci. Total Env.* **584–585** 355–62
[23] Yaqoob A A, Ibrahim M N and Yaakop A S 2021 Application of oil palm lignocellulosic derived material as an efficient anode to boost the toxic metal remediation trend and energy generation through microbial fuel cells *J. Clean. Prod.* **314** 128062
[24] Du Z, Li H and Gu T 2007 A state of the art review on microbial fuel cells: a promising technology for wastewater treatment and bioenergy *Biotechnol. Adv.* **25** 464–82
[25] Fan Y, Sharbrough E and Liu H 2008 Quantification of the internal resistance distribution of microbial fuel cells *Env. Sci. Technol.* **42** 8101–7
[26] Erbay C, Carreon-Bautista S, Sanchez-Sinencio E and Han A 2014 High performance monolithic power management system with dynamic maximum power point tracking for microbial fuel cells *Env. Sci. Technol.* **48** 13992–9
[27] Fornero J J, Rosenbaum M, Cotta M A and Angenent L T 2008 Microbial fuel cell performance with a pressurized cathode chamber *Env. Sci. Technol.* **42** 8578–84
[28] Ghasemi M, Daud W R W, Rahimnejad M, Rezayi M, Fatemi A, Jafari Y, Somalu M and Manzour A 2013 Copper-phthalocyanine and nickel nanoparticles as novel cathode catalysts in microbial fuel cells *Int. J. Hydrogen Energy* **38** 9533–40
[29] Moqsud M A, Omine K, Yasufuku N, Bushra Q S, Hyodo M and Nakata Y 2015 Bioelectricity from kitchen garbage and bamboo wastes *Waste Manage. Res.* **32** 124–30
[30] Habermann W and Pommer E H 1991 Biological fuel cells with sulfide storage capacity *Appl. Microbiol. Biotechnol.* **35** 128–33
[31] He Z, Minteer S D and Angenent L T 2005 Electricity generation from artificial wastewater using an upflow microbial fuel cell *Env. Sci. Technol.* **39** 5262–7
[32] Hu Z 2008 Electricity generation by a baffle-chamber membraneless microbial fuel cell *J. Power Sources* **179** 27–33
[33] Khaled F, Ondel O and Allard B 2016 Microbial fuel cells as power supply of a lowpower temperature sensor *J. Power Sources* **306** 354–60
[34] Yaqoob A A, Guerrero–Barajas C, Ibrahim M N, Umar K and Yaakop A S 2022 Local fruit wastes driven benthic microbial fuel cell: a sustainable approach to toxic metal removal and bioelectricity generation *Env. Sci. Pollut. Res.* **12** 1–6
[35] Kim I S, Chae K J, Choi M J and Verstraete W 2008 Microbial fuel cells: recent advances, bacterial communities and application beyond electricity generation *Env. Eng. Res.* **13** 51–65
[36] Kumar M and Singh R 2020 Sewage water treatment with energy recovery using constructed wetlands integrated with a bioelectrochemical system *Env. Sci.: Water Res. Technol.* **6** 795–808
[37] Li Y, Wu Y, Liu B, Luan H, Vadas T, Guo W and Li B 2015 Self-sustained reduction of multiple metals in a microbial fuel cell–microbial electrolysis cell hybrid system *Bioresour. Technol.* **192** 238–46
[38] Liu F, Sun L, Wan J, Shen L, Yu Y, Hu L and Zhou Y 2020 Performance of different macrophytes in the decontamination of and electricity generation from swine wastewater via an integrated constructed wetland-microbial fuel cell process *J. Environ. Sci.* **89** 252–63

[39] Liu S, Cheng D, Sun H, Huang J, Chen and Cen K Inhibition of microbial growth on air cathodes of single chamber microbial fuel cells by incorporating enrofloxacin into the catalyst layer *Biosens. Bioelectron.* **72** 44–50

[40] Luke W T *et al* 2013 A compact versatile microbial fuel cell from paper *Am. Soc. Mech. Eng.* **55522** V001T01A015

[41] Manohar A K and Mansfeld F 2009 The internal resistance of a microbial fuel cell and its dependence on cell design and operating conditions *Electrochim. Acta* **54** 1664–70

[42] Marco-Urrea E, Perez-Trujillo M, Vicent T and Caminal G 2009 Ability of white-rot fungi to remove selected pharmaceuticals and identification of degradation products of ibuprofen by *Trametes versicolor Chemosphere* **74** 765–72

[43] Oh S E and Logan B E 2005 Hydrogen and electricity production from a food processing wastewater using fermentation and microbial fuel cell technologies *Water Res.* **39** 4673–82

[44] Pamintuan K S and Sanchez K M 2019 Power generation in a plant-microbial fuel cell assembly with graphite and stainless steel electrodes growing *Vigna radiata IOP Conf. Ser.: Mater. Sci. Eng.* **703** 012037

[45] Potter M C 1910 On the difference of potential due to the vital activity of microorganisms *Proc. Univ. Durham Phil. Soc.* **3** 245–9

[46] Rabaey K and Verstraete W 2005 Microbial fuel cells: novel biotechnology for energy generation *Trends Biotechnol.* **23** 291–8

[47] Qing W *et al* 2010 Electricity generation and brewery wastewater treatment from sequential anode–cathode microbial fuel cell *Construct. Build. Mater.* **11** 87–93

[48] Rabaey K, Boon N, Siciliano S D, Verhaege M and Verstraete W 2004 Biofuel cells select for microbial consortia that self-mediate electron transfer *Appl. Env. Microbiol.* **70** 5373–82

[49] Rosenbaum M, Zhao F, Schroder U and Scholz F 2006 Interfacing electrocatalysis and biocatalysis with tungsten carbide: a high performance noble-metal-free microbial fuel cell *Angew. Chem.* **45** 6658–61

[50] Srivastava P, Abbassi R, Yadav A K, Garaniya V and Asadnia M 2020 A review on the contribution of electron flow in electroactive wetlands: electricity generation and enhanced wastewater treatment *Chemosphere* **254** 126926

[51] Taiki Y *et al* 2021 Estimation of total energy requirement for sewage treatment by a microbial fuel cell with a one-meter air-cathode assuming Michaelis–Menten COD degradation *RSC Adv.* **11** 20036–45

[52] Varanasi J L, Roy S, Pandit S and Das D 2015 Improvement of energy recovery from cellobiose by thermophillic dark fermentative hydrogen production followed by microbial fuel cell *Int. J. Hydrogen Energy* **40** 8311–21

[53] Varanasi J L, Nayak A K, Sohn Y, Pradhan D and Das D 2016 Improvement of power generation of microbial fuel cell by integrating tungsten oxide electrocatalyst with pure or mixed culture biocatalysts *Electrochim. Acta* **199** 154–63

[54] Walter X A, Stinchcombe A, Greenman J and Ieropoulos I 2017 Urine transduction to useable energy: a modular MFC approach for smartphone and remote system charging *Appl. Energy* **192** 575–81

[55] Wang Y X, Ye Z Z, Si C Y and Ying Y B 2012 Application of aptamer based biosensors for detection of pathogenic microorganisms *Chin. J. Anal. Chem.* **40** 634–42

[56] Yaqoob A A, Khatoon A, Mohd Setapar S H, Umar K, Parveen T, Mohamad Ibrahim M N, Ahmad A and Rafatullah M 2020 Outlook on the role of microbial fuel cells in remediation of environmental pollutants with electricity generation *Catalysts* **10** 819

[57] Yaqoob A A, Ibrahim M N and Guerrero-Barajas C 2021 Modern trend of anodes in microbial fuel cells (MFCs): an overview *Env. Technol. Innov.* **23** 101579

[58] Zhang X, Cheng S, Wang X, Huang X and Logan B E 2009 Separator characteristics for increasing performance of microbial fuel cells *Env. Sci. Technol.* **43** 8456–61

[59] Moqsud M A, Hannan M A and Omine K 2015 Global assessment of factors influencing bioelectricity generation in paddy plant microbial fuel cells *Adv. Res. J. Agricult. Sci.* **4** 840–50

[60] Moqsud M A and Omine K 2010 Bioelectricity generation by using organic waste in Bangladesh *Proc. of Int. Conf. on Environmental Aspects of Bangladesh*

[61] Moqsud M A and Rahman M H 2004 Biochemical quality of compost from kitchen garbage in Bangladesh *Env. Inform. Arch.* **2** 635–40

Chapter 2

Basic principles and working mechanisms of microbial fuel cells

Asim Ali Yaqoob, Mohamad Nasir Mohamad Ibrahim, Akil Ahmad and Mohammed B Alshammari

Microbial fuel cells (MFCs) are a bio-electrochemical approach where bacteria serve as biocatalysts to oxidize a carbohydrate based organic substrate in order to generate energy. Due to its ecofriendly nature, it is the most promising emerging method at the present time. This chapter summarizes the basic principles and working mechanisms of MFCs in terms of energy generation. Based on a literature survey, we found that the long range electron transfer mechanism is the most dominant in the field of MFCs. A large number of bacteria species are discussed along with their energy efficiency. This chapter will be very useful for new researchers seeking to upgrade their knowledge about MFCs and their basic components. Future perspectives are also included for advanced level researchers in order to bring their research to a more productive level.

2.1 Introduction

Fossil fuels are major energy sources for the entire world and there continues to be an increase in the consumption of these resources due to a rise in energy demands. However, the consumption of these energy sources has further resulted in major issues such as global warming and environmental pollution [1]. Therefore, a lack of energy and clean water are becoming serious problems that need to be addressed at the present time [2]. Organic wastewater is considered to be a source that has the potential to become a renewable feedstock for producing numerous forms of bioenergy as well as helping in the remediation of wastewater through the regulation of biological operations. Bioenergy is receiving attention as an alternative source of energy to fossil fuels as the former can be sustainable [3]. Remediation of wastewater provides a new way to generate bioenergy that is long-lasting and renewable and thus can be advantageous for the environment. In the fields of biotechnology and

doi:10.1088/978-0-7503-4791-4ch2

science, both the fields of energy substitutes and the management of wastewater are two of the least explored. A new technology that can provide simultaneous energy production and wastewater remediation, called microbial fuel cells (MFCs), is receiving attention as it is a method that can subsequently reduce pollution [4, 5]. In MFCs it is possible to produce carbon dioxide and water through the breakdown of an organic substrate, that is usually only achievable by several enzymatic reaction steps [6]. Therefore, ongoing research on MFCs can help in the further development of this technology. This chapter discusses briefly the basic principles and energy generation mechanisms of MFCs. The roles of different bacterial species as catalysts are discussed in detail. Finally, future perspectives and concluding remarks are provided.

2.2 Fuel cells and a brief introduction to the development of MFCs

Direct conversion of the energy in fuel to produce electricity can be achieved easily in a single step using fuel cells. A constant source of fuel enables the fuel cell to generate energy continuously, in contrast to using rechargeable battery. Oxidation at the anode occurs using a commonly used organic substrate, which will come across a catalyst which will cause further conversion to a proton and an electron [7]. The transportation of protons occurs through a proton exchange membrane toward the cathode electrode, while the electrons are transported through the outer circuit. Production of the final product, which is water, occurs through the combination of the transported proton with oxygen molecules through a reduction process on the cathode for current generation in a single step [8]. Research on fuel cells is gaining increasing interest due to its qualities such as high thermodynamic efficiency, ecofriendliness, and sustainability. There are two categories of fuel cells: abiotic and biotic. The standard fuel cell is of the abiotic type, and contains an inorganic catalyst for oxygen reduction and hydrogen oxidation at the cathode, such as the proton exchange membrane fuel cell and the solid oxide fuel cell. An abiotic fuel cell produces a higher magnitude of energy than a biotic fuel cell [9]. Compared to an abiotic cell, a biotic cell contains living organisms, enzymes, or derivatives as the main catalyst. Low-cost and inexpensive substrates and catalysts are used in biological fuel cells (BFCs). BFCs can also be divided into two categories: microbial fuel cells (MFCs) and enzymatic fuel cells (EFCs) [10]. In EFCs the enzyme is either adsorbed or placed on the anode on which oxidation will occur in the presence of an organic substrate. When the microbe acts as a catalyst, the cell is known as an MFC. The utilization of microbes in energy generation was introduced in the early twentieth century [11]. However, prior to that, Luigi Galvani already demonstrated the connection between electricity and biology in 1780. In 1911 Michael Cresse Potter used *Escherichia coli* for current generation [12]. Twenty years later, microbial half fuel cells were prepared and connected together in series by Cohen. The cells generated approximately 35 V and 2 mA current. In the late 1970s the working principle of MFCs using bacteria was described by both Allen and Benetto [8]. Their method required an external supplemented electron mediator. Fuel cells receive more attention as interesting devices in the early 1990s, which further resulted in increasing attention for MFCs as a new sustainable technology [13]. The biggest advancement of MFCs occurred in the late 1990s when MFCs were developed that ran using bacteria and resulted in the direct transfer of electrons to the anode during the oxidation process of waste without the need for a mediator [14]. The benefits of MFCs over other traditional fuel cells are that they enable the use of inexpensive renewable sources to convert current, rather than using fuel cells that require expensive hydrogen or methanol to be oxidized to produce the current. In addition to that, such fuel cells also requires costly chemical substances for the catalyst, such as platinum. By using MFCs various bacterial species can be used for the oxidation at the anode electrode that will make the process more effective. This operation can also be conducted at room temperature and environmental pressure, unlike for some other fuel cells.

2.3 Advantages of MFCs over other bioenergy processes

The benefits of utilizing the technology of MFCs has resulted in their gaining popularity in the past few years. Comparing MFCs with other conventional wastewater treatments, for example the activated sludge process (ASP) and anaerobic digesters (ADs), MFCs enable the direct conversion of waste to electricity in only one step that will provide a high efficiency of conversion [15]. According to a study by Logan *et al* [16], biohydrogen, bioethanol, and biogas showed around 15%–30%, 10%–25%, and 25%–38% conversion efficiency, respectively. In contrast, the efficiency of conversion using MFCs is as high as 80%. Unlike anaerobic digestion, which requires the formation of methane, hydrogen gas, CO_2, or a small volume of hydrogen sulfide, no gas treatment is needed for MFCs. Input energy is needed for the aeration process in the ASP while MFCs are capable of working without the need for aeration because of the passively aerated cathode. Compared to the conventional technologies mentioned previously, the production of sludge is lower using MFCs and a stack of MFCs can also be used as a dispersed source of energy.

2.4 Basic principles and working mechanisms of MFCs

MFCs are made up of an anode electrode, cathode electrode, and electrolyte(s). The cathode and anode chambers can be separated by a proton exchange membrane. The oxidation of substrates can be completed using living microbes in a planktonic state through the formation of a biofilm on the anode. Protons, electrons, and other metabolites will be produced as end-products resulting from the oxidation process. Electrons are released from the microbes and move towards the anode and, with the help of an external load, they are transported to the cathode [17]. In contrast, the protons percolate through the proton exchange membrane or diffuse to the cathode to undergo reduction via electrons in order to complete the circuit. The movement of electrons via the external load produces an electric current. The oxygen present at the anode aids in the reduction of protons forming water as a by-product, resulting in an environmentally friendly process. The scale of MFCs can be both small (microliter-scale or milliliter-scale) and large (liter-scale). The crucial parts of an

MFC are the cathode, anode, and separator, also called a membrane. Carbon materials such as graphite plates, carbon foam, carbon paper, carbon cloth, carbon rods, glassy carbon, carbon felt, stainless steel mesh, Pt, and Pt black have been used as materials for the anode in MFCs. The materials used for the anode must be capable of allowing the formation of biofilm by the bacteria and be both non-corrosive and conductive, whereas the cathode is commonly made up of catalyst materials such as Pt black, MnO_2, Pt, and polyaniline for an improved oxygen reduction reaction (ORR) [18]. The materials used for the cathode are similar to those for the anode, such as carbon felt, graphites, carbon foam, carbon cloth, carbon paper, stainless steel meshes, glassy carbon, and titanium mesh. The proton exchange membrane is used in a dual-chambered MFC. The proton exchange membrane in a single chambered MFC is either hot-pressed or chemically fixed to the cathode so that the cathode membrane is assembled for passive aeration to provide oxygen sources directly for the ORR. Initially, the separator used was usually a salt-bridge between the chambers of the cathode and anode [19]. As time passed it was gradually substituted by a proton exchange membrane. Both the cation exchange membrane and anion exchange membrane (AEM) were reported as membranes. However, the AEM is the more efficient membrane in producing power due to its capability to generate consistent voltage and ability to reduce salt deposition.

2.5 Energy generation through MFCs

The energy generation via MFCs depends on the bacterial community. Distinct varieties of bacterial species are proficient at transferring electrons along with the MFC electrodes, allowing for a varied range of applications [20]. There are several groups of bacteria that have been proven to generate electricity in MFCs, which include *Proteobacteria phyla, Acidobacteria, Firmicutes,* yeast, microalgae, and fungi. Typical bacterial species that are capable of exchanging electrons with electrodes are *Geobacter* spp., *Shewanella* spp., *Clostridium butyricum, Aeromas hydrophila, Geobacter* spp., *Rhodoferax ferrireducens,* and *Enterococcus gallinarum,* etc [21–25]. Specifically, *Geobacter* biofilms are able to serve as transistors and supercapacitors due to the conductivity of the bacterium. The microbes can generate current by producing and passing electrons to electrodes, which they use to regulate their respiration. Since electrodes are not a natural part of a marine system, normally micro-organisms would pass their electrons to natural electron acceptors which are insoluble. Organic substrates of mostly carbohydrate based type are often utilized in MFCs in bioelectrogenesis and bacterial metabolism. By entering glycolysis pathways, these carbohydrate molecules provide acetyl co-enzyme A. This acetyl co-enzyme A goes into the Krebs cycle, which is also known as the tricarboxylic acid and citric acid cycle. One cycle of the Krebs cycle provides three reduced nicotinamide adenine dinucleotide (NADH) molecules and one reduced flavin adenosine dinucleotide ($FADH_2$) [26]. The entire mechanism is carried out in the bacterial cytoplasm. NADH and $FADH_2$ act as electron carriers for the

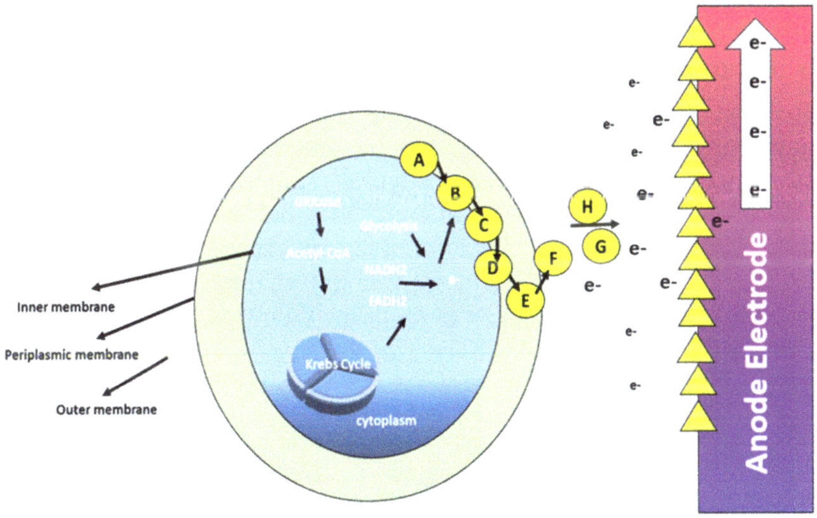

Figure 2.1. Graphic demonstration of the generation of electrons within bacterial cells.

synthesis of adenosine triphosphate and then transmit toward the electron transport chain (ETC). However, the ETC is present in the bacterial cell membrane which consists of internal and external membranes together with periplasm. The produced electrons travel via ETC protein channels such as co-enzyme Q, NADH dehydrogenase, ubiquinone, and cytochromes [6, 27, 28]. Next, it is transferred by pumping out from the bacterial cell, as demonstrated in figure 2.1. The entire cycle generates 34 adenosine triphosphate (ATP) molecules. The main source producing the electrons and protons is the organic substrate, which later follow the Krebs cycle in order to produce ATP. Different organic substrates are used in MFCs and the reactions can be written as follows:

1. If acetate is used as the organic substrate
   Oxidation reaction: $CH_3COOH + 2H_2O \rightarrow 2CO_2 + 8H^+ + 8e^-$
   Reduction reaction: $8e^- + 8H^+ + 2O_2 \rightarrow 4H_2O$
   Overall reaction: $CH_3COOH + 2O_2 \rightarrow 2CO_2 + 2H_2O$ + Electricity + Biomass.
2. If glucose is used as the organic substrate
   Oxidation reaction: $C_6H_{12}O_6 + 6H_2O \rightarrow 6CO_2 + 24H^+ + 24e^-$
   Reduction reaction: $24e^- + 24H^+ + \rightarrow 6O_2 \rightarrow 12H_2O$
   Overall reaction: $C_6H_{12}O_6 + 6O_2 \rightarrow 6CO_2 + 6H_2O$ + Electricity + Biomass.
3. If sucrose is used as the organic substrate
   Oxidation reaction: $C_{12}H_{22}O_{11} + 13H_2O \rightarrow 12CO_2 + 48H^+ + 48e^-$
   Reduction reaction: $4H^+ + 4e^- + O_2 \rightarrow 2H_2O$
   Overall reaction: $C_{12}H_{22}O_{11} + 12O_2 \rightarrow 12CO_2 + 11H_2O$ + Electricity + Biomass.

2.6 The mechanism of electron transfer from the bacterial cell to the anode electrode

The generation of electrons occurs from bacterial respiration during the oxidation of the organic substrate. There are some dominant mechanisms which are mostly followed by bacterial species to transfer the electrons from the bacterial cell to the anode electrode. Table 2.1 shows different bacterial species with their electron transfer mechanism and further the level of energy generation. Figure 2.2 shows a graphical presentation of electron transportation from the bacterial cell to the anode electrode and a detail explanation is given in the following.

Table 2.1. Different bacterial species with their known/unknown electron transfer intermediates and their energy efficiency.

Microbes	Proteins/electron transfer intermediates	Power density ($mA\ m^{-2}$)	Reference
Exoelectrogens with known intermediates of electron transfer			
Geobacter sulfurreducens	c-cytochrome z, type IV pili	3147	[59]
Shewanella oneidensis DsP10	Riboflavin, flavins	5000	[60]
Rhodopseudomonas palustris DX-1	c-type cytochromes	2720	[61]
Desulfovibrio desulfuricans ATTC	c-type cytochromes	1580	[62]
Geobacter metalliriducens	c-type cytochromes, OmcE and OmcB	450	[63]
Desulfuromonas acetoxidans	c-type cytochromes	20000	[64]
Klebsiella pneumonia	2,6-di-tert-butyl-p-benzoquinone	199	[65]
Desulfovibrio alaskensis	Transmembrane complexes, tetraheme cytochrome C3	–	[58]
Pseudomonas aeruginosa	Phenazine-1-carboxamide, pyocyanin	4300	[66]
Thermincola ferriacetica	Anthraquinone-2,6-disulfonate	12000	[67]
Shewanella putrefaciens	c-type cytochromes, OmcA, MtrC, FAD transporters	492	[68]
Dechlorospirillum anomalous strain WD	Anthraquinone-2,6-disulfonate hydrogen	30	[69]
Geobacter lovleyi	Methyl viologen	480	[27]
Chrlorella vulgaris	Methyl viologen, methylene blue	30	[70]
Exoelectrogens with unknown intermediates of electron transfer			
Leptothrix discophora SP-6		70	[71]
Acinetobacter calcoaceticus		110	[72]
Sphingobacterium multivorum		103	[73]

Escherichia coli	3390	[74]
Winogradskyella poriferorum	40	[75]
Acinetobacter johnsonii	20	[75]
Saccharomyces cerevisiae	282	[75]
Pseudomonas fluorescens	210	[75]
Brevundimonas diminuta	90	[75]
Burkholderia cepacia	120	[75]
Branhamella catarrhalis	200	[75]
Enterobacter cloacae	240	[75]
Shigella flexneri	170	[75]
Kingella kingae	250	[75]
Kingella denitrificans	250	[75]
Micrococcus luteus	220	[75]
Bacilus subtilis	250	[75]
Staphylococcus carnosus	280	[75]
Citrobacter sp.	205	[76]
Dechloromonas sp.	300	[77]
Lysinibacillus sphaericus	850	[78]
Arthrospira maxima	100	[79]
Coriolus versicolor	3200	[80]

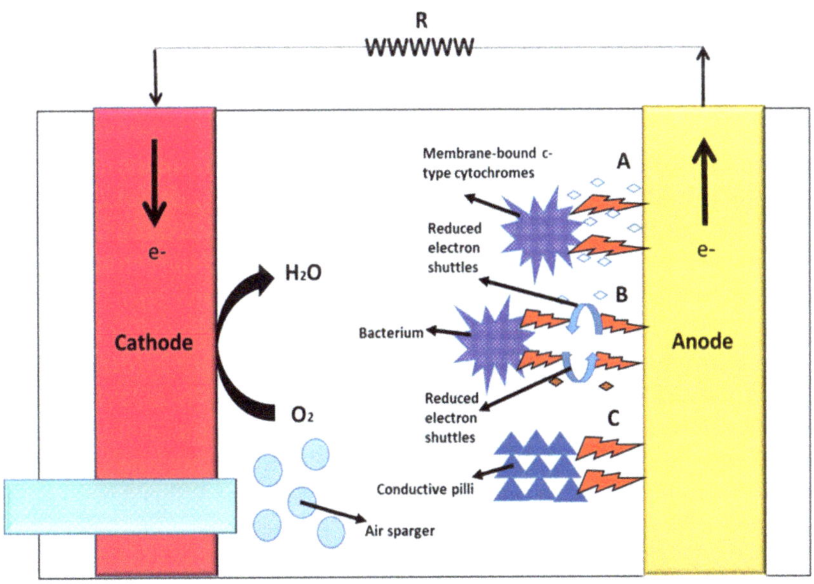

Figure 2.2. Different electron transfer mechanisms of the bacteria cell to the anode electrode.

2.6.1 Electron transportation via electron-shuttle molecules

Several microbiology studies have shown that a variety of bacteria, including *G. fermentans*, *S. oneidensis*, *Lactococcus lactics*, and *P. aeruginosa*, are capable of transporting electrons to anode electrodes through self-producing shuttle molecules [29]. The most popular families of buried electrodes are *Desulfuromonas* and *Geobacteraceae* [30]. Since the former bacteria favor high salinity, the *Geobacter* species predominates on anodes immersed in freshwater sediments, while marine buried sediment anodes are predominated by *Desulfuromonas* species [31]. Bond *et al* [32] were the first to investigate the idea of micro-organisms transferring electrons to electrodes using self-produced soluble electron shuttles. According to the researchers, *G. fermentans* generates soluble electronic shuttles. Two forms of soluble electron shuttles with different redox possibilities were present. Riboflavin is a soluble electron shuttle with a redox potential value of −0.2 V, while the unknown species showed a 0.3 V redox potential [33]. Comparatively, *Shewanella* spp. have the capacity to minimize the poor crystallinity of Fe(III) oxides via the production of riboflavin and flavin mononucleotides functioning as extracellular soluble electron shuttles in order to promote the transfer of electrons to electrodes [34]. Furthermore, intact cells can accomplish direct electron transfer if one deliberately places the electrodes at significantly higher potentials than normal MFCs. In this reaction, the rate of transfer of electron is faster. *S. oneidensis* MR-1 and *Shewanella* sp. MR4 biofilms cultured at poised electron acceptors released a mixture of riboflavin-5-phosphate and riboflavin electron shuttles that were soluble [35]. The moment that the riboflavin was detached from the biofilms, the rate of transfer of electrons to the electrodes decreased by approximately 70%. According to Xu *et al* [36], the region of the carbon cloth electrodes enables them to function as cytochrome-bound cofactors capable of speeding up extracellular electron transfer in *S. oneidensis* MR-1 by endogenous flavins, which represents a model of extracellular electron transfer. Differences in electrode materials and how flavins behave at cytochrome–electrode interfaces are also discussed. Flavin adenine dinucleotide serves as an electron shuttle to transport flavins to the outer membrane, where it is reduced by c-type cytochrome in *S. oneidensis*. Electron transportation is carried out by flavins from the inner parts of the cells to insoluble electrodes by approximately 70% [37]. In *S. oneidensis* MR-1, flavins also control the rate and direction of extracellular transfer of electrons. Yang *et al* [38] discovered a flavin that was secreted endogenously, restricting the extracellular transfer of electrons, in contrast to the synthetic biosynthesis of flavins in *S. oneidensis* MR-1. The flavin concentration increases due to the heterogeneous state by 25.7-fold in contrast to the wild-type strain. MtrA (MtrD, SO4360 and DmsE), MtrC (MtrF and OmcA), Decaheme Cym A, OmcA, and MtrB (MtrE, DmsF and SO4359), which are in *S. oneidensis* MR-1, are parts of a complex of multiproteins that is found at the outer and inner membranes. It allows the electrons to 'hop' to the membrane's exterior surface from the periplasm [39]. In *S. oneidensis* MR-1, Mtr/OmcA facilitates extracellular electron transfer via flavins. MtrD, MtrE, and MtrF are expected to form MtrFDE, a MtrCAB-like outer-membrane-spanning complex. The OmcA to MtrC ratio was found to be 2:1 in both *in vitro* and *in vivo* studies. Extracellular electron transfer is needed for the DMSO reduction through subunits of extracellular catalytic DmsAB, which requires the porin cyctochrome DmsEF. It is unclear what the function is of the porin cytochrome SO4359_60 complex and its related subunit SO4361_62 [40]. According to Wei *et al* [41] in *S. oneidensis* MR-1, the MtrF cytochrome is mobile and has a good affinity for peptide aromatic groups, and the extracellular electron transfer rate across this heme group is very high. When compared to an MtrF monolayer on a flat gold substrate, Pirbadian [42] discovered that in the recombinant genome, MtrF has a better potential for extracellular transfer of electrons. Since this complex is needed for the transfer and attachment of electrons to electrodes, OmcA/MtrC mutants provided little current. As little flavin was produced, mutants without the Mtr/OmcA-coupled protein were capable of transferring only half as many electrons to the electrodes [43].

2.6.2 Redox-active proteins

Since it includes enzymes used in core anaerobic metabolism, *Geobacter sulfurreducens* has also been thoroughly researched for direct electron transfer to electrodes. Electrons are passed to other electron acceptors, including fumarate, metal ions, and elemental sulfur, by oxidizing carbon to carbon dioxide. Another essential feature of *G. sulfurreducens* is that it is directly linked to other *Geobacter* species. By passing electrons to electrodes, it can produce current from the conversion of organic substrates when oxygen intrusion is not artificially opposed and very low. *G. sulfurreducens* contains genes for various c-type cytochromes that have heme as the central molecule and protrude from the outer membrane [44].

G. sulfurreducens cells grown on graphite, were compared in terms of gene translation. When compared to several other cytochromes, OmcZ was found to be more abundant, for example OmcB, OmcE, OmcT, and OmcS, according to microarray analysis, in a situation that encourages current harvesting. OmcZ is often used in two forms: the large form (OmcZL) and short form (OmcZs), which are hydrophobic with a molecular weight of 50 and 30 kDa, respectively [45]. OmcZ contains eight groups of heme that consist of a wide range of redox potentials (−60 to −420 mV). As a result, OmcZ is critical for the transfer of electrons to electrodes and biofilm production. Immunogold studies showed that biofilms from *G. sulfurreduens* grow well with fumarate graphite as a result of high OmcZ production [46]. Furthermore, the deletion of OmcF genes reduced the existing productivity since OmcF participates in the transcription of several genes required in current processing in MFCs, as well as in exoelectrogen systems [47]. In addition to the c-type cytochromes of the external membrane, OmpC, OmpB, and multicopper proteins were found in *G. sulfurreducens.*

The proteins mentioned above are needed to reduce Fe(III) oxides because they consist of two dimensions. The first is type III fibronectin with an Fe(III) binding site [48]. Some Gram-positive bacteria, such as *T. potens*, have recently been reported to provide direct electron transfer to electrodes with the help of c-type cytochromes. *Desulfovibrio alaskensis* G20, a sulfate-reducing bacterium, is made up of a type I tetraheme cytochrome C_3 and transmembrane complexes, both which are needed for the mechanism of exoelectrogens [49].

2.6.3 Electron transportation via bacterial conductive pili

G. sulfurreducens was the first bacterium to be described in biofilm conduction [80]. Biofilm conduction involves a dense network of pili and their metallic characters. Protein type IV pili, which can be configured easily, are encoded by the gene PilA, and are 10–20 μm in length, 3–5 nm in width, and have 7–20 kDa molecular mass [50]. The pili have a conductivity that is similar to that of metal, which includes the following features: (i) increased conductivity with low temperature, which can be seen with metals; (ii) an increase in the number of protons will also increase pilus conduction as observed with polyaniline; and (iii) the PilA protein of *G. sulfurreducens* consists of an N-terminal that has an α-helix, and a C-terminal with a conserved amino acid chain (Phe, Tyr, Met, Trp, and His) which, as in metals, are aligned to provide π–π stacking [23, 51].

When nonsoluble electron acceptors such as Fe(III) and Mn(IV) are present, the pili can be expressed by *Geobacter metallireducens* but this does not occur in the presence of soluble acceptors of electrons, for example Fe(III) citrate. PilA gene deletion also ceased Fe(III) oxide reduction, while the reduction of Fe(III) citrate was still possible. These discoveries were supported with further research: (i) the pili have a unique relationship with Fe(III) oxides at the nanoscale; (ii) the PilA gene is only needed for the reduction of Fe(III) oxides and is not needed for the transfer of electrons; and (iii) *G. sulfurreducens* pili that provide long range transfer of electrons to Fe (III) oxides indicate that pili behave as microbial nanowires and, due to this, a thicker biofilm is produced by *G. sulfurreducens* on the anode than for other microbes [52, 53]. The factors mentioned previously give the most probable explanation that the *Geobacter* species create the largest densities of current compared to other types of microbes. *Geobacter sulfurreducens* also results in a linear relationship between the production of current density and conduction as a biofilm [53]. Current output and biofilm conduction were also linked in the strain KN400, which generated the most electricity. Gene deletion of PilA in KN400 stopped the Fe(III) oxide reduction, but after an incubation period of 1 month this strain adapted for the reduction of Fe(III) oxides with a low input of current, which was linked with the c-Cys Pgc A production [54]. Fluorescence microscopy of the enveloping membrane of *S. oneidensis* showed conductive extensions regarded as pili previously. The diameters of filaments range from 50 to 150 nm, and in certain cases they are larger compared to the cell itself. The filaments' diameter was decreased more than ten-fold to a value of 10 nm after they were dried for conduction [55, 56]. Subramanian *et al* [57] discovered that a similar type of cytochrome tunneling could occur in *S. oneidensis* in hydrated environments, however, the exact cytochromes' structure on the filaments and functions in the transfer of electrons remains unknown. According to Leung *et al* [58] the nanowires of *S. oneidensis* MR-1 display tunable, p-type electronic behavior for extracellular transfer of electrons.

A lot of exoelectrogens also create conductive filaments, such as the thermophilic fermentative bacterium *Pelotomaculum thermopropionicum*, strains of *Synechocystis*, and *Methanothermobacter thermautotrophicus* [23].

2.7 The bacterial community as a biocatalyst in MFCs

2.7.1 Anodic bacterial biofilm

In MFCs, bacterial species have received the most research attention. There exist categories of bacterial species in MFCs. *E. coli* and *Bacillus* sp. belong to the indirect MFCs. Here, the cell needs an external mediator as the microbes cannot directly give an electron to the anode. Examples of the next category are *Shewanella putrefaciens* and *Geobacter sulfurreducens* pure cultures, that have been used the most because they can donate the electrons to the anode directly [23]. These bacteria are called electroactive and are usually known as androphiles or exoelectrogens because they produce a conductive biofilm on the anode of the cell. A biocatalyst that can be utilized in MFCs at the anode chamber is a mixed culture. The formation of biofilm on the surface of the anode has been observed for different phylogenetic groups. Further research reported that, under anaerobic conditions, *Hansenula anomala* and *Sacchaomyces cerevisiae* yeast strains were able to generate electricity [4]. Various types of pure cultures that can generate current in MFCs can be found in table 2.2. The significant contributors to the bioanode are the dissimilatory sulfate-reducing bacteria (SRB) and metal-reducing bacteria (DMRBs). The removal of DMRBs can be achieved by using selective media consisting of either solid Mn (IV) or Fe (III) oxides as insoluble acceptors of electrons. When a conductive anode is absent in the natural condition, oxidized metal ions (Mn (IV) or Fe (III)) will be used as terminal acceptors of electrons for the continuous process of respiration. As such, this probably explains the mechanisms of extracellular electron transfer (EET) for both the anode and metal reduction, which are identified to be similar to each other. Ideally, bacterial isolation should be performed from the formation of biofilm on the anode in the MFC supplemented with an inoculum used as the seed [12].

Table 2.2. Utilization of pure cultures on the anode as biocatalysts for MFCs.

Biocatalysts	Fuel	Power per volume ($W\ m^{-3}$)	Power density ($mW\ m^{-3}$)	References
Rhodoferax ferrireducens	Glucose		33.4	[81]
Shewanella oneindensis	Lactate	500	3000	[82]
Escheria coli, acclimated	Complex medium		600	[83]
Clostridial isolate	Complex medium	5.62	–	[84]
Geobacter sulfurreducens	Acetate	0.356	13	[85]

2.7.2 Exoelectrogen isolation

With regard to electroactive direct isolation of electroactive bacteria (EABs) from the MFC anode, there are a few studies available. In MFCs a wide range of EABs and non-EABs can be found on the anode surface. One study reported the culture and isolation from anolyte suspension of an Fe(III)-reducing *Bacteroides* sp. W7 [86]. *Clostridium butyricum* EG3 is recognized as an obligatory anaerobe for MFCs processing wastewater anolyte that contains starch. However, it was found that direct interactions with the anode may not occur for all micro-organisms related to biofilm, but in the presence of an electron mediator released by members of the anodophiles, indirect interaction could occur. For example, *Brevibacillus* sp. PTH1 were found predominantly in the anode even though it is not capable of producing electricity on its own. This may be due to a compound excreted by *Pseudomonas* sp., such as phenoxine or pyocyanin, that allowed electron donation to the anode from *Brevibacillus* sp. PTH1 when it was cocultured with *Pseudomonas* sp. Without the coculture or supernatant with the *Pseudomonas* sp. in MFCs, the current produced was very low [12].

2.7.3 Mixed consortia

The simultaneous wastewater treatment and generation of energy using an MFC system for industrial applications requires the utilization of a bacterial culture consortium such as anaerobic digester sludge or activated sludge [87]. When it is adjusted to new climate conditions, MFC technology is anticipated to be selective and enhance its electrochemically active consortia from a mixed culture over a period of time. The substrate availability, conditions of operation, microbial communities, and system architectures of MFCs differ based on the inoculum source. Table 2.3 shows a variety of mixed bacterial consortia enriched from different sources in MFCs.

Table 2.3. Mixed culture bacterial species in MFCs.

Type of mixed culture	Organic substrate	Power density	Volumetric power density	References
Activated sludge	Glucose, glutamate	560	102	[88]
Activated sludge	Glucose	115.6	–	[83, 89]
	Landfill leachate	–	2.06	
	Glucose	3987	–	
	Landfill leachate	–	6.82	
Thermophilic effluent from anaerobic digestion of brewery wastewater	Acetate	1030	–	[90]
Granular anaerobic sludge	Sucrose	–	29.3	[91]
Domestic wastewater biofilm	Glucose	766–1540	51	[92, 93]
	Domestic wastewater	464	15.5	

2.8 Future perspectives

Overall, MFCs' potential to improve bioremediation and energy efficiency has been determined, but further achievements in future research will be needed for practical application in the field.

2.8.1 Modeling of the MFC set-up

Modeling MFCs will be possible based on comprehensive findings from biology, electrochemistry, geochemistry, and hydrology in deep ecosystems. Prior to field implementation, the finest MFC models will accurately predict energy output and bioremediation. When designing MFCs models, other factors, such as the depth of the water, the depth of the anode, and the distance between electrodes, will need to be taken into consideration. If the separation between the electrodes increases, it will also increase the ohmic losses. As constant space electrodes are used, a floating MFC model could be an option for the future to minimize ohmic losses. Increasing anode depth and adding more water increases ohmic losses, and at different depths, only certain micro-organisms are active. As a consequence, determination of the depth of the anode should be done on the site [23].

2.8.2 Sensing and monitoring

A heterogeneous and deep ecosystem is the biggest area of concern in gathering more knowledge regarding the interaction between microbial ecology and MFC electrodes and other elements; thus more regulated conditions can be created. The area-sensitive monitoring of microbial community behaviors, along with structure and growth parameters at sequential scales, may be part of this environment [4]. Some molecular biology techniques that can be used to learn more about the microbial consortia of MFCs include deposition-fluorescence *in situ* hybridization, metagenomics, and proteomic analysis.

2.8.3 Electrode materials for MFCs

Since electrodes act as acceptors of electrons and also affect the development of exoelectrogens, the substance used for the electrodes is crucial. As a result, the electrodes must resist corrosion and biodegradation, possess a wide contact area, have a large porosity, be physically durable, be naturally nonfouling, have excellent electron discharging performance, and be practical and biocompatible. Owing to their high mechanical strength, the most often used electrodes are rods made of stainless steel. However, they have a poor specific surface area and corrode quickly in anoxic conditions. Therefore, carbon-based materials can be appealing for utilization as electrodes [5]. Carbon mesh, reticulated vitreous carbon, graphite fiber brushes, graphite plates, carbon cloth electrodes, graphite flakes, carbon paper, graphite foils, and graphite felts can be utilized as cathodes. However, they cannot be utilized as anodes because burying graphite plates in sediments is difficult and expensive. Compared to other graphite plates, graphite rods are easier to insert into sediments. Carbon or graphite can be used in a variety of ways, including fibers, foam, granules, felt, and nanotubes to maximize the anodic surface area. As H_2 gas is emitted in the chamber of the cathode, the rate of the oxidation reduction reaction should be increased by the catalyst. On the cathode, the catalysts used can be polyaniline and platinum. Graphite rods and felts have lower strength than carbon electrodes with platinum coatings. A conductive polymer containing carbon nanotubes can also be efficient as an electrode material. Because of their synergistic relationship with the microbial population, NR-woven, Fe^{3+}-woven, and Mn (IV)-woven graphite could be excellent electrode materials in the future [7].

2.8.4 The concept of electromicrobiology

Electromicrobiology presents a significant opportunity for the future. Since little is known about the mechanisms, further research into how mixed cultures or monoculture bacteria pass electrons to electrodes and from electrodes to cells is necessary. There are a lot of species of bacteria that have been identified in MFCs, such as *Desulfocapsa*, Epsilonproteobacteria, *Syntrophus*, Gammaproteobacteria, *Desulfuromonas acetoxidans*, and *Alcaligenes faecalis*. More analysis has been conducted on *S. oneidensis*, *G. sulfurreducens*, and *T. potens* in earlier studies as their mechanisms aided their development and these bacteria have different electron transport routes than most micro-organisms [23]. The conductivities of the pili and filaments of various micro-organisms should also be studied further [6]. Since the time it takes to move electrons to electron shuttles from the electrodes and then to the micro-organisms would be reduced, a mechanism for direct electron transfer to bacteria from electrodes is being developed, resulting in a very quick reduction reaction of inorganic and organic compounds. In addition, thick biofilms hinder the transfer of electrons to the electrodes, thus research into the function and formation of biofilms in a bio-catalyzed system is needed to better understand electron transfer processes and optimize bio-electrogenic operation. It is essential to discover the ideal electroactive biofilms as they will be beneficial for a variety of uses, including bioremediation, biosensors, metal removal, biocorrosion, and possibly the human–microbe interface.

2.8.5 Scaling up

There are no records of MFCs generating power at the watt level owing to the high cost and time requirements of microcosm analyses. However, it is sensible to investigate how microbes evolve and are continuous to develop MFCs. The manipulation of optimum MFCs is extremely challenging, due to the rapid changes of marine microbiology and ecological factors; however, promising findings have been reported in laboratories. The poor energy production can be resolved by using a control scheme that includes a discharge and cyclic charge of a capacitor, as well as increasing the output. The surface area of the MFCs can be increased by joining several electrodes in series or by joining various MFCs in cycles. Ewing *et al* [94] used multiple electrodes in MFCs with a power control scheme, and the electrodes in this system were electronically isolated. Later, Ewing *et al* [95] used 3D-micro-emulsion-doped anodes to study the long-term efficiency of MFCs that slowly discharge lactate and its fermented derivatives. According to Abazarian *et al* [60], three MFCs in a stacking series outperformed four MFCs in a stacking series in terms of current density. The generation of electricity was found to be based on the configuration of MFCs and the number of cells used in these experiments, which may be beneficial for future research. Scaled-up MFCs and different numbers of cells will be the subject of future study. Upcoming scaled-up MFC studies will place a greater emphasis on factors such as material selection, reactor configuration, reliability, operating mode, and economic and environmental sustainability. When scaling up MFCs, the electrode material chosen has a significant impact on the interactions between bacteria, electrodes, and pollutants, as well as the system's cost and long-term reliability. The selection of appropriate MFC configuration sets, as well as their potential synergistic uses with other remediation techniques, would be significant aims for optimizing MFCs.

2.9 Conclusions

MFCs have triggered a field of study that can be demonstrated to be ecologically balanced, controllable, and practical for electricity production. The mechanism of electron transfer in bacterial species has primarily been studied in *Geobacter* spp. and *Shewanella* spp. Further studies into the mechanisms of electron transfer are required. In exoelectrogens and electrotrophs, genetic modification may also increase electron transfer or receiving efficiency. There is currently no optimal MFC model for power production. MFC research has so far been conducted on a laboratory scale. Researchers face several challenges, such as including the disruption of MFCs by the flow of current, electrode electrochemical performance, MFC settings, and the entire electrochemical reaction performance. Thus, joint efforts by researchers in a number of areas, such as environmental sciences, electrochemistry, biotechnology, materials science, and microbiology, can address the modern complexities of MFCs.

Conflicts of interest

The authors declare that they have no known competing financial interests or personal relationships that could have appeared to influence the work reported in this paper.

Acknowledgements

This research article was supported by Universiti Sains Malaysia, Malaysia.

References

[1] Bose B K 2010 Global warming: energy, environmental pollution, and the impact of power electronics *IEEE Ind. Electron. Mag.* **4** 6–17

[2] Martins F, Felgueiras C, Smitkova M and Caetano N 2019 Analysis of fossil fuel energy consumption and environmental impacts in European countries *Energies* **12** 964

[3] Nicoletti G, Arcuri N, Nicoletti G and Bruno R 2015 A technical and environmental comparison between hydrogen and some fossil fuels *Energy Convers. Manag.* **89** 205–13

[4] Yaqoob A A, Khatoon A, Mohd Setapar S H, Umar K, Parveen T, Mohamad Ibrahim M N, Ahmad A and Rafatullah M 2020 Outlook on the role of microbial fuel cells in remediation of environmental pollutants with electricity generation *Catalysts* **10** 819

[5] Yaqoob A A, Ibrahim M N M and Rodríguez-Couto S 2020 Development and modification of materials to build cost-effective anodes for microbial fuel cells (MFCs): an overview *Biochem. Eng. J.* **164** 107779

[6] Yaqoob A A, Ibrahim M N M, Yaakop A S, Umar K and Ahmad A 2020 Modified graphene oxide anode: a bioinspired waste material for bioremediation of Pb^{2+} with energy generation through microbial fuel cells *Chem. Eng. J.* **417** 128052

[7] Yaqoob A A, Mohamad Ibrahim M N, Rafatullah M, Chua Y S, Ahmad A and Umar K 2020 Recent advances in anodes for microbial fuel cells: an overview *Materials* **13** 2078

[8] Pandit S and Das D 2018 Principles of microbial fuel cell for the power generation *Microbial Fuel Cell* (Berlin: Springer) pp 21–41

[9] Hemalatha M, Sravan J S and Mohan S V 2020 Self-induced bioelectro-potential influence on sulfate removal and desalination in microbial fuel cell *Bioresour. Technol.* **309** 123326

[10] Li M, Zhang H, Xiao T, Wang S, Zhang B, Chen D, Su M and Tang J 2018 Low-cost biochar derived from corncob as oxygen reduction catalyst in air cathode microbial fuel cells *Electrochim. Acta* **283** 780–8

[11] Das I, Das S and Ghangrekar M 2020 Application of bimetallic low-cost CuZn as oxygen reduction cathode catalyst in lab-scale and field-scale microbial fuel cell *Chem. Phys. Lett.* **751** 137536

[12] Pandit S, Chandrasekhar K, Kakarla R, Kadier A and Jeevitha V 2017 Basic principles of microbial fuel cell: technical challenges and economic feasibility *Microbial Applications* vol 1 (Berlin: Springer) pp 165–88

[13] Santoro C, Arbizzani C, Erable B and Ieropoulos I 2017 Microbial fuel cells: from fundamentals to applications. a review *J. Power Sources* **356** 225–44

[14] Yaqoob A A, Ibrahim M N M, Yaakop A S and Ahmad A 2021 Application of microbial fuel cells energized by oil palm trunk sap (OPTS) to remove the toxic metal from synthetic wastewater with generation of electricity *Appl. Nanosci.* **11** 194–61

[15] Logan B E 2008 *Microbial Fuel Cells* (New York: Wiley)

[16] Logan B E, Hamelers B, Rozendal R, Schröder U, Keller J, Freguia S, Aelterman P, Verstraete W and Rabaey K 2006 Microbial fuel cells: methodology and technology *Env. Sci. Technol.* **40** 5181–92

[17] Rabaey K and Verstraete W 2005 Microbial fuel cells: novel biotechnology for energy generation *Trends Biotechnol.* **23** 291–8

[18] Varanasi J L, Nayak A K, Sohn Y, Pradhan D and Das D 2016 Improvement of power generation of microbial fuel cell by integrating tungsten oxide electrocatalyst with pure or mixed culture biocatalysts *Electrochim. Acta* **199** 154–63

[19] Khilari S, Pandit S, Ghangrekar M, Das D and Pradhan D 2013 Graphene supported α-MnO_2 nanotubes as a cathode catalyst for improved power generation and wastewater treatment in single-chambered microbial fuel cells *RSC Adv.* **3** 7902–11

[20] Bond D R 2010 Electrodes as electron acceptors, and the bacteria who love them *Geomicrobiology: Molecular and Environmental Perspective* (Berlin: Springer) pp 385–99

[21] Kumar R, Singh L, Wahid Z A and Din M F M 2015 Exoelectrogens in microbial fuel cells toward bioelectricity generation: a review *Int. J. Energy Res.* **39** 1048–67

[22] Gorby Y A, Yanina S, McLean J S, Rosso K M, Moyles D, Dohnalkova A, Beveridge T J, Chang I S, Kim B H and Kim K S 2006 Electrically conductive bacterial nanowires

produced by *Shewanella oneidensis* strain MR-1 and other microorganisms *Proc. Natl Acad. Sci.* **103** 11358–63
[23] Abbas S Z, Rafatullah M, Ismail N and Syakir M I 2017 A review on sediment microbial fuel cells as a new source of sustainable energy and heavy metal remediation: mechanisms and future prospective *Int. J. Energy Res.* **41** 1242–64
[24] Chuo S C, Mohamed S F, Mohd Setapar S H, Ahmad A, Jawaid M, Wani W A, Yaqoob A A and Mohamad Ibrahim M N 2020 Insights into the current trends in the utilization of bacteria for microbially induced calcium carbonate precipitation *Materials* **13** 4993
[25] Yaqoob A A, Mohamad Ibrahim M N, Umar K, Bhawani S A, Khan A, Asiri A M, Khan M R, Azam M and AlAmmari A M 2021 Cellulose derived graphene/polyaniline nanocomposite anode for energy generation and bioremediation of toxic metals via benthic microbial fuel cells *Polymers* **13** 135
[26] Puntel R L, Nogueira C W and Rocha J B 2005 Krebs cycle intermediates modulate thiobarbituric acid reactive species (TBARS) production in rat brain *in vitro Neurochem. Res.* **30** 225–35
[27] Kalathil S and Pant D 2016 Nanotechnology to rescue bacterial bidirectional extracellular electron transfer in bioelectrochemical systems *RSC Adv.* **6** 30582–97
[28] Titov D V, Cracan V, Goodman R P, Peng J, Grabarek Z and Mootha V K 2016 Complementation of mitochondrial electron transport chain by manipulation of the NAD +/NADH ratio *Science* **352** 231–5
[29] Bond D R and Lovley D R 2005 Evidence for involvement of an electron shuttle in electricity generation by geothrix fermentans *Appl. Environ. Microbiol.* **71** 2186–9
[30] Liu B, Williams I, Li Y, Wang L, Bagtzoglou A, McCutcheon J and Li B 2016 Towards high power output of scaled-up benthic microbial fuel cells (BMFCs) using multiple electron collectors *Biosens. Bioelectron.* **79** 435–41
[31] Wang H, Song H, Yu R, Cao X, Fang Z and Li X 2016 New process for copper migration by bioelectricity generation in soil microbial fuel cells *Env. Sci. Pollut. Res.* **23** 13147–54
[32] Bond D R, Holmes D E, Tender L M and Lovley D R 2002 Electrode-reducing microorganisms that harvest energy from marine sediments *Science* **295** 483–5
[33] Arunasri K, Modestra J A, Yeruva D K, Krishna K V and Mohan S V 2016 Polarized potential and electrode materials implication on electro-fermentative di-hydrogen production: microbial assemblages and hydrogenase gene copy variation *Bioresour. Technol.* **200** 691–8
[34] Von Canstein H, Ogawa J, Shimizu S and Lloyd J R 2008 Secretion of flavins by *Shewanella* species and their role in extracellular electron transfer *Appl. Env. Microbiol.* **74** 615–23
[35] Reiche A, Sivell J-l and Kirkwood K M 2016 Electricity generation by *Propionibacterium freudenreichii* in a mediatorless microbial fuel cell *Biotechnol. Lett.* **38** 51–5
[36] Xu S, Jangir Y and El-Naggar M Y 2016 Disentangling the roles of free and cytochrome-bound flavins in extracellular electron transport from *Shewanella oneidensis* MR-1 *Electrochim. Acta* **198** 49–55
[37] Xu Y-S, Zheng T, Yong X-Y, Zhai D-D, Si R-W, Li B, Yu Y-Y and Yong Y-C 2016 Trace heavy metal ions promoted extracellular electron transfer and power generation by *Shewanella* in microbial fuel cells *Bioresour. Technol.* **211** 542–7

[38] Yang Y, Ding Y, Hu Y, Cao B, Rice S A, Kjelleberg S and Song H 2015 Enhancing bidirectional electron transfer of *Shewanella oneidensis* by a synthetic flavin pathway *ACS Synth. Biol.* **4** 815–23

[39] Sanchez D V, Jacobs D, Gregory K, Huang J, Hu Y, Vidic R and Yun M 2015 Changes in carbon electrode morphology affect microbial fuel cell performance with *Shewanella oneidensis* MR-1 *Energies* **8** 1817–29

[40] Magazù S, Migliardo F and Ramirez-Cuesta A 2005 Inelastic neutron scattering study on bioprotectant systems *J. R. Soc. Interface* **2** 527–32

[41] Wei T, Ma H and Nakano A 2016 Decaheme cytochrome MtrF adsorption and electron transfer on gold surface *J. Phys. Chem. Lett.* **7** 929–36

[42] Pirbadian S 2015 Bacterial nanowires of *Shewanella oneidensis* MR-1: electron transport mechanism, composition, and role of multiheme cytochromes *Thesis* University of Southern California

[43] Chenot C C, Robiette R L and Collin S 2019 First evidence of the cysteine and glutathione conjugates of 3-sulfanylpentan-1-ol in hop (*Humulus lupulus* L.) *J. Agric. Food Chem.* **67** 4002–10

[44] Yan Z, Jiang H, Cai H, Zhou Y and Krumholz L R 2015 Complex interactions between the macrophyte *Acorus calamus* and microbial fuel cells during pyrene and benzo[a]pyrene degradation in sediments *Sci. Rep.* **5** 1–11

[45] Estevez-Canales M, Kuzume A, Borjas Z, Füeg M, Lovley D, Wandlowski T and Esteve-Núñez A 2015 A severe reduction in the cytochrome C content of *Geobacter sulfurreducens* eliminates its capacity for extracellular electron transfer *Env. Microbiol. Rep.* **7** 219–26

[46] Zhang B, Tian C, Liu Y, Hao L, Liu Y, Feng C, Liu Y and Wang Z 2015 Simultaneous microbial and electrochemical reductions of vanadium (V) with bioelectricity generation in microbial fuel cells *Bioresour. Technol.* **179** 91–7

[47] Kokhan O, Ponomarenko N S, Pokkuluri P R, Schiffer M, Mulfort K L and Tiede D M 2015 Bidirectional photoinduced electron transfer in ruthenium (II)-tris-bipyridyl-modified PpcA, a multi-heme c-type cytochrome from *Geobacter sulfurreducens J. Phys. Chem.* B **119** 7612–24

[48] Alves M N, Fernandes A P, Salgueiro C A and Paquete C M 2016 Unraveling the electron transfer processes of a nanowire protein from *Geobacter sulfurreducens Biochim. Biophys. Acta Bioener.* **1857** 7–13

[49] Santos T C, Silva M A, Morgado L, Dantas J M and Salgueiro C A 2015 Diving into the redox properties of *Geobacter sulfurreducens* cytochromes: a model for extracellular electron transfer *Dalton Trans.* **44** 9335–44

[50] Hernández-Eligio A, Andrade Á, Soto L, Morett E and Juárez K 2017 The unphosphorylated form of the PilR two-component system regulates pilA gene expression in *Geobacter sulfurreducens Env. Sci. Pollut. Res.* **24** 25693–701

[51] Yaqoob A A, Serrá A, Bhawani S A, Ibrahim M N M, Khan A, Alorfi H S, Asiri A M, Hussein M A, Khan I and Umar K 2022 Utilizing biomass-based graphene oxide-polyaniline-Ag electrodes in microbial fuel cells to boost energy generation and heavy metal removal *Polymers* **14** 845–59

[52] Nair R R, Silveira C M, Diniz M S, Almeida M G, Moura J J and Rivas M G 2015 Changes in metabolic pathways of *Desulfovibrio alaskensis* G20 cells induced by molybdate excess *J. Biol. Inorg. Chem.* **20** 311–22

[53] Ghoreishi K B, Ghasemi M, Rahimnejad M, Yarmo M A, Daud W R W, Asim N and Ismail M 2014 Development and application of vanadium oxide/polyaniline composite as a novel cathode catalyst in microbial fuel cell *Int. J. Energy Res.* **38** 70–7
[54] Adhikari R Y, Malvankar N S, Tuominen M T and Lovley D R 2016 Conductivity of individual *Geobacter pili RSC Adv.* **6** 8354–7
[55] Liu X, Tremblay P-L, Malvankar N S, Nevin K P, Lovley D R and Vargas M A 2014 *Geobacter sulfurreducens* strain expressing *Pseudomonas aeruginosa* type IV pili localizes OmcS on pili but is deficient in Fe (III) oxide reduction and current production *Appl. Env. Microbiol.* **80** 1219–24
[56] Pirbadian S, Barchinger S E, Leung K M, Byun H S, Jangir Y, Bouhenni R A, Reed S B, Romine M F, Saffarini D A and Shi L 2014 *Shewanella oneidensis* MR-1 nanowires are outer membrane and periplasmic extensions of the extracellular electron transport components *Proc. Natl Acad. Sci.* **111** 12883–8
[57] Subramanian P, Pirbadian S, El-Naggar M Y and Jensen G J 2016 Electron cryo-tomography of nanowires in *Shewanella oneidensis* MR-1 *Biophys. J.* **110** 23a
[58] Leung K M, Wanger G, El-Naggar M Y, Gorby Y, Southam G, Lau W M and Yang J 2013 *Shewanella oneidensis* MR-1 bacterial nanowires exhibit p-type, tunable electronic behavior *Nano Lett.* **13** 2407–11
[59] Lovley D R 2011 Powering microbes with electricity: direct electron transfer from electrodes to microbes *Env. Microbiol. Rep.* **3** 27–35
[60] Abazarian E, Gheshlaghi R and Mahdavi M A 2016 The effect of number and configuration of sediment microbial fuel cells on their performance in an open channel architecture *J. Power Sources* **325** 739–44
[61] Nevin K P, Kim B-C, Glaven R H, Johnson J P, Woodard T L, Methé B A, DiDonato R J Jr, Covalla S F, Franks A E and Liu A 2009 Anode biofilm transcriptomics reveals outer surface components essential for high density current production in *Geobacter sulfurreducens* fuel cells *PLoS One* **4** e5628
[62] Ringeisen B R, Ray R and Little B 2007 A miniature microbial fuel cell operating with an aerobic anode chamber *J. Power Sources* **165** 591–7
[63] Cordas C M, Guerra L T, Xavier C and Moura J J 2008 Electroactive biofilms of sulphate reducing bacteria *Electrochim. Acta* **54** 29–34
[64] Reimers C E, Tender L M, Fertig S and Wang W 2001 Harvesting energy from the marine sediment–water interface *Env. Sci. Technol.* **35** 192–5
[65] Logan B E 2005 Simultaneous wastewater treatment and biological electricity generation *Water Sci. Technol.* **52** 31–7
[66] Keller K L, Rapp-Giles B J, Semkiw E S, Porat I, Brown S D and Wall J D 2014 New model for electron flow for sulfate reduction in *Desulfovibrio alaskensis* G20 *Appl. Env. Microbiol.* **80** 855–68
[67] Qiao Y, Li C M, Bao S-J, Lu Z and Hong Y 2008 Direct electrochemistry and electrocatalytic mechanism of evolved *Escherichia coli* cells in microbial fuel cells *Chem. Commun.* **2008** 1290–2
[68] Marshall C W and May H D 2009 Electrochemical evidence of direct electrode reduction by a thermophilic gram-positive bacterium, *Thermincola ferriacetica Energy Env. Sci.* **2** 699–705
[69] Carmona-Martínez A A, Harnisch F, Kuhlicke U, Neu T R and Schröder U 2013 Electron transfer and biofilm formation of *Shewanella putrefaciens* as function of anode potential *Bioelectrochemistry* **93** 23–9

[70] Thrash J C, Van Trump J I, Weber K A, Miller E, Achenbach L A and Coates J D 2007 Electrochemical stimulation of microbial perchlorate reduction *Env. Sci. Technol.* **41** 1740–6
[71] Powell E E, Mapiour M L, Evitts R W and Hill G A 2009 Growth kinetics of *Chlorella vulgaris* and its use as a cathodic half cell *Bioresour. Technol.* **100** 269–74
[72] Xing D, Zuo Y, Cheng S, Regan J M and Logan B E 2008 Electricity generation by *Rhodopseudomonas palustris* DX-1 *Env. Sci. Technol.* **42** 4146–51
[73] Nguyen T A, Lu Y, Yang X and Shi X 2007 Carbon and steel surfaces modified by *Leptothrix discophora* SP-6: characterization and implications *Env. Sci. Technol.* **41** 7987–96
[74] Raghavulu S V, Goud R K, Sarma P and Mohan S V 2011 *Saccharomyces cerevisiae* as anodic biocatalyst for power generation in biofuel cell: influence of redox condition and substrate load *Bioresour. Technol.* **102** 2751–7
[75] Erable B, Vandecandelaere I, Faimali M, Delia M-L, Etcheverry L, Vandamme P and Bergel A 2010 Marine aerobic biofilm as biocathode catalyst *Bioelectrochemistry* **78** 51–6
[76] Cournet A, Délia M-L, Bergel A, Roques C and Bergé M 2010 Electrochemical reduction of oxygen catalyzed by a wide range of bacteria including gram-positive *Electrochem. Commun.* **12** 505–8
[77] Xu S and Liu H 2011 New exoelectrogen *Citrobacter* sp. SX-1 isolated from a microbial fuel cell *J. Appl. Microbiol.* **111** 1108–15
[78] Wu S J, Hu Z H, Zhang L L, Yu X and Chen J M 2009 A novel dichloromethane-degrading *Lysinibacillus sphaericus* strain wh22 and its degradative plasmid *Appl. Microbiol. Biotechnol.* **82** 731–40
[79] Inglesby A E, Beatty D A and Fisher A C 2012 *Rhodopseudomonas palustris* purple bacteria fed *Arthrospira maxima* cyanobacteria: demonstration of application in microbial fuel cells *RSC Adv.* **2** 4829–38
[80] Wu C, Liu X-W, Li W-W, Sheng G-P, Zang G-L, Cheng Y-Y, Shen N, Yang Y-P and Yu H-Q 2012 A white-rot fungus is used as a biocathode to improve electricity production of a microbial fuel cell *Appl. Energy* **98** 594–6
[81] Chaudhuri S K and Lovley D R 2003 Electricity generation by direct oxidation of glucose in mediatorless microbial fuel cells *Nat. Biotechnol.* **21** 1229–32
[82] Ringeisen B R, Henderson E, Wu P K, Pietron J, Ray R, Little B, Biffinger J C and Jones-Meehan J M 2006 High power density from a miniature microbial fuel cell using *Shewanella oneidensis* DSP10 *Env. Sci. Technol.* **40** 2629–34
[83] You S, Zhao Q, Zhang J, Jiang J and Zhao S 2006 A microbial fuel cell using permanganate as the cathodic electron acceptor *J. Power Sources* **162** 1409–15
[84] Prasad D, Sivaram T, Berchmans S and Yegnaraman V 2006 Microbial fuel cell constructed with a micro-organism isolated from sugar industry effluent *J. Power Sources* **160** 991–6
[85] Bond D R and Lovley D R 2003 Electricity production by *Geobacter sulfurreducens* attached to electrodes *Appl. Env. Microbiol.* **69** 1548–55
[86] Wang H and Ren Z J 2013 A comprehensive review of microbial electrochemical systems as a platform technology *Biotechnol. Adv.* **31** 1796–807
[87] Li Y, Marshall A and Gostomski P A 2014 Gaseous pollutant treatment and electricity generation in microbial fuel cells (MFCs) utilising redox mediators *Rev. Env. Sci. Biol Technol.* **13** 35–51
[88] Moon H, Chang I S and Kim B H 2006 Continuous electricity production from artificial wastewater using a mediator-less microbial fuel cell *Bioresour. Technol.* **97** 621–7

[89] You S J, Zhao Q L, Jiang J Q, Zhang J N and Zhao S Q 2010 Sustainable approach for leachate treatment: electricity generation in microbial fuel cell *J. Env. Sci. Health* A **41** 2721–34

[90] Jong B C, Kim B H, Chang I S, Liew P W Y, Choo Y F and Kang G S 2006 Enrichment, performance, and microbial diversity of a thermophilic mediatorless microbial fuel cell *Env. Sci. Technol.* **40** 6449–54

[91] He Z, Wagner N, Minteer S D and Angenent L T 2006 An upflow microbial fuel cell with an interior cathode: assessment of the internal resistance by impedance spectroscopy *Env. Sci. Technol.* **40** 5212–7

[92] Cheng S, Liu H and Logan B E 2006 Increased performance of single-chamber microbial fuel cells using an improved cathode structure *Electrochem. Commun.* **8** 489–94

[93] Cheng S, Liu H and Logan B E 2006 Increased power generation in a continuous flow MFC with advective flow through the porous anode and reduced electrode spacing *Env. Sci. Technol.* **40** 2426–32

[94] Ewing T, Ha P T, Babauta J T, Tang N T, Heo D and Beyenal H 2014 Scale-up of sediment microbial fuel cells *J. Power Sources* **272** 311–9

[95] Ewing T, Ha P T and Beyenal H 2017 Evaluation of long-term performance of sediment microbial fuel cells and the role of natural resources *Appl. Energy* **192** 490–7

IOP Publishing

Microbial Fuel Cells
Emerging trends in electrochemical applications
Mohamad Nasir Mohamad Ibrahim, Asim Ali Yaqoob and Akil Ahmad

Chapter 3

Electrochemical characterization of microbial fuel cells

Claudia Guerrero Barajas, María del Carmen Monterrubio Badillo and Elsa M Arce Estrada

Microbial fuel cells (MFCs) are a biotechnology to which considerable research efforts have been devoted to reach the stage at which it can be broadly applied at large scale, particularly in wastewater treatment. Due to the complex nature of the bioprocesses that take place in these engineered systems, it is necessary to know in depth the methods and techniques that better serve to analyse the performance of MFCs. In addition to the basis of thermodynamics and microbiology that inevitably need to be considered in the process, it is important to know the electrochemical techniques that can be utilized to elucidate the phenomena that occur in an MFC. As has been experienced within the scientific community devoted to the study of MFCs, collecting different efforts and expertise is useful in order to select the appropriate electrochemical technique, achieve a comprehensive analysis of the data obtained from it, and determine the correct interpretation and implications. The current chapter presents a description of the most utilized electrochemical techniques in MFC analysis and recommends the most recent research and review documents that provide excellent examples of the application and interpretation of data acquired with these techniques.

3.1 General aspects of microbial fuel cells

Research on microbial fuel cells (MFCs) requires the in-depth involvement of several disciplines. In order to understand these systems, an integrated approach has been required that includes experts in the following areas: environmental, biochemical, biotechnological, and chemical engineering, microbiology, mathematics (to solve the system equations describing the transport phenomena in MFCs and to accurately describe the mathematical modeling), and electrochemistry.

doi:10.1088/978-0-7503-4791-4ch3

The latter, along with a deep knowledge of the biological processes in which the redox reactions are carried out, is known as bioelectrochemistry. To date, the journey of MFCs has undergone several stages, from the laboratory scale, of a few milliliter (100–2000 ml) single or dual chamber MFCs, assembled as a single unit or in stacked arrangements [1], to the large scale, i.e. 1500 l [2, 3], well controlled and adjusted processes (i.e. bioelectrochemical toilet MFCs). In both the laboratory and large-scale MFCs, wastewater treatment has been assayed, focusing on optimizing the start-up process of the system to treat brewery wastewater [4], for example, and to a large scale to treat human waste, as described in extensive reviews on the applications and scaling up of MFCs [2, 3]. However, MFCs are a technology that is still being investigated to reach a wider large-scale application, particularly in wastewater treatment in regard to the removal of chemical oxygen demand (COD), nitrogen, phosphate, and sulfide. Recent literature surveys demonstrate that ongoing research into this technology is being conducted to further extend the field of application of MFCs. This includes the removal of metals such as Cu^{2+}, Cd^{2+}, Pb^{2+}, U^{6+}, Cr^{6+}, Hg^{2+}, Zn^{2+}, Ag^{+}, Co^{2+}, and Au^{3+} [5, 6] and biodegradation of organic pollutants such as BTEX [7], phenol, nitro and chlorophenols [8, 9], and azo dyes [5, 10, 11]. Figure 3.1 presents a schematic of a MFC and the main applications of this technology that have been studied so far.

It has been stated that in an MFC the main protagonist is the microbial community utilized to process the organic matter or organic compound, that could be a toxic water pollutant, and therefore catalyze the electron transfer process. Pure strains and consortia of bacteria belonging to the domains Alphaproteobacteria (genera *Acidiphilum* and *Ochrobactrum*), Betaproteobacteria (genera *Rhodoferax* and *Comamonas*), Deltaproteobacteria (genera *Geobacter* and *Desulforomonas*), Epsilonproteobacteria (i.e. the genus *Arcobacter*), Gammaproteobacteria (genera *Shewanella*, *Klebsiella*, *Citrobacter*, and *Pseudomonas*), and Firmicutes have been

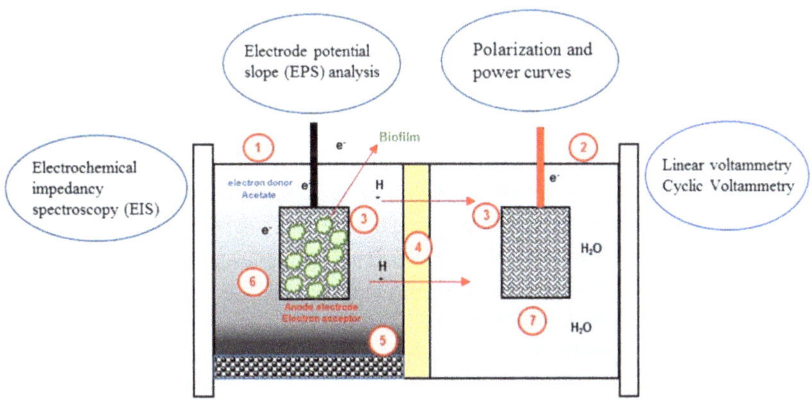

Figure 3.1. Electrochemical techniques utilized to characterize MFCs. Example of an MFC. 1: Anodic chamber (anaerobic); 2: cathodic chamber (aerobic); 3: graphite cloth electrodes; 4: cation exchange membrane; 5: electrogenic bacteria; 6: mineral medium; 7: distilled water. The redox reactions that occur in the MFC are as follows:

$$CH_3COO^- + 4H_2O \rightarrow 2HCO_3^- + 9H^+ + 8\ e^- \quad \text{anode, acetate oxidation reaction}$$
$$8e^- + 8H^+ + 2O_2 \rightarrow 4H_2O \quad \text{cathode, oxygen reduction reaction.}$$

identified as electrogenic and have been utilized to generate energy in MFCs. Some Archaea and Eukarya have also been shown to exhibit the electrogenic feature to be utilized as a catalyst in MFCs [12]. The active biofilm formed by the bacteria on the electrodes of the MFC can be formed either on the anode or the cathode depending on the purpose of the MFC. In an anodic chamber, the anode, to which the biofilm is attached (bioanode) can promote the oxidization of the organic matter from, for example, a wastewater stream. In the case of a biocathode, particularly an anaerobic biocathode in which the electron acceptors can be sulfate (SO_4^{2-}), nitrate (NO_3^-), manganese (Mn^{4+}), CO_2, etc, the microbial communities attached to it have been investigated for their capability of catalyzing the reduction of either inorganic (such as the ions just mentioned) or organic compounds, with the purpose of obtaining compounds of industrial interest. For example, the bioelectrochemical reduction of acetate and butyrate has been reported in a biocathode on which a biofilm of sulfate reducing bacteria was the biological catalyst [13, 14]. As a biocatalyst, electrogenic bacteria can enhance the effectiveness of electron transfer and interactions with the corresponding electrolyte in the MFC chambers through an electrode that combines a suitable shape, size (effective active area), and material. Therefore, the second protagonist in the MFC that is frequently the focus of attention is the electrode. Electrodes play an important role in MFCs and a suitable design will contribute to the improvement of the biofilm–electrode interaction by promoting a better attachment of the biofilm and also by promoting an efficient interfacial transport of electrons, that is, the charge transfer reaction that occurs between the interface of the electrode and the electrolyte [15]. Advancements in electrode design have been focused on reducing the cost while increasing the efficacy. The efforts to improve the efficiency of the anodes have considered the following aspects: (i) material surface area; (ii) biocompatibility of the material with the bacteria that will form the biofilm; (iii) electrical conductivity; (iv) material stability and durability; and (v) material availability and cost [16]. The advancements in research regarding anode materials have been notable, and anode materials can now be classified as conventional materials, which are carbon based materials, i.e. carbon paper, cloth, felt, brushes, etc, and modern anode materials, which can be natural source based carbon anodes, which are made of natural waste from sugarcane, coconut shell, corn stem, and bamboo charcoal, just to mention a few, or composite based material anodes, which can be made of metal/metal oxides, and combinations of carbon based materials with conductive polymers, i.e. MnO_2–carbon felt, PANI–carbon cloth, Fe_3O_4–carbon cloth, among several others [16]. In designing the cathodes it is important to consider the catalyst (i.e. platinum or ferricyanide) and the material that will be used to increase the rate of the oxygen reduction, in the case of an aerobic biocathode (i.e. graphene based cathodes) [17] or the combination of photosensitive enzymes with nanocatalysts to enhance the reduction efficiency of CO_2, in an anaerobic biocathode as recently reported by Sahoo *et al* [18].

In order to take advantage of the MFCs and other types of bioelectrochemical systems, for example microbial electrolysis cells (MECs), a new approach is starting to couple the MFC system to an MEC to take advantage of the energy generated in the MFC as a supply of energy for the MEC to promote the generation of hydrogen without any other external power source other than that provided by the MFC [19]. Another approach to coupling an MFC to a wastewater treatment process has been assembling constructed wetlands to an MFC configuration in a variety of designs that have been recently reviewed by Gupta *et al* [20]. As the approaches for the study of MFC applications become more sophisticated, the evaluation of the different processes that take place in these systems, such as the redox reactions, sustainability of the biofilms formed, side reactions among components of the electrolytes and bacteria, and any other physical or chemical processes that are involved in these systems, needs to become more detailed. Therefore, it is of utmost importance to implement multidisciplinary participation in this research area, otherwise a proper electrochemical characterization of these systems may be unaffordable in terms of time and economic investment. As in the description of the microbial communities involved in these systems, the molecular biology techniques that best suit the analysis are needed and require skills that range from DNA extraction to bioinformatics, the input of an expert or experts on electrochemical analysis is useful to reach a better understanding of the electrochemistry of the MFC in a range that covers knowledge of the foundations of the electrochemical analysis technique, from the theoretical reactions and theoretical calculations to the skills to properly manage the equipment and achieve an accurate analysis of the data obtained, which is not always a straightforward endeavor.

The current chapter presents a description of the most utilized electrochemical techniques to characterize MFCs and suggests some of the most relevant and, in most cases, recent reviews on MFCs with updates on the application of the electrochemical analytical techniques and research documents in which these techniques have been used, all of them used as representative examples of the utilization of these analytical procedures. It is highly recommended that the reader follow some of the studies cited in this chapter in which these techniques have been used to further understand the factors that influence the MFC performance. Figure 3.1 mentions some of the electrochemical techniques that will be reviewed in the chapter.

3.2 Fundamentals of the thermodynamics and calculations for describing potential in MFCs

In electrochemical and bioelectrochemical systems, as for example in an MFC, the driving force that causes the electrons to flow from an external circuit from the anode to the cathode is the electromotive force (EMF). The EMF of a cell is therefore E_{cell} and is known as the cell potential and is measured in volts (V), therefore, sometimes it is called cell voltage. For spontaneous reactions in the cell the cell potential will be positive.

The EMF of the cell depends on the reactions that occur in the anode and in the cathode, the concentrations of the reactants, the products, and temperature. If the cell is operated at standard conditions, the temperature is assumed to be 25 °C at 1 atm pressure and 1 M for the concentrations of the chemical compounds involved. Under these standard conditions the EMF is the standard EMF or the standard cell potential, E°_{cell}. The E°_{cell} is obtained from the standard reduction potential of the cathode reaction (E°_{red} cathode) minus the standard reduction potential of the anode reaction (E°_{red} anode) (some reduction potentials of environmental interest frequently utilized in MFC studies are presented in table 3.1):

$$E^{\circ}_{cell} = E^{\circ}_{red}\,(\text{cathode}) - E^{\circ}_{red}\,(\text{anode}). \tag{3.1}$$

The dependence of the EMF on concentration can be obtained from the dependence of the free-energy change (ΔG_r). ΔG_r is related to the standard free-energy change ΔG°_r:

$$\Delta G_r = \Delta G_r + RT \ln Q \tag{3.2}$$

$$\Delta G^{o}_r = \sum \Delta G^{o}_f\,(\text{products}) - \sum \Delta G^{o}_f\,(\text{reactants}). \tag{3.3}$$

Taking into account that the work produced by the cell is given by the exchange of electrons,

$\Delta G_r = -nFE$, the equation gives the Nernst equation

$$-nFE = -nFE + RT \ln Q, \tag{3.4}$$

where Q is given by $\left(\dfrac{[\text{products}]^p}{[\text{reactants}]^r}\right)$, the reaction quotient, which considers the concentrations that exist in the reaction mixture at any given moment, n is the number of exchanged electrons, and F is the Faraday constant (96 485 C mol^{-1}). The reactions that need to be considered in the system are the oxidation reaction that occurs in the anode and the reduction reaction that occurs in the cathode.

Therefore, the common Nernst equation representation is as follows:

$$E\,(\text{EMF}) = E^{\circ}_{cell} - \frac{RT}{nF} \ln\left(\frac{[\text{products}]^p}{[\text{reactants}]^r}\right), \tag{3.5}$$

where E°_{cell} is the standard potential of the cell, R is the gas constant (8.314 J mol^{-1} K^{-1}), T is the temperature (K), n is the number of electrons, and F is Faraday's constant (96 485 C mol^{-1}). This value of EMF is the maximum of cell voltage that can be obtained from the cell.

If the reactions that are carried out in the cell are as follows:

$CH_3COO^- + 4H_2O \rightarrow 2HCO_3^- + 9H^+ + 8e^-$ anode, acetate oxidation reaction,

$8e^- + 8H^+ + 2O_2 \rightarrow 4H_2O$ cathode, oxygen reduction reaction,

Then for the anode: $E_{anode} = E^{\circ}_{anode} - RT/8F \ln([CH_3COO^-]/[HCO_3^-]^2[H^+]^9)$

And for the cathode: $E_{cathode} = E^{\circ}_{cathode} - RT/4F \ln(1/p\,O_2\,[H^+]^4)$.

Once the theoretical anode and cathode potentials are obtained through the Nernst equation, equation (3.6) is used to calculate the theoretical EMF that is the maximum theoretical voltage attained from the MFC [23]:

$$\text{EMF} = E_{cathode} - E_{anode}. \tag{3.6}$$

In the actual operation of the MFC, the electrode at which the reaction that is studied occurs is known as the working electrode or indicator. The actual potential

Table 3.1. Some reduction potentials of environmental interest and frequently utilized in MFCs.

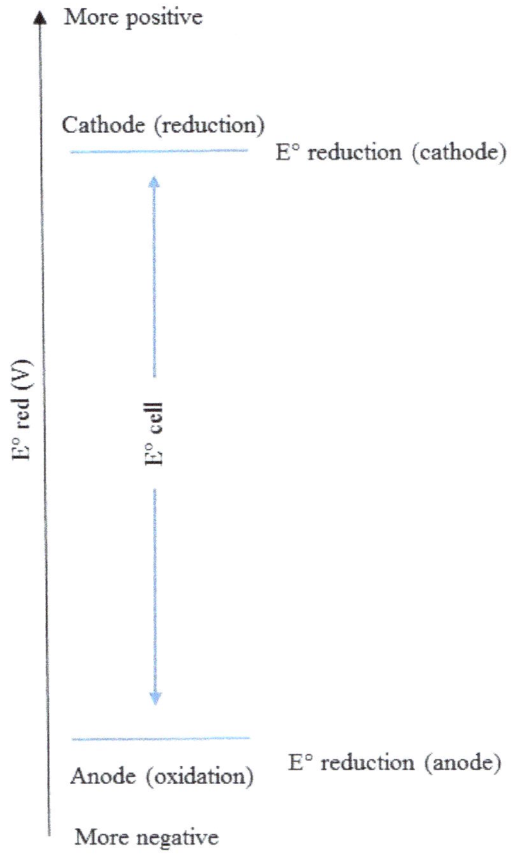

Standard potential E°(V)	Reduction half reaction
1.51	$MnO_{4(aq)} + 8H^+_{(aq)} + 5e^- \rightarrow Mn^{2+}_{(aq)} + 4H_2O_{(l)}$
1.36	$Cl_{2(g)} + 2e^- \rightarrow 2Cl^-_{(aq)}$
0.82	$O_{2\,(g)} + 4H^+ + 4e^- \rightarrow 2H_2O$
0.77	$Fe^{3+} + e^- \rightarrow Fe^{2+}$
0.74	$NO_3^-{}_{(aq)} + 6H^+ + 5e^- \rightarrow N_2 + 6H_2O$
0.59	$MnO_{4(aq)} + 2H_2O + 3e^- \rightarrow MnO_{2(s)} + 4OH^-$
0.36	$NO_3^-{}_{(aq)} + 10H^+ + 8e^- \rightarrow NH_4^+{}_{(aq)} + 3H_2O$
0.42	$NO_3^- + + 2H^+ + 2e^- \rightarrow NO_2^- + H_2O$
0.139	AsO_4^{3-}/AsO_3^{3-}
0.035	Cytochrome ox/red
0	$2H^+_{(aq)} + 2e^- \rightarrow H_{2(g)}$
-0.075	Menaquinone ox/red
-0.12	Flavodoxin ox/red*
-0.18	AQDS ox/AH_2QDS^- red
-0.19	Piruvate$^-$ /Lactate $^-$
-0.22	$SO_4^{2-} + 9H^+ + 8e^- \rightarrow HS^- + 4H_2O$
-0.24	$S_{(s)} + 2H^+ + 2e^- \rightarrow H_2S_{(g)}$
-0.24	$CO_{2\,(g)} + 8H^+ + 8e^- \rightarrow CH_{4(g)} + 2H_2O$
-0.27	$S_{(s)} + 2H^+ + 2e^- \rightarrow H_2S_{(g)}$
-0.29	$CO_2 + HCO_3^- + 8H^+ + 8e^- \rightarrow CH_3COO^- + 3H_2O$
-0.32	$NAD^+ + 2H^+ + 2e^- \rightarrow NADH + H^+$
-0.37	Flavodoxin ox/red*
-0.38	$CO_2 + 6H^+ + 6e^- \rightarrow CH_3OH + H_2O$
-0.43	$6CO_2 + 24H^+ + 24e^- \rightarrow C_6H_{12}O_6 + 6H_2O$
-0.44	$Fe^{2+}{}_{(aq)} + 2e^- \rightarrow Fe_{(s)}$

Data recompiled from Madigan *et al* [21] and Eweis *et al* [22]. The half reactions * refer to one e^- each.

of the cell E_{cell} is always lower than the theoretical potential (voltage) of the cell at equilibrium due to irreversible losses.

Open circuit voltage (OCV)
The open circuit voltage (OCV) refers to the voltage of the cell that can be measured in the absence of current. The theoretical OCV, which should be close to the cell EMF, is actually lower than the cell EMF due to potential losses. In many cases, the measured potential of each electrode is much lower than the theoretical value, and the energetic loss is known as the overpotential, that is, the difference between the theoretical potential (at equilibrium) and the actual potential.

The overall potential of the cell
The actual overall potential of the cell (or cell voltage) can be presented as a function of the EMF of the cell, the overpotential of the electrodes, and the ohmic losses of the circuit as follows:

$$E_{\text{cell}} = E_{\text{EMF}} - \left(\sum \eta_{\text{An}} + \left| \sum \eta_{\text{Cat}} \right| + R_{\Omega} I \right), \tag{3.7}$$

where $\Sigma\eta_{An}$ and $|\Sigma\eta_{Cat}|$ are the overpotential of the anode and cathode, respectively, $R_{\Omega}I$ is the sum of all ohmic losses in the cell as a function of the current (I) and the ohmic resistance of the system (R_{Ω}). The overpotential of the electrodes is current dependent and due to the following categories of losses, activation losses, bacteria metabolic losses, and mass transport losses (or concentration losses). In practice, the overall cell potential described in equation (3.7) could be presented by the following equation:

$$E_{\text{cell}} = \text{OCV} - R_{\Omega} I, \tag{3.8}$$

where the OCV includes the overpotentials of the electrodes and $R_{\Omega}I$ includes the current-dependent overpotentials and the ohmic resistance. The ohmic losses, which include the resistance of the electrons to flow and the flow of cations through the cation exchange membrane (CEM) in the case of an MFC including a membrane. The other losses in the MFC can be described briefly as follows. Activation losses, which are due to the energy required for the redox reaction to occur, can be diminished by increasing the electrode surface area. Bacteria metabolic losses depend on the choice of electron donor that the bacteria can use in the MFC and favor their growth but do not necessarily favor the difference of potential in the MFC to generate an electron transfer to the anode that may in turn generate voltage. In some cases, if any other compound can serve as an electron acceptor in the anodic chamber, and the reaction is thermodynamically favored for the bacteria, they would prefer to take advantage of the energy of that electron acceptor and the electrons would not be received in the electrode, therefore the MFC generation of energy would decrease due to the metabolism of the bacteria. Concentration losses are due to hindrances in mass transfer in the MFC and occur mainly at high current densities due to low diffusion of the chemical species to the electrode surface.

Reactions in an MFC and basic measurements and calculations for electrochemical characterization

The reactions that take place in an MFC, for example an MFC designed for oxidation of organic compounds from wastewater treatment, are as follows. The model of the organic compound that is derived from the fermentation of organic matter in this case is acetate, this is taken as an example from among the multiple organic compounds that could be oxidized in the anodic chamber of an MFC coupled to the reduction of oxygen in the cathodic chamber while the electrons flow from the anode to cathode. A scheme of an MFC in which acetate is oxidized is shown in figure 3.2; the reaction takes place at pH 7:

The voltage measurements in an MFC are taken with a digital multimeter. Once the voltage measurements are obtained, the current and power densities are calculated using Ohm's law:

$$I = V/R_{\text{ext}}, \tag{3.9}$$

where V is the closed circuit voltage (V), I is the current (A), and R_{ext} represents each one of the resistances used (kΩ). The current density is obtained by dividing the current

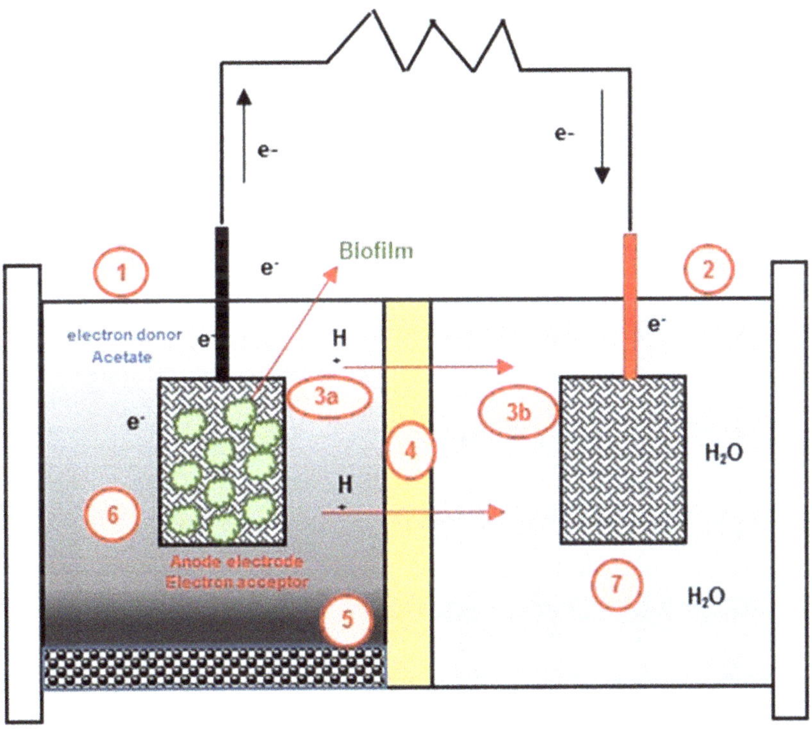

Figure 3.2. MFC type 'H'. 1: Anodic chamber (anaerobic) for the evaluation of acetate oxidation; 2: cathodic chamber (aerobic); 3: graphite cloth electrodes (3a anode and 3b cathode); 4: cation exchange membrane; 5: electrogenic bacteria; 6: mineral medium; 7: distilled water. The redox reactions that occur in the MFC are as follows:

$CH_3COO^- + 4H_2O \rightarrow 2HCO_3^- + 9H^+ + 8e^-$ anode, acetate oxidation reaction

$8e^- + 8H^+ + 2O_2 \rightarrow 4H_2O$ cathode, oxygen reduction reaction.

obtained with the resistance by the anode area. The power is calculated with the following equation and the result is divided by the area to obtain the power density:

$$P = I^{*}V. \tag{3.10}$$

Coulombic efficiency (CE)
The CE represents the ratio of coulombs that are actually transferred from the substrate to the anode to the maximum coulombs that could possibly be obtained if one hundred per cent of the substrate consumed was utilized to produce current. The CE is calculated using the following equation taking into account the change of the COD (ΔCOD) as a measure of the substrate consumption and integrating the current measured in the MFC when this is operated in a fed batch mode:

$$\mathrm{CE} = \frac{8\int_0^t I\,dt}{FV_{\mathrm{an}}\Delta\mathrm{COD}}, \tag{3.11}$$

where 8 is derived from taking the oxygen reduction as the reaction taking place in the cathode (this is 32 g mol^{-1} of oxygen divided by the 4 electrons that 1 mol of oxygen can accept), thus it is a constant used for COD, F is Faraday's constant, and V_{an} is the volume of liquid in the anode compartment.

Ohmic resistance determination using the current interruption technique
The determination of the ohmic resistance (R_Ω) in an MFC can be done by using the current interruption technique that consists of operating the MFC at a current at which no concentration losses occur. At some point, the circuit is opened and then an increase (steep increase) in the initial potential is observed followed by a slower further increase to the OCV value. The ohmic losses are proportional to the produced current and ohmic resistance. At the moment of current interruption, the ohmic losses disappear immediately. Therefore, the steep potential is E_{R}, which is proportional to the ohmic resistance (R_Ω) and the current (I) produced just before the interruption. The estimation of R_Ω can be achieved by means of Ohm's law, $R_\Omega = E_{\mathrm{R}}/I$. It is assumed that the slower potential that takes place after the steep

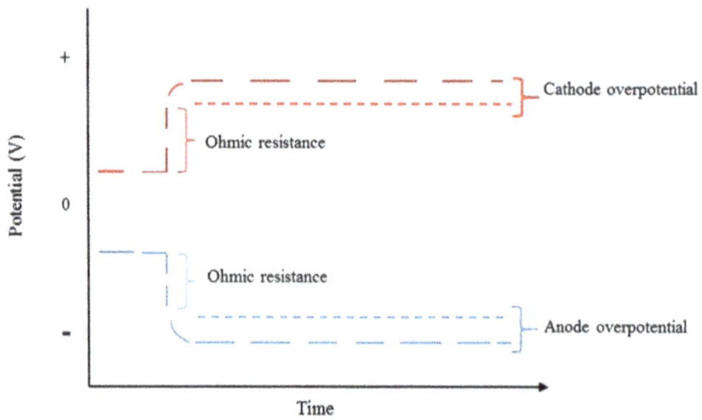

Figure 3.3. Electrode potentials over time during the current interruption technique.

increase to the OCV gives the electrode overpotentials that occurred during current generation. Figure 3.3 shows the voltage over time when applying the current interruption technique.

Polarization and power curves

The polarization curves present the voltage as a function of current, frequency, and current density.

These curves can be constructed for the anode, the cathode, or for the whole MFC if the equipment is available (potentiostat). The polarization curve can be constructed without using a potentiostat by varying the external load with a resistor box. The load is gradually increased or decreased and the voltage is measured. The current is calculated using Ohm's law. The anode or cathode can be evaluated separately using a reference electrode (RE). When constructing the polarization curve, the variation of the external resistance has to be gradual and for each resistance it is important to allow time for the system to reach a pseudo steady state. Therefore, this procedure may take some time since the pseudo steady state may take several minutes to be reached in the system and it needs to be reached every time that the external resistance is changed. As discussed in the literature [23], the polarization curves are divided in three regions: (a) the activation losses region (a steep decrease of voltage from the OCV), (b) the ohmic losses region (the voltage drop is almost linear with current), and (c) the concentration losses region, at high currents (a rapid drop of voltage), see figure 3.4(a). Power (power density) as a function of current (current density) can be plotted in a power curve, the data of which are calculated from the polarization curve (see figure 3.4(b)).

Improving the accuracy of the electrode potential measurements

The measurements of the electrode potential need to be as accurate as possible in order to know the influence of the material and the distance between the electrodes on cell performance. In this regard, in a series of documents, Logan and his research group have reported in-depth work on the improvement of the accuracy of electrodes potential measurements. In [24] the effect of the solution ohmic resistance on the measured electrode potential and power production in an MFC was explained through a series of equations developed to correct the measured potential of the electrode and that include the effect of the solution ohmic resistance on the power production of the cell. The experimental work used to explain the importance of the material and distance of the electrodes was carried out with conventional anode materials and shapes, such as graphite fiber brush and flat carbon felt, whereas the cathodes were made of a carbon–metal combination and the adjustments to the measurements of the electrode potential were conducted with the equations explained in the study, for which the resistance measurement needed was obtained by means of the electrochemical impedance spectroscopy (EIS) technique. However, if EIS is not available for the resistance measurement, the calculations to obtain an estimate based on the solution conductivity and distance between the electrodes was also explained. At this point, the reader may want to refer to the publication [24] in which the complete experimental work is detailed in order to apply the method to a current electrode measurement. In this chapter just the main equations will be summarized.

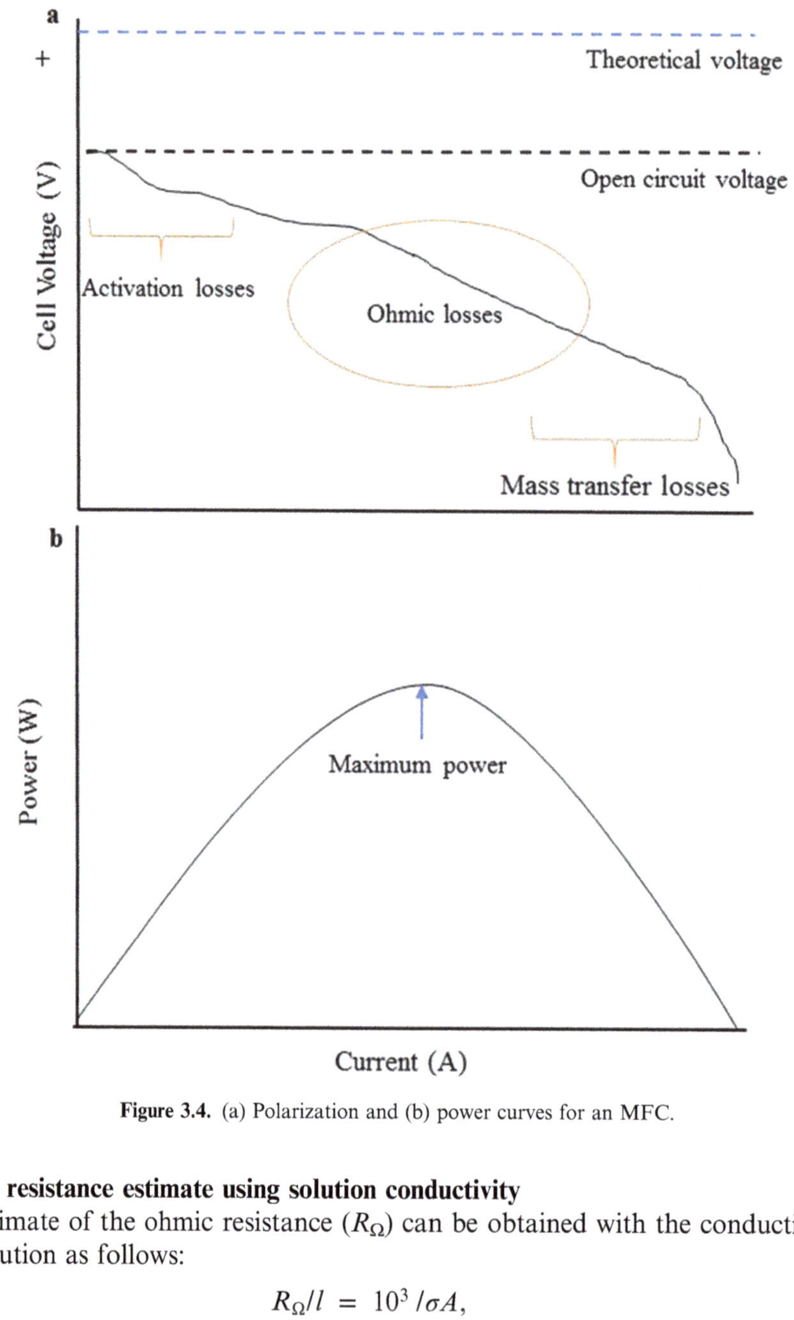

Figure 3.4. (a) Polarization and (b) power curves for an MFC.

Ohmic resistance estimate using solution conductivity

An estimate of the ohmic resistance (R_Ω) can be obtained with the conductivity of the solution as follows:

$$R_\Omega/l = 10^3/\sigma A, \tag{3.12}$$

where R_Ω is in Ω cm^{-1}, σ is conductivity (mS cm^{-1}), l is length, A is the cross sectional area (exposed surface of the electrode), and 10^3 is used to convert the conductivity in mS into S (1 S = Ω^{-1}). It was suggested that this value needs to be compared with the actual resistance value that can be obtained from EIS since the calculation underestimates other resistances at the electrode surface.

Electrode potentials using one reference electrode

In those cases in which one RE is used to measure the electrode potential, the distance between the working electrode and the RE, an ohmic drop will be present due to ohmic resistance and this is a function of current. In case that the working electrode (WE) is the anode, the correction for the anode potential due to ohmic resistance using one RE (E_{AnRE}, in mV), can be estimated as follows:

$$E_{AnRE}\,(\mathrm{mV}) = E_{An} - (10^3\,R_\Omega\,d_{An\text{-}RE}/l)I, \tag{3.13}$$

where d_{An-RE} (cm) is the distance between the anode (WE) and the RE and I (A) is the current. In the case that this can be done with a potentiostat, the software may conduct the calculation for the correction of the anode potential due to ohmic resistance and then $E_{An} = E_{An'}$. The usual calculation for E_{cat} is sometimes inaccurately done taking into account the uncorrected anode potential and the overall cell voltage as follows:

$$E_{cat,\,U}(\mathrm{mV}) = E_{An} + U. \tag{3.14}$$

At this point Logan *et al* [24] explain that the equation contains two errors that are not taken into account in the reporting of the cathode potential and may lead to wrong values in the reports involving these measurements. The first error is to utilize an uncorrected anode potential and the second error is to utilize the overall cell voltage (U) since it contains the potential loss due to the ohmic resistance of the solution between the anode and cathode. In order to obtain a better estimate of the cathode potential, the first error may be amended by using the following equation which, as explained, can be used for the estimation of a corrected anode potential:

$$E_{cat,\,U+\,AnRE}\,(\mathrm{mV}) = E_{AnRE} + U. \tag{3.15}$$

This cathode potential will be equal or more positive than that calculated with equation (3.14). The second error that usually occurs in the calculation of the cathode potential has to do with the overall cell potential U. This error can be amended by adding the potential lost due to the current-dependent potential whole cell ohmic drop due to ohmic resistance based on the distance between the cathode and anode (d_{An-Cat}). Therefore, a better estimate for the cathode potential using one RE will be as follows:

$$E_{Cat,U+\,AnRE\,+IP} = E_{AnRE} + U + (10^3 R_\Omega d_{An\text{-}Cat}/l)\,I. \tag{3.16}$$

It is suggested that the cathode potential can be corrected by using the same equation for the correction of the anode potential but in this case the cathode potential would be the working electrode and the distance would be between the cathode and the RE, d_{Cat-RE}. In using these equation for the corrections, it is important to take the measurement of the distance of the electrodes as the measure of the space between the closest edges of each electrode.

Electrode potentials using two reference electrodes

Accuracy in the estimation of electrode potentials may be higher using the following two equations. This is possible if two REs are used, one close to the anode and another close to the cathode (if distance between anode and cathode makes it possible):

$$E_{AnREAn} = E_{An} - (10^3 R_\Omega d_{An\text{-}REAn}/l)I, \tag{3.17}$$

$$E_{AnRECat} = E_{Cat} + (10^3 R_\Omega d_{Cat\text{-}RECat}/l)I, \tag{3.18}$$

where $d_{An-REAn}$ is the distance (cm) between the anode and the RE used for the anode, $d_{Cat-RECat}$ is the distance (cm) between the cathode and the RE used for the cathode, and E_{Cat} is the measured cathode potential. It is important to note that if the solution conductivity and the distances between the electrodes are all correct, the estimate of E_{cat} obtained using one RE with equation (3.16) would be the same as the E_{cat} estimated with equation (3.18) for two REs.

Calculation of power densities accounting for the impact of ohmic losses

Following the same reasoning, in order to determine the power of the cell taking into account the impact of the ohmic losses, Logan *et al* [24] proposed a series of calculations that would serve this purpose. The power produced in the MFC (P_p) is normalized by the cathode area (cathode projected area) where A_{Cat} is assumed to be the cross sectional area of the cathode and is obtained by the whole cell voltage (U, mV) and an external resistor R_{ex} used in the circuit:

$$P_p = 10UI/A_{Cat} = U^2/100A_{Cat}R_{ex}. \tag{3.19}$$

The plot of the power as a function of the current will show the maximum power at the highest point of the curve (P_{max}). The power that is lost due to the ohmic resistance of the solution between the electrodes (P_Ω), that is the distance due to the distance between the anode and cathode as a function of the current, can be estimated by the following equation:

$$P_\Omega = 10^4 R_\Omega d_{An\text{-}Cat} I^2/lA_{Cat}\ . \tag{3.20}$$

The sum of the power produced by the cell and that lost due to the ohmic resistance ($P_{p+\Omega}$) is as follows:

$$P_{p+\Omega} = P_p + P_\Omega. \tag{3.21}$$

This equation is recommended to evaluate the impact of ohmic losses on the power generation at given conditions. The maximum power density production in an MFC with high ohmic resistance generally occurs at half of the maximum current ($\frac{1}{2}I_{max}$) and a voltage lower than half of the OCV ($\frac{1}{2}U_{OCV}$, mV) for a power density curve. If the internal losses are mainly due to ohmic resistances, then the maximum current ($I_{max,\Omega}$) can be estimated by the following equation:

$$I_{max,\Omega} = U_{ocv} l/10^3 R_\Omega d_{An\text{-}Cat}. \tag{3.22}$$

The maximum power ($P_{max,\Omega}$) that can be obtained taking into account the losses, can be calculated by $\frac{1}{2}I_{max,\Omega}$ and $\frac{1}{2}U_{ocv}$, given by

$$P_{max,\Omega} = 10U_{OCV}I_{max,\Omega}/4A_{Cat}. \tag{3.23}$$

When the anode and cathode electrode potentials are measured, the power can be calculated as follows without taking into account the ohmic losses between electrodes, therefore this can be taken as a rough estimate:

$$P_{\text{Cat-An}} = 10(E_{\text{Cat}}\text{-}E_{\text{An}})I/A_{\text{Cat}}. \tag{3.24}$$

In the case in which the corrected potentials are used and the power due to the ohmic losses for this potential difference and current density are included in this calculated power, then the total power can be estimated as follows:

$$P_{\text{Cat REc-AnREa-IR}} = [E_{\text{CatREc}} - E_{\text{AnREa}} - (R_{\Omega}d_{\text{An-Cat}}/l)I]10^{*}I/A_{\text{Cat}}. \tag{3.25}$$

Therefore, if the corrections for ohmic losses were estimated correctly, this last equation (3.25) would give the same result as equation (3.19) in which power was determined by the voltage of the cell over an external resistor.

It is suggested to consider the series of corrections presented by Logan *et al* [24] in order to improve the accuracy for the electrode potentials and the power of cells. In their report, the authors mention that in several cases the distances between the RE and the working electrode are not reported and the method to obtain the cathode potential is also missing, thus making comparisons of the limits of power generation in MFCs difficult, particularly in those cases in which the conductivity of the solution is low, as for example in wastewater. The authors pointed out that the EIS is a complementary technique that in this case was used to obtain the value of the ohmic resistance, which value was in agreement with the estimate of ohmic resistance made with the solution conductivity and the distance between the RE and working electrode and the distance between the anode and cathode.

The electrode potential slope analysis

In order to improve the validity of the comparisons of power generation among MFCs, Rossi *et al* [25] proposed a method to calculate electrode and solution area based resistances. The electrode potential slope (EPS) analysis was based on the behavior of the potential of the anode and cathode as a function of current. Therefore, the slopes of the electrode polarization data, near peak power, based on the linear region were used to calculate the electrode area based resistance ($m\Omega\ m^2$) whereas the *y*-intercepts of the potential versus current density data were utilized to calculate effective half-cell potentials. This was proposed since in MFCs the polarization data are linear in a wide range of current densities near the peak power. Anode potentials are linear until high current densities before they drop due to several phenomena occurring in the MFCs, on the other hand, cathode potentials typically have an initial drop due to the activation losses of the oxygen reduction reaction at low current densities, afterwards the cathode potential tends to be linear at high current densities near the peak power. Therefore, it is proposed that the slopes of the polarization curve for the whole cell or the individual electrodes can be used to obtain area normalized resistances and the *y*-intercepts of the potentials can be a good estimate of the actual working open circuit voltages (OCVs). The report in which Rossi *et al* [25] proposed this EPS analysis is very interesting due to the demonstration of the application of the calculations of this method to their experiments, which were included in the same publication, and the application of this same method to the data of previous published studies to compare results for electrodes of different sizes and materials.

A summary of the equations of the EPS method is provided here but it is highly recommended to see the original publication [25] for discussion of the experiments in which it was applied and the comparisons that were made using the method to data for electrodes reported in other works.

The overall potential of the cell is calculated as a function of the EMF of the cell calculated from the cathode and anode half-cell potentials E_{An} and E_{Cat} and resistances in the circuit. The electrode overpotentials are a function of the current and they decrease the voltage as a result of the variations in the MFC due to microbial changes, pH modifications, activation losses, and mass transfer limitations. A convenient form to present the current density data takes into account the size of the electrode and it is presented per unit of area of the electrode, thus the resistance should be normalized as well, for example in mΩ m^2 for any given voltage (mV):

$$E_{cell} = E_{EMF} - \left(\sum \eta_{An} + |\sum \eta_{Cat}| + R\Omega I \right). \tag{3.26}$$

In terms of the electrode potentials, equation (3.26) would be as follows:

$$E_{cell} = E_{Cat} - E_{An} - \left(\sum \eta_{An} + |\sum \eta_{Cat}| + R\Omega I \right), \tag{3.27}$$

where the theoretical E_{EMF} calculated from the half-cell potentials for the cathode and anode are E_{Cat} and E_{An}, respectively, η_{An} is the anode overpotential, η_{Cat} is the cathode overpotential, R_{Ω} is the ohmic resistance, and I is the current. It is important to note that the measured $E_{EMF,m}$ of the cell under open circuit conditions ($E_{EMF,m0}$) does not have a unique value because the oxygen reduction reaction can present variations. These variations in the oxygen reduction reaction are widely explained in Rossi *et al* [25]. The measured cell voltages decrease due to several losses that impact the operational conditions near the electrode at low current densities. As the voltage declines, the electrode potentials become a linear function of current.

For operation past the low current densities, the electrode potentials can be replaced by electrode potentials that fit the linear region of the measured cathode and anode potentials, producing operational potentials that better describe the working electrode potentials in terms of experimental electrode potential conditions for the anode ($E_{An,e0}$) and cathode ($E_{Cat,e0}$) potentials. Since all the changes in the electrodes and solution potentials are now linear as a function of current, then equation (3.27) becomes

$$E_{cell} = E_{Cat,e0} - E_{An,e0} - I(R_{An} + R_{Cat} + R\Omega), \tag{3.28}$$

where the electrode overpotentials are now quantified in terms of individual resistances, anode resistance (R_{An}), cathode resistance (R_{Cat}), and electrolyte (solution) resistance (R_{Ω}).

In an MFC experiment, the load of the circuit is varied from an external resistance, and the operational current of the MFC is calculated from the cell voltage, thus E_{cell} can be re-employed by the external resistance in the circuit as a function of current, IR_{ext}. For this EPS analysis, the electrode open circuit potentials and resistances were obtained from the polarization test by changing the external resistance in the circuit (see figure 3.5):

$$IR_{ext} = E_{Cat,e0} - E_{An,e0} - I(R_{An} + R_{Cat} + R\Omega), \tag{3.29}$$

where $E_{Cat,e0}$ and $E_{An,e0}$ are the cathode and anode potentials at the experimental electrode potential conditions, R_{An} is the anode resistance, R_{Cat} is the cathode resistance, and R_{Ω} is the electrolyte (solution) resistance. R_{Ω} includes the membrane or separator resistance if there is one in the MFC. The membrane or separator resistance can also be determined separately from the electrolyte resistance.

Then the electrode potentials are plotted on the y-axis versus the current density in the x-axis, and the linear portion of the data corresponds to the following expression:

$$E = mI + b, \tag{3.30}$$

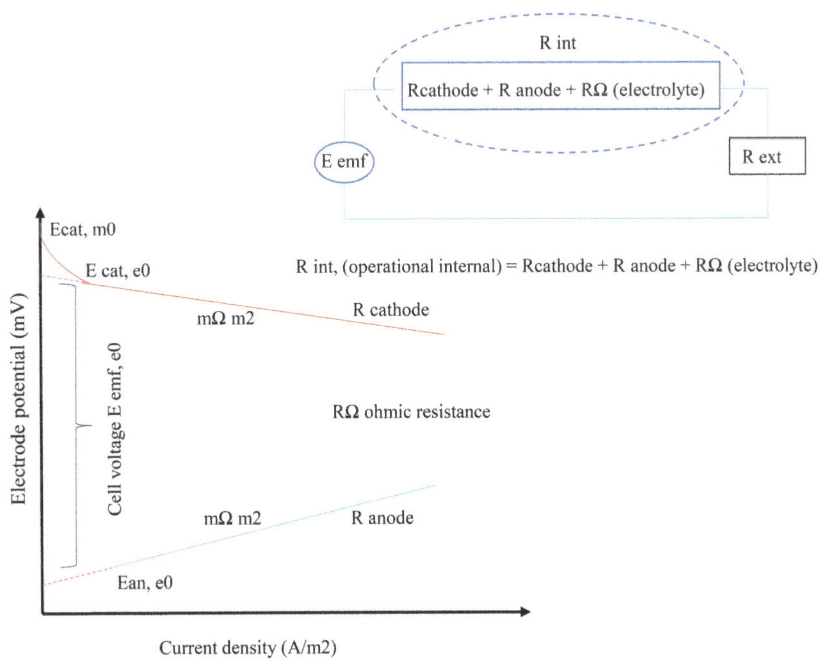

Figure 3.5. Electrode slope potential representation for an MFC. The dashed lines represent the linearization that is obtained from the polarization test. The continuous lines represent the linear part of the polarization data that serve to calculate the cathode ($R_{cathode}$) and anode (R_{anode}) resistances. (Adapted from [25].)

where the slope m is the specific resistance of each electrode (R_{cat} or R_{An}, mΩ m^2) and the y-intercepts are used to calculate the experimental open circuit potential of the cathode and anode electrodes (cathode $E_{cat,e0}$, anode $E_{An,e0}$), see figure 3.5.

The solution resistance (R_Ω, mΩ m^2) can be obtained from the conductivity of the solution (σ mS cm^{-1}) and the distance between electrodes (l, cm):

$$R_\Omega = 100l/\sigma, \tag{3.31}$$

where 100 is used to convert mS to S (1 S = 1 Ω^{-1}) and cm^2 to m^2, l is the electrode spacing in cm, and σ is the solution conductivity. In the case that a membrane is used, the electrochemical impedance spectroscopy (EIS) technique can be used to determine the resistance of the membrane from the combined solution and membrane resistance. The reader may want to refer to [25] for further details on the utilization of this method to compare the data reported in literature or in the case of application of the method to the reader's own existing data or ongoing work.

A combination of the EPS analysis and the EIS technique has been utilized by Rossi and Logan [26] in a two chamber H-type MFC to determine the contributions of the components of the internal resistances of the cell involving different projected areas of electrodes. A combination of EPS, EIS, and linear sweep voltammetry was presented by the same research group [27] to investigate the behavior of an activated carbon commercial cathode and brush anode in an MFC fed with acetate and phosphate buffer solution in synthetic wastewater. These authors demonstrated the practical value of the EPS in combination with electrochemical techniques such as EIS and linear sweep voltammetry (LSV) to characterize anode and cathode performance in MFCs. The understanding and application of the methodology presented in this series of reports can contribute to the improvement of the MFC design and/or to identify the key factors that need more attention in the case of a scaling up of the system or the study of different solutions (electrolytes) in the MFC chambers, such as different sources of wastewater, different buffer solutions, pH, etc.

3.3 Linear and cyclic voltammetry for the characterization of MFCs

Linear sweep (LSV) and cyclic voltammetry (CV)

Linear sweep and cyclic voltammetry are two electrochemical techniques that are used widely to characterize the role of the microbial community in the redox reactions that take place in an MFC. In LSV the potential of the working electrode (WE, i.e. anode) is varied linearly versus the reference electrode (RE, i.e. Ag/AgCl) and this can be used to measure the current flow between the WE and the counter electrode (CE, i.e. Pt) as a function of the potential applied. The data obtained are plotted, i.e. potential versus current (current density) or logarithm of current versus overpotential, in which case they are known as Tafel plots. The Tafel plots are a representation of the equilibrium potential on the surface of the electrode when the oxidation current is equal to the reduction. The Tafel plots may also be utilized to estimate the number of electrons that are transferred in the electrochemical reaction. Figure 3.6 shows an example of a Tafel plot.

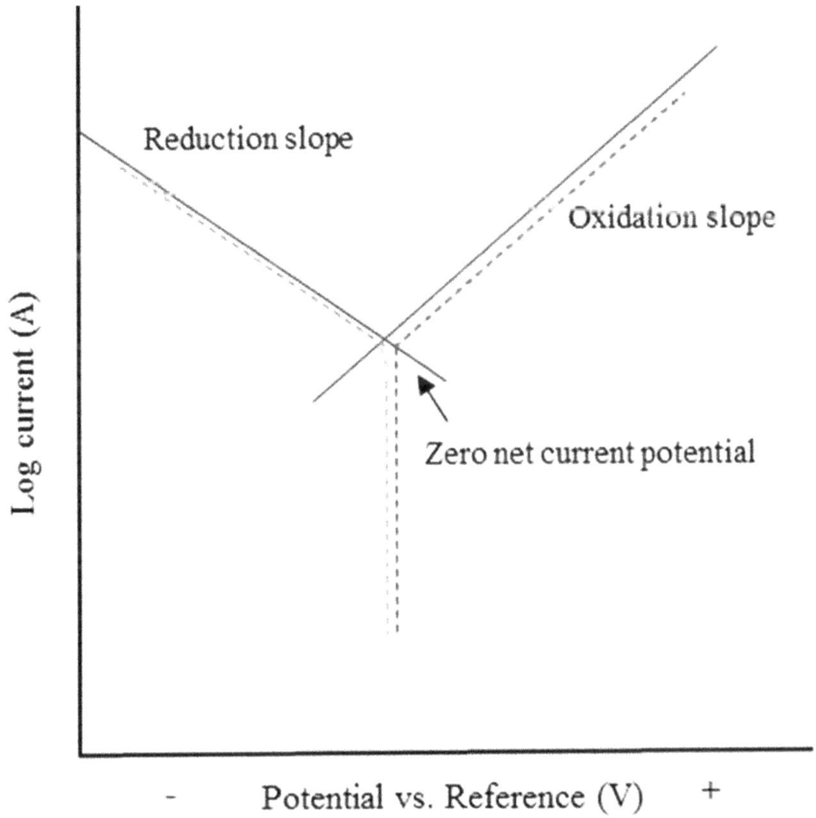

Figure 3.6. Tafel plot. It presents the oxidation and reduction slopes of the working electrode (WE) in a three electrode system.

Cyclic voltammetry is used to investigate the electron transfer mechanisms in the MFC and therefore the role of the micro-organisms in the redox process. This technique is used to characterize biofilms and also electrode materials. Technically, LSV and CV are similar but in CV the potential is increased and decreased cyclically, which is known as forward and reverse scan from two set potentials. Usually, the potential is varied ±1 V and this variation is chosen based on the specific experiment that is carried out. Relatively low speed scans are used (i.e. 1–50 mV s^{-1}). The redox active components that are involved in the experiment (the MFC experiment) are shown as peaks in the voltammagrams (current versus voltage plots) and, according to the system, the oxidation–reduction of the components can be observed in the forward and reverse scans. It is important to mention that if a redox active compound shows a peak in the forward and the reverse scan, the equidistant point from the two peaks (midpoint potential) points out the redox potential of the compound, which compared to a reference value will give information to identify the compound in the system. CV can be conducted several

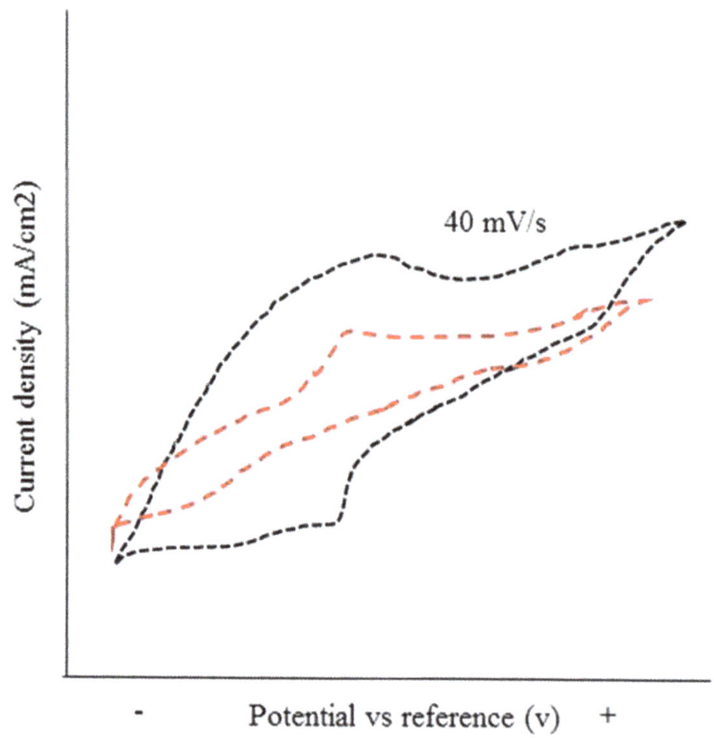

Figure 3.7. Cyclic voltammetry at a 40 mV s^{-1} scan rate to identify redox activity in MFCs for biofilm (black) and planktonic cells (red).

times depending on the stability of the system and it is used to determine the several redox reactions that take place in the MFC. Figure 3.7 shows an example of CV.

3.4 Electrochemical impedance spectroscopy applied to MFC characterization

Electrochemical impedance spectroscopy (EIS) is a non-destructive technique for the study of very complex systems in many fields of science and technology. EIS is based on the measurement of the complex impedance in a wide range of frequencies, separating the effects of relaxation and allowing more precise, or inclusive, measurements of quantities that for other techniques are not possible. EIS has been used widely in fields such as corrosion, coatings, batteries, sensors, biosensors, polymer-based membranes, fuel cells, electrocatalysis, capacitors, analytical chemistry studies, and more recently in microbial and enzymatic fuel cells (MFCs), among others.

Classical electrochemical techniques such as CV, chronoamperometry, potentiometry, and coulometry, based on direct current (DC), analyze the relationship between currents and electrical charges as a function of time and electrode potential. On the other hand, EIS studies the signal as a function of frequency at a constant potential, providing information on the relaxation processes in the frequency range from 10^{-8} to 10^{-4} Hz. The analysis of the real and imaginary parts of the complex impedance is used to obtain information on phenomena that predominate in a given range of frequencies.

With regard to research on MFCs, the EIS technique has been utilized to characterize electrode properties, the formation of biofilm, and interactions of the biofilm with the electrolyte. As described by Domínguez-Benetton *et al* [28], EIS has been largely utilized for determining the internal resistance (R_{int}) in bioelectrochem ical systems, either single or dual chamber MFCs or in arrangements of several MFCs or stacked MFCs. The three components of the R_{int} taken into consideration in these systems are the ohmic resistance, or solution resistance (R_{Ω}), diffusion resistance, that involves mass transfer phenomena (R_d or R_{diff}), and charge transfer resistance (R_{ct}). The identification of these resistances is important in order to know the contribution of each of them in the overall performance of an MFC. The contribution of the internal resistance on the performance of an MFC can be described as follows:

$$P = E_b * I - R_{int} * I^2, \tag{3.32}$$

where P is the power density of the cell, E_b is the linear extrapolation of the open circuit voltage, which is the intercept of the linear curve with the voltage axis, I^2 is the current density (A/surface unit), and R_{int} is the internal resistance for which is desirable to know the magnitude of its components. Figure 3.8 depicts a polarization curve for an MFC in which ohmic losses, charge transfer, and mass transfer limitation regions are shown. EIS has been used to calculate R_{int} in the MFC by fitting the data into equivalent electronic circuit EECs from which an explanation will be given in the following sections and which is not a straightforward method.

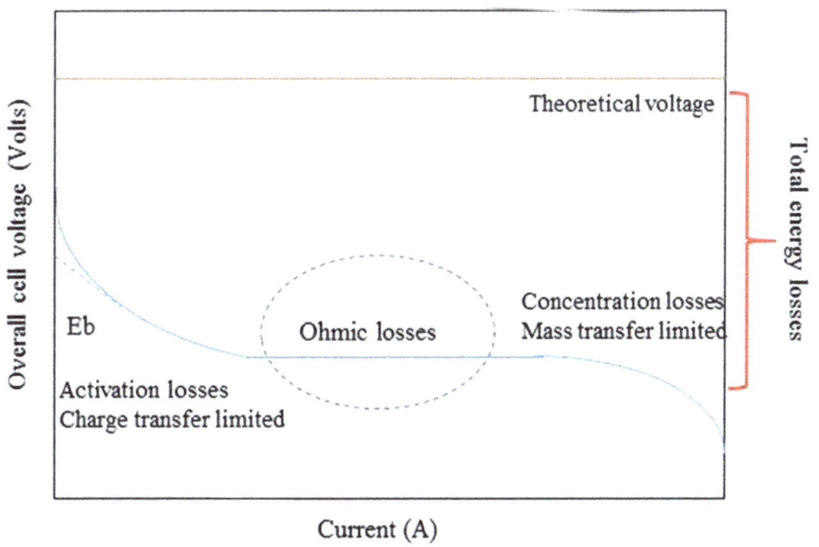

Figure 3.8. Current versus voltage polarization curve of an MFC. The ohmic, activation, and charge transfer losses are presented, which describe the $P = E_b * I - R_{int} * I^2$ power equation. (Adapted from [28].)

Theoretical aspects

Electrical resistance (R), according to Ohm's law, represents the opposition of an electrical circuit to the current flow. The mathematical expression is

$$R = \frac{E}{I}, \tag{3.33}$$

where R is the resistance (in Ω), E is the voltage (in V), and I is the current (in A). This expression is only valid for an ideal resistor. However, real electrochemical devices are more complex than a simple resistive device, requiring a more complex analysis such as EIS. The EIS technique is based on applying a small amplitude sinusoidal excitation signal (potential or current) to the system. The response being measured is an AC current or voltage signal. When a sinusoidal potential excitation is applied with an amplitude E_0, and a radial frequency ω (radians s^{-1}) excitation signal, at time t (s), the potential at time t is given by

$$E_t = E_o \sin(wt), \tag{3.34}$$

where $\omega = 2\pi f$ and f is the frequency in Hz. The generated current at time t shows a shift in phase with an amplitude I_0. The current at time t is given by the following expression:

$$I_t = I_0 \sin(\omega t + \phi), \tag{3.35}$$

where ϕ is the shift in phase. Figure 3.9 shows that current response to a sinusoidal potential at the same frequency but shifted in phase.

The resistance also exhibits complex behavior and the term impedance (Z) is used to represent the resistance to the flow of an AC. The expression for impedance at time t considering Ohm's law is

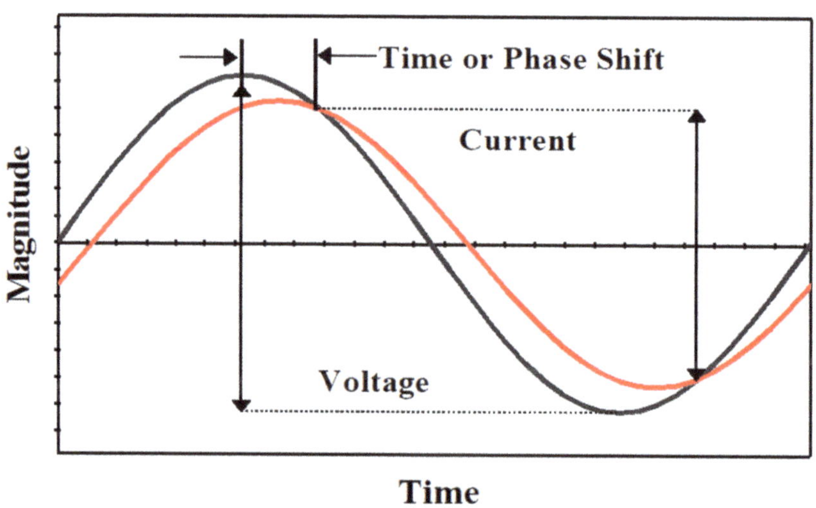

Figure 3.9. Current and potential as a function of time.

$$Z_t = \frac{E_t}{I_t} = \frac{E_0 \sin(wt)}{I_0 \sin(wt + \varnothing)} = Z_0 \frac{\sin(wt)}{\sin(wt + \varnothing)}. \tag{3.36}$$

Potential, current, and impedance represented as complex functions by the Euler relationship, are expressed as

$$E_t = E_0 \exp(jwt), \tag{3.37}$$

$$I_t = I_0 \exp(jwt - \varnothing), \tag{3.38}$$

$$Z_w = Z_0 \exp(j\varnothing) = Z_0(\cos\varnothing + j\sin\varnothing), \tag{3.39}$$

where j (−1) is the imaginary number. Z_w contains two contributions, the real and imaginary parts, represented as Z' and Z'', respectively.

Commonly, the EIS spectrum is described as a complex impedance plane plot, $-Z''$ versus Z' from each frequency, known as Nyquist plot. Figure 3.10 shows a Nyquist plot of the EIS spectrum of a system with a single time constant. The experimental data, Z_i', Z_i'', and ω_i ($I = 1, 2, \ldots, n$) of n measurements taken in any frequency range. The low frequency data are on the right side of the plot, and higher frequencies are on the left. The frequency values are not visible in the Nyquist plot.

It is observed in figure 3.10 that the magnitude of the impedance (Z) (modulus of the impedance) is represented as a vector, and the angle between this vector and the x-axis is the phase angle, $\phi = \arg Z$. (Z) is calculated with the following expression:

$$|Z| = \sqrt{Z'^2 + Z''^2}. \tag{3.40}$$

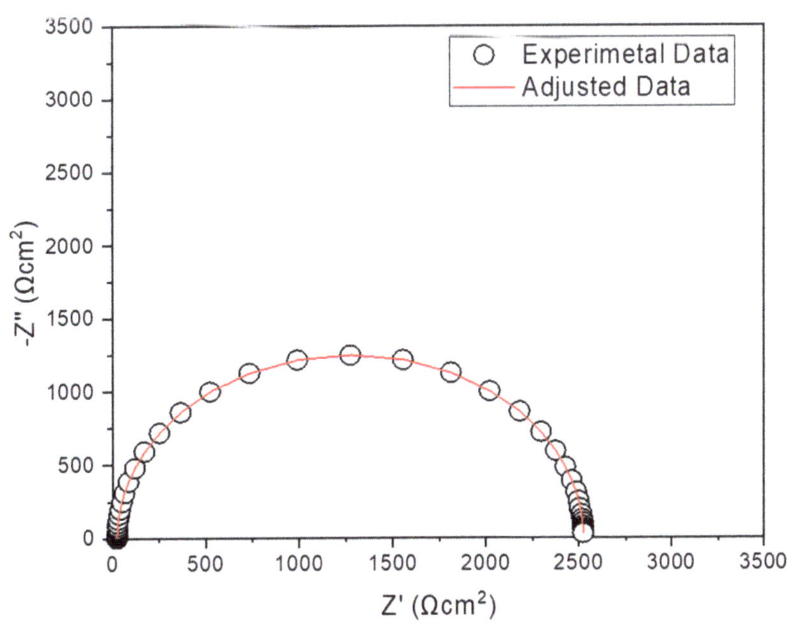

Figure 3.10. Representation of EIS data in a Nyquist plot.

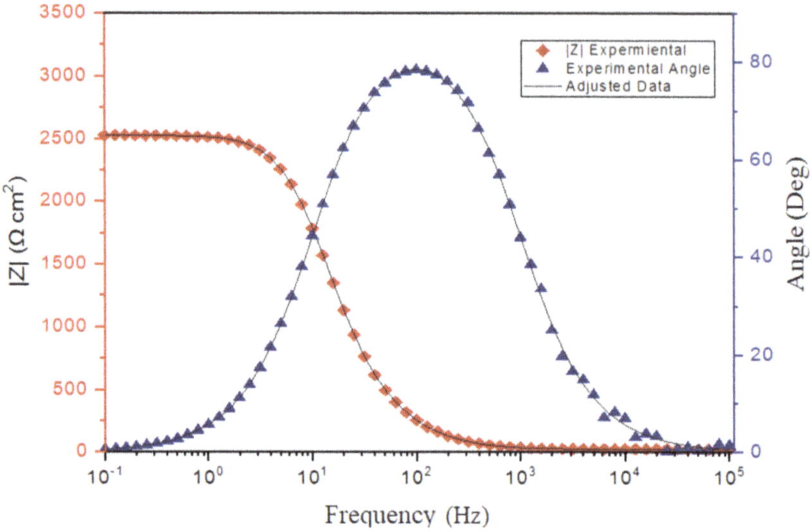

Figure 3.11. Representation of EIS data in Bode plots.

The Bode plots in figure 3.11 represent the same EIS data but in another format, in which the impedance magnitude and phase angle ϕ are plotted against frequency. The phase angle is given by

$$\phi = \tan^{-1}\left(\frac{Z''}{Z'}\right). \tag{3.41}$$

Unlike the Nyquist plot, the Bode plots show the frequency, and the observed small and large impedance values.

Equivalent circuit models

Equivalent circuit models are used to analyze the EIS data. The models include electrical elements such as resistors (R), inductors (L), and capacitors (C), and others such as a constant phase element and Warburg impedance. These elements are connected in series and/or parallel combinations. The impedance expressions for each one are given by

$$Z_R = R, \tag{3.42}$$

$$Z_L = j\omega L, \tag{3.43}$$

$$Z_C = \frac{1}{j\omega C}. \tag{3.44}$$

It is observed that Z_R of an ideal resistor does not depend on the frequency and only has a real component. The current through the resistor is in phase with the voltage, while the impedance of an ideal inductor and capacitor only have an imaginary component. Z_L increases as the frequency increases, while Z_C decreases as the frequency increases. The current through the inductor and capacitor are phase-shifted 90° and −90°, respectively, with respect to voltage. An equivalent circuit must contain electrolyte resistance, a double-layer capacity, and the impedance of the Faradaic or non-Faradaic processes. Each one of these elements has a physical meaning, and some of them are described below.

Electrolyte resistance. The resistance of an ionic solution depends on the type of ions, ionic concentration, temperature, and the geometry of the area in which the current is transported. The electrolyte resistance (also ohmic resistance, denoted R_s or R_Ω) is given by

$$R_\Omega = \rho \frac{l}{A} = \frac{1}{\sigma}\frac{l}{A}, \tag{3.45}$$

where ρ is the resistivity of the solution, l and A are the length and the area across which the current is transported, and σ is the conductivity, reciprocal of the resistivity. When a sine wave voltage is applied to the electrochemical cell, the sine wave current through the resistor is in phase with the voltage, thus $\phi = 0$. For an MFC it is necessary to consider that the solution resistance has two contributions, the resistance of the electrolyte and the resistance due to the presence of micro-organisms in the solution. However, by adjusting the experimental EIS data, it is possible to calculate the R_Ω value.

Double-layer capacitance. At the electrode–electrolyte interface, ions from the solution approach the electrode surface. One of the phases acquires a positive excess charge, which is balanced by a countercharge of the same magnitude but the opposite sign in the other phase. The separation of the charge generates the double-layer capacitor (C_{dl}), whose impedance is inversely on the frequency. Frequently, the double-layer capacitance of a metal varies between 20 and 60 μF cm^{-2}, depending on ion type, concentration, temperature, films on the electrode (oxides, impurities), biofilms, the roughness of the electrode surface, applied potential, etc. The capacitance can be calculated by

$$C = \frac{\varepsilon\varepsilon_o A}{l}, \tag{3.46}$$

where C is the capacitance in farads, A is the plate area, ε is the dielectric constant of the material separating the plates, ε_o is the vacuum permittivity, ($\varepsilon_o = 8.85 \times 10^{-14}$ farads cm^{-1}), and l is the thickness of the material. The adjustment of the experimental EIS data also allows one to obtain this value.

Polarization resistance. At the equilibrium, $I_T = I_{ox} - I_{red} = 0$. The departure from the equilibrium by applying another potential, $\eta = E_{app} - E_{eq}$, polarizes the electrode, favoring the oxidation or the reduction reaction, where η is the over-potential. The generated current is given by the Butler–Volmer equation:

$$I_T = I_o\left[e^{\frac{(1-\alpha)nF\eta}{RT}} - e^{\frac{-\alpha nF\eta}{RT}}\right], \tag{3.47}$$

where I_o is the exchange current density, α is the transfer coefficient, n is the number of electrons involved in the reaction, F is the Faraday constant, R is the ideal gas constant, and T is the temperature (K). When $\eta \to 0$,

$$I_0 = \frac{b_a b_c}{2.303(b_a + b_c)} \frac{1}{R_p}, \tag{3.48}$$

where b_a and b_c are the anodic and cathodic Tafel slopes, and R_p is the polarization resistance, behaving like a resistor. R_p is not ideal resistance and is a function of the frequency. The polarization resistance has two contributions, the charge transfer-limited activation losses (R_{act}) and diffusion-limited concentration losses (R_d or R_{dif}).

Charge transfer resistance. The charge transfer through the interface of the metal–solution depends on the type of reaction, temperature, concentration of the reactive species, and potential. The current is given by

$$I = I_o\left[\frac{C_O}{C_O^*}\exp\left(\frac{(1-\alpha)nF}{RT}\eta\right) - \frac{C_R}{C_R^*}\exp\left(\frac{\alpha nF}{RT}\eta\right)\right], \tag{3.49}$$

where C_O and C_R are the concentration of the oxidant and the reductant at the surface electrode, C_O^* and C_R^* are the concentration of the oxidant and the reductant in the bulk. When $\frac{C_O}{C_O^*} = 1$ and $\frac{C_R}{C_R^*} = 1$, the current is given by

$$I = I_o\left[\exp\left(\frac{(1-\alpha)nF}{RT}\eta\right) - \exp\left(\frac{\alpha nF}{RT}\eta\right)\right]. \tag{3.50}$$

If the overpotential $\eta \to 0$ the system is at equilibrium, and the following equation gives the charge transfer resistance:

$$R_{ct} = \frac{RT}{nF} I_o. \tag{3.51}$$

The kinetics of reactions in an MFC depends on the charge transfer resistance of both the anode (R_P^a) and the cathode (R_P^c), and affects the mass transport by diffusion to and from electrodes.

Diffusion control. When the charge transfer is influenced by the mass transport of the electroactive species to and from the electrode, the Warburg impedance must be included in the electrical circuit. Figures 3.12 and 3.13 show the Nyquist and the Bode plots of a system where the polarization is controlled by a combination of kinetic and diffusion processes and show a line with a slope of 45° in the lower frequency region and a phase shift of 45°, respectively. The Warburg impedance (Z_W) depends on frequency; at high frequency Z_W is small as the electroactive species do not need to be transported very far; in contrast, at low frequency the species have to be transported far away by diffusion and Z_W increases.

The Warburg impedance is only valid if the diffusion layer has an infinite thickness and is given by

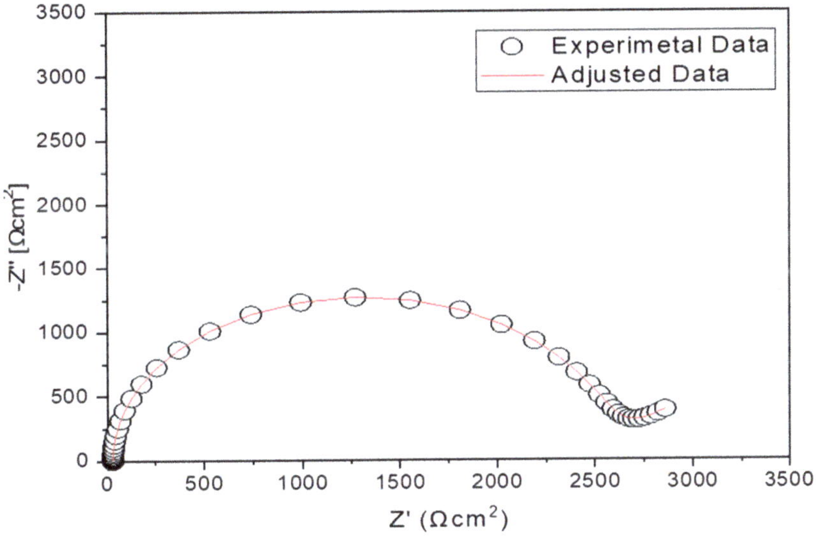

Figure 3.12. Representation of EIS data in a Nyquist plot with a Warburg element.

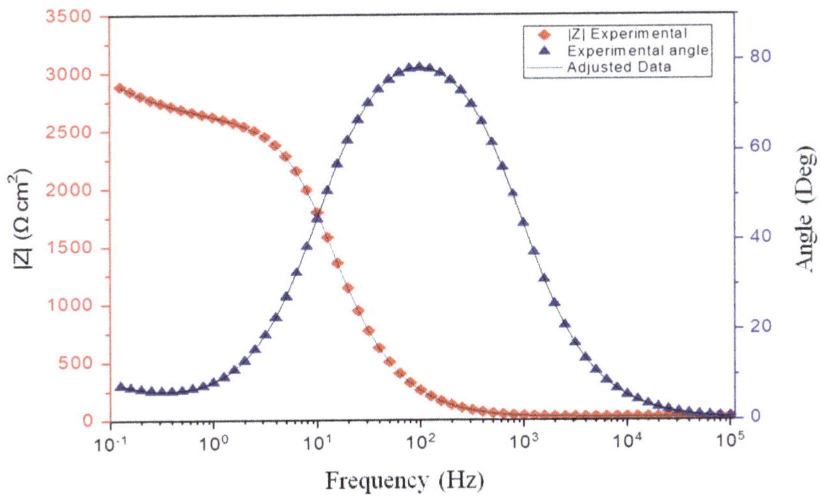

Figure 3.13. Representation of EIS data in Bode plots with a Warburg element.

$$Z_W = \sigma\omega^{-1/2}(1 - j), \tag{3.52}$$

where σ is the Warburg coefficient expressed by the following equation:

$$\sigma = \frac{RT}{n^2F^2A\sqrt{2}}\left(\frac{1}{C_O^+\sqrt{D_O}} + \frac{1}{C_R^*\sqrt{D_R}}\right), \tag{3.53}$$

where D_O and D_R are the diffusion coefficients of the oxidant and the reductant, respectively. If the diffusion layer is not infinite, at lower frequencies, the impedance is given by

$$Z_W = \sigma\omega^{-1/2}(1 - j)\tanh \delta\left(\frac{j\omega}{D}\right)^{1/2}, \tag{3.54}$$

where δ is the Nernst diffusion layer thickness and D is the average value of the diffusion coefficients of the electroactive species.

In an MFC the internal resistance, R_{int}, is given by the sum of the ohmic or solution resistance (R_Ω or R_s), anodic polarization resistance R_P^a, cathodic polarization resistance R_P^c, anodic concentration polarization R_C^a, and cathodic concentration polarization R_C^c:

$$R_{\text{int}} = R_s + R_P^a + R_P^c + R_C^a + R_C^c. \tag{3.55}$$

Constant phase element. A constant phase element (CPE or Q) is considered when the capacitor does not behave ideally. The impedance of the CPE is given by

$$Z_{\text{CPE}} = A(j\omega)^{-\alpha}, \tag{3.56}$$

where A is the admittance (inverse of the capacitance) and α is an exponent. When the surface of the electrode is homogeneous, it is possible to consider an ideal capacitor, then $\alpha = 1$, whereas at $\alpha = 0$ the CPE behaves as a pure resistor. CPE is used to model the double-layer capacitance and coating capacitance; in an MFC these are due to the presence of the biofilm.

Equivalent circuit models. Through the Nyquist plot is possible to determine the solution resistance (R_S) and the charge transfer resistance (R_{ct}) that correspond to the intersections of the curve to the real axis. However, using equivalent circuits, it is possible to analyze the EIS data, employing various electrical components such as resistors, capacitors, and inductors related to chemical, physical, and biological processes. The Randles circuit, figure 3.14, is an equivalent electrical circuit widely

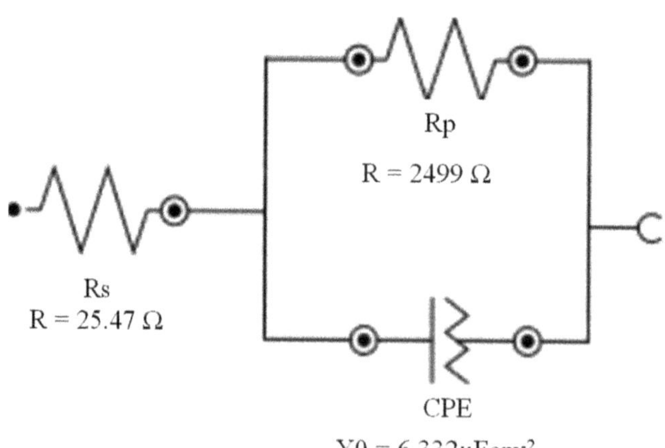

Figure 3.14. Representation of a Randles equivalent circuit.

used in the analysis of the corrosion of metals or coated metals and in also in MFCs. It is often used as a starting point for other more complex models. A Randles circuit comprises two resistors, R_S and R_{ct}, and a capacitance associated to C_{dl}. If there are limitations due to mass transport, a Warburg element is included.

It must be considered that experimental data can be modeled with several equivalent circuits and that an equivalent circuit can be used to model various experimental data, so it is important to consider that a proper assignment of the circuit elements is possible only if sufficient information is available on the various chemical, physical, and electrochemical processes that take place in the system being analyzed. By adjusting the experimental data with the appropriate equivalent circuit model, it is possible through computational algorithms to obtain the values of the impedances of the circuit elements.

3.5 Examples of applications of electrochemical techniques in MFCs

At this point, we want to respectfully recall and quote Professor Bruce E Logan and his research team [12]: 'a specific reason for a high power density—in a[n] MFC study—needs to be identified to publish such a finding'. A specific reason for a high power density is not only the presence of electrogenic bacteria, the metabolism of which certainly promotes and favors the process, but also depends on other factors such as, for example, viability over time and interactions with the components in the medium (wastewater, organic pollutants, metal ions, etc). In order to identify the reasons for a high or low power production beyond the electrogenic character of the micro-organisms, the utilization of one or several electrochemical characterization techniques needs to be conducted in combination with a systematic procedure to assess the MFC performance. As the interest in the development of efficient MFCs systems increases, different approaches to characterize these bioreactors are being developed and it is important to introduce the study of electronic circuitry fully in the bioelectrochemistry research team. It has been observed that electronic circuitry is not always taught as part of a bioelectrochemistry course and one of the reasons may be the cost of the equipment, since potentiostats are expensive, and another reason is that not all microbiologists, environmental or biochemical engineers, and biochemists who are developing work on MFCs have the necessary knowledge for managing the equipment and interpreting the results. The more basic equipment to conduct research on MFCs can be used easily, but the utilization of a potentiostat is not easy, taking into account that it can be used to obtain a great deal of information but only if the experiment is designed correctly. There is a large amount of recent work in the literature that presents a combination of several electrochemical techniques and some of the most recent are presented briefly here.

The study presented by Littfinski *et al* [29] focused on the comparison of six electrochemical techniques to evaluate the ohmic resistance of MFC during wastewater treatment. The comparison was meant to elucidate the compatibility and validity of the methods in order to facilitate the comparison of results among different research works. The methods evaluated were varying circuit resistance (VCR), linear sweep voltammetry (LSV), current interruption (CI), EIS, pulse-width modulation (R-PWM), and the perturbation observation (P/O) algorithm. The authors reported that the ohmic resistances obtained with the CI, EIS, and R-PWM methods are in high agreement. The investigation of the activation resistance through the EIS and R-PWM methods led to identical results. In their work, it was demonstrated that the real time method R-PWM can be applied by complementing with the calculations of the maximum power densities obtained from the methods VCR and LSV along with the adjusted resistance from the P/O algorithm. Particularly useful for researchers interested in applying these methods is the scheme of the model conception that involves the different methods proposed in that work which can be analyzed for further application in different MFC studies.

In order to distinguish between the internal and ohmic resistance in an MFC, Kim *et al* [30] proposed the scanning EIS. The method proposed the evaluation of the internal resistance and the charge resistance of each electrode of an MFC through a variation of the external resistance, in their case in the range of 0.03–100 kΩ selected from a discharge test. The recommendation made by this research group is to establish a systematic scanning EIS in order to obtain an accurate value for the internal resistance of the MFC and identify the phenomena that limit the MFC performance. The specific proposal presented by these authors is to replace the calculation for the internal resistance of the cell from the V–I and peak power calculation method with their methodology. In their method, R_{int} (internal resistance) and R_{ct} (charge resistance) are calculated. The R_{ohm}, R_{ct}, and R_{d} regions can be distinguished in the V–I curve and can be calculated by ohm's law but this is not possible for the internal resistance of the cell under different external resistances, therefore a measurement of the R_{int} needs to be made for each external resistor through the scanning EIS test and then the R–I curve can be interpreted to elucidate the internal resistance as a function of current. Although time consuming, this method could be assayed in other research groups in order to make more comparisons.

The work presented by López Zavala *et al* [31] includes the application of cyclic voltammetry to characterize the electrochemical performance of a dual MFC applied to the treatment of wastewater with regard to the effect of the electrolyte, pressure, reaction time, and sweep potentials. For this evaluation the potentiostat was used under three different settings and five experimental conditions. Since dual MFCs are often evaluated for wastewater treatment, this publication provides information on the utilization of CV that can serve as an example for the utilization of the technique in other cases for the evaluation of dual MFCs intended for wastewater treatment.

A comprehensive review of the background and applications of the electrochemical techniques presented in this chapter was recently published by Sánchez *et al* [32]. In that review, the reader will encounter an explanation of the principles and utilization of the electrochemical techniques that can serve, for example, as an introduction to a course on electronic circuitry. Research on MFCs does not always include the utilization of all the electrochemical techniques that are available, a frequent approach is to start with a basic MFC design and an inoculum with moderate or high electrogenic character to remove chemical oxygen demand (COD), phosphate, or nitrate from a wastewater stream and measure the power generated by the cell [33]. This approach has been followed since the beginning of this biotechnology to date, in view of the importance of deriving energy from biological processes [33]. Based on the results obtained from these first experimentations, the MFC designs are modified to improve the MFC performance, sometimes at a larger scale than in the laboratory (i.e. pilot scale) and then one or several of the electrochemical techniques available are used to gain insight into the factors that influence the MFC performance [34]. The composition of wastewater, either municipal, industrial, or specifically contaminated with organic toxic compounds or metals, frequently involves a combination of one or more of the electrochemical techniques in the analysis of the MFC performance. The design of the electrodes, the utilization of new materials for the electrodes, and the utilization of membranes in the MFC also require the application of one or several techniques. In table 3.2 a brief list of applications of electrochemical characterization techniques is presented depending on the purpose of the study involving the MFCs.

3.6 Conclusions and recommendations

In addition to the information available on the electrochemical characterization techniques of MFCs in the form of review articles and research reports in which several of these techniques are applied, it is important to include in-depth professional training in electronics and the utilization of the relevant equipment for at least one of the research members in the MFC laboratory team. While this may be obvious for most laboratories in the world, it has been the case for several authors to which this chapter refers, that the systematic application of the characterization techniques is not always carried out. In some instances the lack of equipment may be the reason for this, due to the cost, and in others it is due to the lack of sufficient knowledge on electronics that would allow the correct application of the techniques and interpretation of the results. A lack of adequate training with regard to which technique may be the best for each case may lead to difficulties in measuring the real advancement of MFC research in some cases. In addition to the observations and comments provided by several of the review articles cited in the current chapter, we could mention that interdisciplinary teamwork and maybe the development of tutorials to communicate the essentials for each technique could contribute to disseminate the correct utilization of the techniques, although it may be necessary to always count on the close support of an expert in the field of electronics.

Table 3.2. Recent reviews on the principles, applications, and advances on the most utilized electrochemical techniques to characterize MFCs.

Some relevant reviews on electrochemical characterization of MFCs	Electrochemical characterization technique that is revised and explained	Scope of the review	Highlights on the application of the techniques
Dave *et al* [15] Review chapter	Cyclic voltammetry (CV) Electrochemical impedance spectroscopy (EIS)	Fundamentals of voltammetry, descriptions of linear sweep voltammetry (LSV), differential pulse voltammetry (DPV), and cyclic voltammetry (CV). Fundamentals of impedance.	Detection of formation of biofilm. Characterization of electrodes. Characterization of electrode–electrolyte interaction. Characterization of bacteria–electrolyte interface. Determination of the electrode polarization resistance. Determination of the rate of corrosion and rate of consumption of metallic substrates.
Kretzschmar and Harnisch [35] Critical review article	EIS focused on biofilm–electrode interactions.	Discussion of EIT application for biofilm–electrode characterization and the effect of electrode material, size, and shape on the EIT measurements. Emphasis on the requirements for a meaningful application of EIT: (a) linearity, (b) stability, (c) guaranteed causality, and (d) finiteness of Z of the applied frequencies.	Understanding electrochemical mechanisms in biofilm electrodes. Biofilm capacitance. Charge transfer resistance. Diffusion at biofilm electrodes. Monitoring biofilm electrodes. Monitoring membrane fouling.

(*Continued*)

Table 3.2. (*Continued*)

Some relevant reviews on electrochemical characterization of MFCs	Electrochemical characterization technique that is revised and explained	Scope of the review	Highlights on the application of the techniques
Sánchez *et al* [32] Review article	Background and fundamentals on the electrochemical techniques focused on bioelectrochemical system characterization: direct current techniques; chronoamperometry and chronopotentiometry; fundamentals and applications for LSV and CV; fundamentals on alternating current techniques; EIS.	Description of the working principles of the potentiostat. Explains potentiostat circuitry, use, and limitations. Polarization analysis. Electrode potential control and the effect of the potential on the microbial metabolism.	Explanation of potential control as a tool to start-up an MFC that promotes the biodegradation of certain compounds in the anodic chamber and the synthesis of others in the cathodic chamber.
Yang *et al* [36] Review article	Characterization of electrodes to evaluate the charge transfer resistance.	Discussion of the factors that improve mass transfer in MFCs and characterization of the electrodes that can promote a higher mass transfer by EIS.	Examples of anode characterization using EIS in order to improve mass transfer.
Kashyap *et al* [37] Review article	Application of electrochemical impedance.	–	Comparison of the advantages and disadvantages among the different electrochemical characterization techniques.
Sekar and Ramasamy [38] Review article	EIS description.	Detailed fundamentals of EIS that allow the different resistances in the MFCs to be identified and guidance of its application in the characterization of MFCs.	Examples for anode and cathode impedance and the effect of biofilm, pH, electrolytes, and substrates. Example of biocathode EIS characterization. Examples of EIS for characterization of complete MFC systems.

References

[1] Mathuriya A S, Jadhav D A and Ghangrekar M M 2018 Architectural adaptations of microbial fuel cells *Appl. Microbiol. Biotechnol.* **102** 9419–32

[2] Jadhav D A, Das I, Ghangrekar M M and Pant D 2020 Moving towards practical applications of microbial fuel cells for sanitation and resource recovery *J. Water Process.* **38** 101566

[3] Jadhav D A, Mungray A K, Arkatkar A and Kumar S S 2021 Recent advancement in scaling-up applications of microbial fuel cells: from reality to practicability *Sustain. Energy Technol. Assess.* **45** 101226

[4] Haavisto J M, Kokko M E, Lakaniemi A M, Sulonen M L K and Puhakka J A 2020 The effect of start-up on energy recovery and compositional changes in brewery wastewater in bioelectrochemical systems *Bioelectrochemistry* **132** 107402

[5] Yaqoob A A, Khatoon A, Setapar S H M, Umar K, Parveen T, Mohamad Ibrahim M N, Ahmad A and Rafatullah M 2020 Outlook on the role of microbial fuel cells in remediation of environmental pollutants with electricity generation *Catalysts* **10** 819

[6] Kakar R, Ashraf Abdullahi A A, Rashid M, Hussain R T, Yaakop A S and Bhawani S A 2021 Bioremediation of pollutants and sustainable energy production through bacterial activities in microbial fuel cells: an overview *Asian J. Chem.* **33** 253–65

[7] Umar M F, Rafatullah M, Mohamad Ibrahim M N and Ismail N 2021 Enhanced benzene bioremediation and power generation by double chamber benthic microbial fuel cells fed with sugarcane waste as a substrate *J. Clean. Prod.* **310** 127583

[8] Umar M F, Rafatullah M, Mohamad Ibrahim M N and Ismail N 2021 Bioelectricity production and xylene biodegradation through double chamber benthic microbial fuel cells fed with sugarcane waste as a substrate *J. Hazard. Mater.* **419** 126469

[9] Miran W, Jang J, Nawaz M, Shahzad A and Lee D S 2018 Sulfate-reducing mixed communities with the ability to generate bioelectricity and degrade textile diazo dye in microbial fuel cells *J. Hazard. Mater.* **325** 70–9

[10] Lu N, Li L, Wang C, Wang Z, Wang Y, Yan Y, Qu J and Guan J 2021 Simultaneous enhancement of power generation and chlorophenol degradation in nonmodified microbial fuel cells using an electroactive biofilm carbon felt anode *Sci. Total Env.* **783** 147045

[11] Fadzli F S, Bhawani S A and Mohammad R E A 2021 Microbial fuel cell: recent developments in organic substrate use and bacterial electrode interaction *J. Chem.* **2021** 4570388

[12] Logan B E, Rossi R, Ragab A and Saikaly P 2019 Electroactive microorganisms in bioelectrochemical systems *Nat. Rev. Microbiol.* **17** 307–19

[13] Sharma M, Sarma P M, Pant D and Domínguez-Benetton X 2015 Optimization of electrochemical parameters for sulfate-reducing bacteria (SRB) based biocathode *RSC Adv.* **5** 39601–11

[14] Sharma M, Varanasi J L, Jain P, Dureja P, Lal B, Domínguez-Benetton X, Pant D and Sarma P M 2014 Influence of headspace composition on product diversity by sulphate reducing bacteria biocathode *Bioresour. Technol.* **165** 365–71

[15] Dave S, Das J, Shivani Dave P, Mohanty G M, Jagtap P P and Pawar R 2021 Contribution of cyclic voltammetry and electrochemical impedance spectroscopy in deciphering the electron transport system in biofilm *Analytical Methodologies for Biofilm Research* (Berlin: Springer) pp 129–52

[16] Yaqoob A A, Mohamad Ibrahim M N and Guerrero-Barajas C 2021 Modern trends on anodes in microbial fuel cells (MFCs): an overview *Env. Sci. Innov.* **23** 101579

[17] Mier A A, Olvera-Vargas H, Mejía-López M, Longoria A, Verea L, Sebastian P J and Arias D M 2021 A review of recent advances in electrode materials for emerging bioelectrochemical systems: from biofilm-bearing anodes to specialized cathodes *Chemosphere* **283** 131138

[18] Sahoo P C, Pant D, Kumar M, Puri S K and Ramakumar S S V 2020 Material–microbe interfaces for solar-driven CO_2 bioelectrosynthesis *Trends Biotechnol.* **38** 1245–61

[19] Wilberforce T, Sayed E T, Abdelkareem M A, Elsaid K and Olabi A G 2021 Value added products from wastewater using bioelectrochemical systems: current trends and perspectives *J. Water Process Eng.* **39** 101737

[20] Gupta S, Srivastava P, Patil S A and Kumar Yadav A 2021 A comprehensive review on emerging constructed wetland coupled microbial fuel cell technology: potential applications and challenges *Bioresour. Technol.* **320** 124376

[21] Madigan M, Martinko J, Dunlap P and Clark D 2009 *Brock Biology of Microorganisms* (London: Pearson Education)

[22] Eweis J B, Ergas S J, Chang D P Y and Schroeder E D 2001 *Bioremediation Principles* (New York: McGraw Hill)

[23] Logan B E, Hamelers B, Rozendal R, Schröder U, Keller J, Freguia S, Aelterman P, Verstraete W and Rabaey K 2006 Microbial fuel cells: methodology and technology *Env. Sci. Technol.* **40** 5181–92

[24] Logan B E, Zikmund E, Yang W, Rossi R, Kim K-Y, Saikaly P E and Zhang F 2018 Impact of ohmic resistance on measured electrode potentials and maximum power production in microbial fuel cells *Env. Sci. Technol.* **52** 8977–85

[25] Rossi R, Cario B P, Santoro C, Yang W, Saikaly P E and Logan B E 2019 Evaluation of electrode and solution area-based resistances enables quantitative comparisons of factors impacting microbial fuel cell performance *Env. Sci. Technol.* **53** 3977–86

[26] Rossi R and Logan B E 2020 Unraveling the contributions of internal resistance components in two-chamber microbial fuel cells using the electrode potential slope analysis *Electrochim. Acta* **348** 136291

[27] Rossi R, Hall D M, Wang X, Regan J M and Logan B E 2020 Quantifying the factors limiting performance and rates in microbial fuel cells using the electrode potential slope analysis combined with electrical impedance spectroscopy *Electrochim. Acta* **348** 136330

[28] Dominguez-Benetton X, Sevda S, Vanbroekhoven K and Pant D 2012 The accurate use of impedance analysis for the study of microbial electrochemical systems *Chem. Soc. Rev.* **41** 7228–46

[29] Littfinski T, Nettmann E, Gehring T, Krimmler S, Heinrichmeier J, Murnleitner E, Lübken M, Pant D and Wichern M 2021 A comparative study of different electrochemical methods to determine cell internal parameters of microbial fuel cells *J. Power Sources* **494** 229707

[30] Kim B, Chang I S, Dinsdale R M and Guwy A J 2021 Accurate measurement of internal resistance in microbial fuel cells by improved scanning electrochemical impedance spectroscopy *Electrochim. Acta* **366** 137388

[31] López Zavala M A, González Peña O I, Cabral Ruelas H, Delgado Mena C and Guizani M 2019 Use of cyclic voltammetry to describe the electrochemical behavior of a dual-chamber microbial fuel cell *Energies* **12** 3532

[32] Sánchez C, Dessi P, Duffy M and Lens P N 2020 Microbial electrochemical technologies: electronic circuitry and characterization tools *Biosens. Bioelectron.* **150** 111884

[33] Negassa L W, Mohiuddin M and Tiruye G A 2021 Treatment of brewery industrial wastewater and generation of sustainable bioelectricity by microbial fuel cell inoculated with locally isolated microorganisms *J. Water Process. Eng.* **41** 102018

[34] Babanova S, Jones J, Phadhke S, Lu M, Angulo C, García J, Carpenter K, Cortese R, Chen S and Phan T 2019 Continuous flow, large-scale, microbial fuel cell system for the sustained treatment of swine waste *Water Env. Feder.* **92** 60–72

[35] Kretzshmar J and Harnisch F 2021 Electrochemical impedance spectroscopy on biofilm electrodes—conclusive or euphonious? *Curr. Opin. Electrochem.* **29** 100757

[36] Yang W, Li J, Fu Q, Zhang L, Wei Z, Liao Q and Zhu X 2021 Minimizing mass transfer losses in microbial fuel cells: Theories, progresses and prospectives *Renew. Sustain. Energy Rev.* **136** 110460

[37] Kashyap D, Dwivedi P K, Pandey J K, Kim Y H, Kim G M, Sharma A and Goel S 2014 Application of electrochemical impedance spectroscopy in bio-fuel cell characterization: a review *Int. J. Hydrogen Energy* **39** 20159–70

[38] Sekar N and Ramasamy R P 2013 Electrochemical impedance spectroscopy for microbial fuel cell characterization *J. Microb. Biochem. Technol.* **S6** 1–14

IOP Publishing

Microbial Fuel Cells
Emerging trends in electrochemical applications
Mohamad Nasir Mohamad Ibrahim, Asim Ali Yaqoob and Akil Ahmad

Chapter 4

The effect of bioelectrochemical reactor design on the electrochemical performance

Tayyaba Kanwal, Akil Ahmad and Mohammed B Alshammari

Energy and water storage have become worldwide problems due to a variety of factors, including environmental changes, increased populations, a rising demand for energy at both the commercial and residential levels, and high material costs. To solve these problems, bioelectrochemical systems are being promoted as a cutting-edge technology that can generate electricity while also enhancing the removal efficiency of different pollutants from wastewater. Bioelectrochemical systems have the potential to generate electricity from natural waste sources and may also be used to clean wastewater. This chapter examines the various designs and configurations of bioelectrochemical systems, with a focus on energy generation and wastewater treatment. This chapter also goes over several different approaches to bioelectrochemical systems, as well as some possible future perspectives. These issues, as well as future recommendations, are incorporated in this chapter.

4.1 Introduction

The effects of electrical currents on glutamic acid have been studied since the late 1970s. Many electrically influenced biosynthesized materials and their applications have been described in the literature, along with the emerging research area of bioelectrical systems [1, 2]. Early research reported by Krieg *et al* [3] was the first time bioelectrochemical systems were used by developing and utilizing a basic experimental set-up. Two glass bottles were used to make these reactors and each of them was considered as a half-cell with a cathode and an anode. The flasks were fused with glass in an H-shaped model, known as H-shaped cells or more commonly as H-cells. In this reported method, a membrane is used for separation of the electrolytes of the two half cells for ionic connection maintenance [4]. These types of reactors are still in use, in particular in the area of electrobiotechnology. However,

doi:10.1088/978-0-7503-4791-4ch4

this field is still progressing with the development of various distinct enzymatic and microbial bioelectrosynthesis techniques. Therefore, different processes, and standard and adaptable reactors can be used to meet progressive microbial and electrochemical requirements [5, 6]. Reactor design mainly depends on the substrates and products used. For membrane and membrane-free techniques, further work is needed. Moreover, several research groups are exploring bioelectrochemical techniques with their own distinctive reactor designs and can provide low-cost, simple, and quick solutions. As a result, the literature contains a wide range of non-optimized reactors [7]. The reactors utilized for microbial and enzymatic electrosynthesis are discussed in this chapter, along with their scope for future application. This chapter discusses a wide range of reactors based on their shape and presentation that can be evaluated from an engineering or technical standpoint.

4.2 General characterization of the systems

The yield and the rate of reactions are important factors that characterize industrial and bio-production processes. Therefore the procedures of reactors must be investigated and an optimization process should be carried out to optimize the parameters. Apart from the common substrates of carbon and nitrogen used in all bioprocesses, electrons are likewise important reactants in electrobiotechnology, and are either added to (through the cathode) or removed from (by the anode) the reactor broth [9].

Several mechanisms can be used to transport electrons which always depend on an electrode which is in contact with the reactor liquid [10, 11]. Bioelectrochemical systems are hybrid because the have to interface electrodes with fluids and separate anode–cathode cavities. These systems must meet the requirements of conventional bioreactors while also possessing the characteristics of an electrochemical reactor.

Electrode engineering may be used for microbial electrosynthesis (MES) reactor engineering by modifying the specific approach [12]. Reactors along with enhanced mass-transfer are needed for microbial electrosynthesis based on facilitated electron transport. In fact, there are currently no off-the-shelf and no gold-standard reactors for bioelectrochemical systems and in order to develop such reactors a variety of resources and geometries have been used [3].

Until now, these vessels have only been employed in particular studies and in specific laboratories [13]. Therefore only a few comparisons can be made between experiments and reactors. This is correct when using microbial synthesis as a baseline for MES without using electrons as reactants. To solve these problems, various publications highlight importance of the standardization of data representation and this will pave the way for MES to be assessed in a systematic manner.

The existing literature provides an overview of the key parameters for bioelectrochemistry, container engineering, electrochemistry, biotechnology, and biofilm categorization in detail. In particular, reactor model factors are under-reported in previous studies [14, 15]. It may be stated in an infinite number of ways, which is useful for characterizing the flow area. We would like to bring in special consideration to the coulombic efficiency, which is frequently analyzed in research, among all the parameters described. However, its definition may differ based on the authors' point of view, so it is critical to properly clarify the mathematics so that there are no misunderstandings when comparing different results [3]. Moreover, a simple and familiar experimental error can skew coulombic efficiency calculations, such as the adding of a chemically ambiguous component to a predetermined growth medium that can possibly function as a carbon source, i.e. yeast isolate or meat hydrolysate. Using the quantity of chemical oxygen demand as the overall carbon equivalent for the coulombic efficiency calculation is a simple solution to solve this problem. We consider that comprehensive development of processes spanning scales varying from fluid changing aspects to microbial extracellular electron transport is required in knowledge-driven engineering [16]. Hence, future research into scaling up variables for electrobiotechnological processes is needed.

4.3 Reactors for microbial electrolysis cells and microbial electrosynthesis

Microbial electrolysis cells (MECs) first originated from microbial fuel cell (MFCs), which were used for the synthesis of hydrogen gas by microbes, electronic energy, and an effluent stream [17]. The formation of hydrogen occurs at the (bio)cathode, where protons produced at the (bio)anode are reduced by electrons. In this anodic process, microbes decompose waste into electrons, protons, and tiny molecules such as CO_2, or an abiotic catalyst converts water into oxygen, electrons, and protons [18]. The voltage produced from a bioanode connected with a cathode used for hydrogen production is insufficient to reduce protons to hydrogen. Thus, if electrical current is generated at the anode through reduction of waste, more energy must be delivered externally to permit hydrogen synthesis. In addition, the voltage must be leveraged, thus, in comparison to the energy from the MFC, a utilized circuit power of at least 110 mV is required. The result at the cathode might be abiotic (mediated by platinum electrodes) or biotic (catalyzed by micro-organisms) [19]. MECs can be utilized to make various fuels and chemicals in a broader sense. If more complicated chemicals are produced, the process is known as microbial electrosynthesis (MES). However, this actually applies to all anodic and cathodic biomanufacturing procedures, irrespective of whether they start with carbon dioxide or organic substrata. The great majority of MEC and MES experiments have been carried out in the research laboratory with a volume scale varying from milliliters to liters [20]. Therefore, these experimental systems are typically suited to a basic investigation subject, however, they are frequently under-described from a manufacturing point of view. Several reactor designs have been created to date, and rest of this section will describe these in detail.

4.3.1 Single-chamber systems

Many reactors utilize membranes to split up anode and cathode chambers, however, these membranes provide additional barriers to H^+ distribution to the cathode, increasing the system's internal resistance [21, 22]. To solve this problem, there have been developments in single-chamber reactors. The anodic response is designed to generate carbon dioxide from the decay of an organic substance while generating hydrogen at the cathode, which can be collected with the hydrogen in the reactor's headspace. The hydrogen generated at the cathode can be used by microbes from the anode which is a major drawback. As a result, if methanogens are not suppressed, most systems designed to produce hydrogen end up producing methane. To boost the production and purity of the product, Fu *et al* [23] proposed cultivating a methane-manufacturing community in a single-chamber reactor and then moving the cathode to a two-chamber system. Membrane and membrane-less systems produce various products according to Xafenias *et al* [24]. Single-chamber systems may be a superior alternative, in particular for scale-up, as layers are prone to biofouling, require preservation, and are hence very expensive. There is only one electrolyte mixture in single-chamber arrangements. The electrolyte mixture is divided into a catholyte and an anolyte in two-chamber systems. Separators can be used to physically separate electrodes in single-chamber systems to prevent short circuits [25].

4.3.2 Single-chamber cube-type reactors

A cube-type system was used to construct the main single-chamber reactors, which was nearly identical to the one illustrated in figure 4.1 except for the absence of the membrane. This mechanism can yield both hydrogen and methane. This model does not inhibit chemical crossover due to its basic and membrane-free design; yet cube-type devices are frequently employed for level-up research due to their ease of

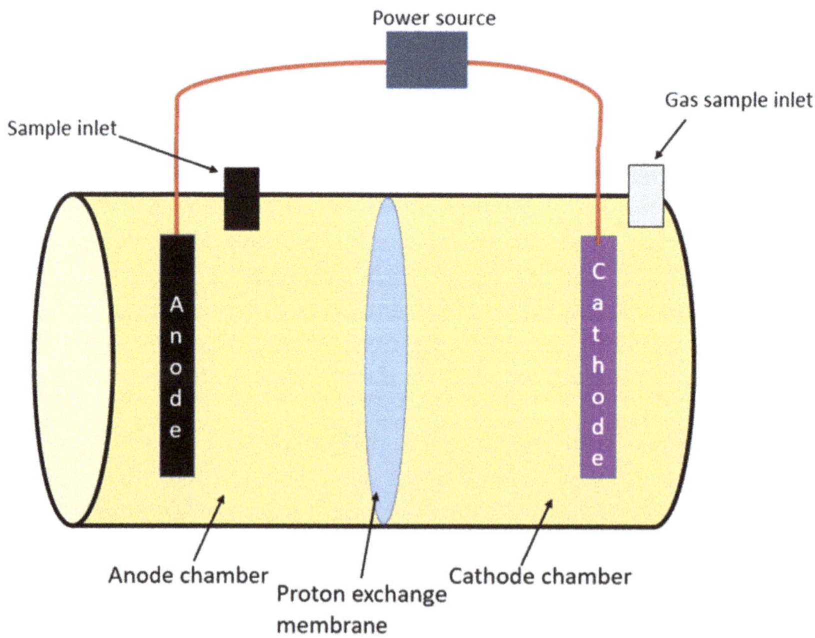

Figure 4.1. Cube-type reactors with a single chamber.

handling [26, 27]. Moreover, an increase in the quantity of conductors or three-dimensional assemblies can be used to change the electrode surface area. Because increasing the quantity of electrodes for scale-up is easier than raising the electrode surface zone, Rader *et al* [26] proposed integrating numerous pairs of electrodes into a single-chamber system. This may also help to avoid ohmic losses caused by internal resistance drops. Thus they devised a system that resembled a number of coupled cube-type systems, with anodes and cathodes positioned alternately. Short circuits were avoided by using barriers between the anodes and cathodes. Hydrogen and methane were formed with a constant substrate flow. Cusick *et al* [28] employed an analogous system in a pilot-scale application to make methane and hydrogen. In a 1000 l reactor, this arrangement had 144 pairs of electrodes.

4.3.3 Cylindrical reactors

Cylindrical single-chamber systems have been employed in the same way as double-chamber spherical cylindrical systems, as shown in figure 4.2. A reactor constructed by Bo *et al* [29], for example, contained a steel pipe that served as the cathode and a circular anode inside the cathode cylinder. This permits constant ability spreading throughout the container. However, compared to the electrode–membrane packs commonly utilized in cube-type reactors, the gap between the anode and cathode is large. One problem is that the cathode material, which is critical for invention creation, is difficult to change for future research.

4.3.4 Bottle-type reactors

The bottle-type method is one system that is particularly well suited to examination and initial research laboratory measurement analyses, as shown in figure 4.3. Equally

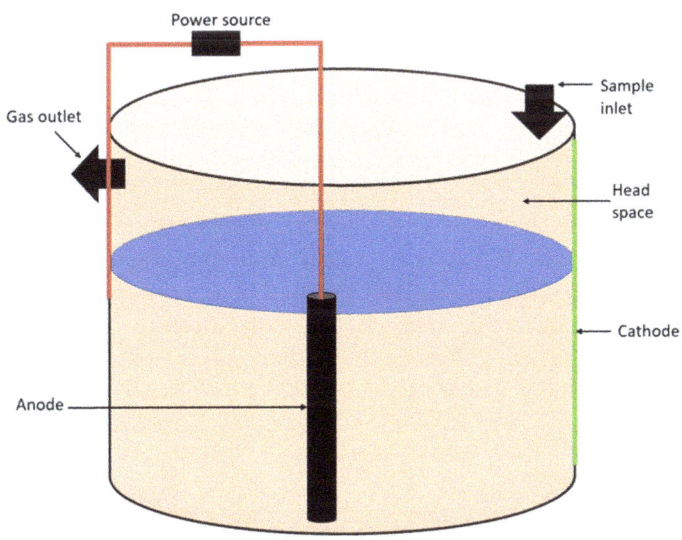

Figure 4.2. Single-chamber cylindrical reactors.

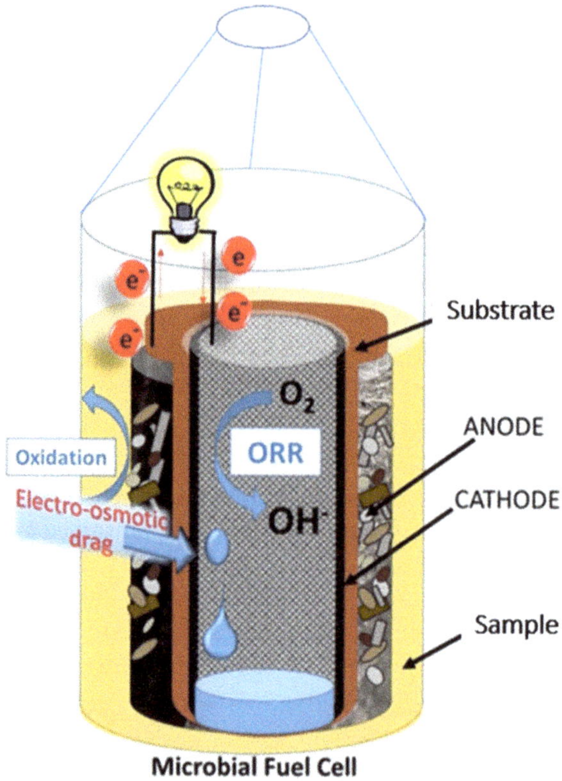

Figure 4.3. Single-chamber bottle-type reactors. (Reproduced with permission from [33]. Springer Nature Open Access.)

probes are positioned in a minor flask, [30, 31], and an isolator can be placed between the electrodes to prevent short circuiting. However, the electrode placement is not stable, reducing the experiment's repeatability and controllability. However, since all of the essential elements are routinely found in laboratories, this system may be set up quickly and inexpensively. These mini-reactors have been used in screening systems, which use a single power supply for multiple tests at once [31]. The high support vector regression (SVR) of the chosen cathodes drives the performance of these systems [32]. Unfortunately, due to the limited area within the septum containers, tests involving three-dimensional electrodes cannot be conducted in a broad range. In comparison to experiments with other reactor designs, the particular electrode area appears to be low in general [30]. As a result, a comparable patter may be extracted in a cylindrical reactor by switching out of the septum bottle.

4.3.5 Column-type reactors

A cathode-on-top strategy has been devised to avoid hydrogen interaction with anodic bacteria capable of metabolizing the product [34]. A schematic presentation is shown in figure 4.4. The cathode is positioned beyond the anode, enabling

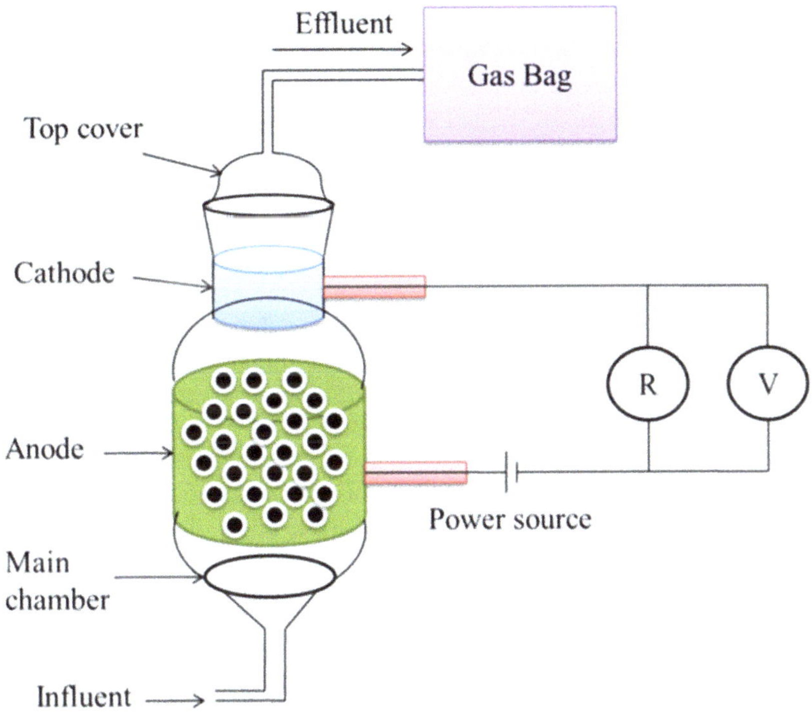

Figure 4.4. Cathode-on-top reactors. (Adapted with permission from [39]. MDPI Open Access.)

hydrogen boils produced at the cathode to leave the culturing process without passing through the anode or associated micro-organisms. The electrodes are not placed horizontally. The inner conformation of the reactors is significant because of the huge gap between the anode and the cathode [34]. Because it provides superior support for microbial attachment, the anode can be built as a permanent substrate. Liu *et al* [35] devised a cathode-on-top system for hydrogen production that uses a liquidized bed anode rather than a fixed bed anode. Carbon particles are liquidized at the lowest side of the apparatus to rise to the anode surface and provide biofilm sustenance in this reactor. Katuri *et al* [36] proposed using resonating-fiber layers as cathodes to merge filtering with electrosynthesis. As a result, an electromicrobial membrane reactor can be described as a unique column-type reactor system. Li *et al* [37] have already proposed mathematical models for this type of reactor architecture. The goal of the cathode-on-top shape is to enhance hydrogen manufacture without the use of a membrane. However, if gases (such as oxygen or carbon dioxide) are created at the anode, it is not acceptable because it reduces the purity of the produced gas or, in the case of methane manufacture at the cathode, inhibits micro-organisms by producing oxygen. To prevent this, an anode-on-top design, in which the anode is situated above the cathode, is recommended. For MES based on CO_2, to prevent the generated oxygen from coming into contact with the strictly anaerobic microbes at the anode without using a membrane is an attractive approach [38]. This sort of reactor was utilized in the initial generation of MFC investigations; however, it was eventually replaced by systems with smaller internal resistances.

4.3.6 Rotating disk reactors

One further fascinating reactor was constructed by Cheng *et al* [40] who employed a tubular single-compartment reactor together with casual rotational half-disks as electrodes. Together the electrode half-disks are immersed in liquid and subjected alternately to gaseous phases. As a result, each disc half acts as both an anode and a cathode, with the cathode acting as an anode when subjected to the vaporous phase and the anode acting as an anode while immersed in the liquid phase. At any given time, a portion of the gas-phase electrode is buried in the liquid form, resulting in an ionic association between the electrodes [40]. As a result of this method, a consistent biofilm forms all around the disc. Methane is major invention, which is discharged immediately into the gas phase. One benefit of this arrangement is that there is no pH increase between the cathode and the anode. Even though the quantity of electrodes can be increased, the scale-up is restricted due to the high energy required to replace the rods in a possibly viscous medium [41, 42].

4.4 Two-chamber systems

Many MECs/MESs have two chambers separated by a membrane: an anode section and a cathode section. Despite the fact that layers can limit a system's working due to their electrical confrontation and ion flux resistance, they can also improve it. As it prohibits contact between the electrodes (chambers) through the distribution of decreased and/or oxidized molecules, a layer can be useful, particularly when high-level coulombic efficiencies are required (chemical short circuits). Membranes also keep products from leaking between chambers. In the case of cathodic methane synthesis, for example, a sheath prevents oxygen from reaching the cathode chamber via water electrolysis, potentially inhibiting the oxygen-sensitive microbial community [43]. Of course, this is only true if the membrane tools are selected based on the desired selectivity. Reactors using layers typically produce purer results or off-gas streams than reactors with no membranes.

4.4.1 H-cell reactors

The H-cells, as previously mentioned, are one of the very popular dual-chamber reactor arrangements. They are made up of two chambers, each of which is normally a glass bottle, divided by a coating. Figure 4.5 depicts the conventional shape of an H-cell. Nevin *et al* [44] originally reported MES in H-cells initially from CO_2 in 2010; since then, many MES studies in H-cells have been conducted. The best part of these studies indicate acetic acid as the main or only result; however, additional products such as methane or butyrate were also observed in some cases. H-cells offer a wide range of applications, are very simple to set up, can accommodate a magnetic stirrer and a gas intake, and can be used with a variety of electrode materials. Two 200 ml bottles are divided by a cation-exchange layer, and graphite rods serve

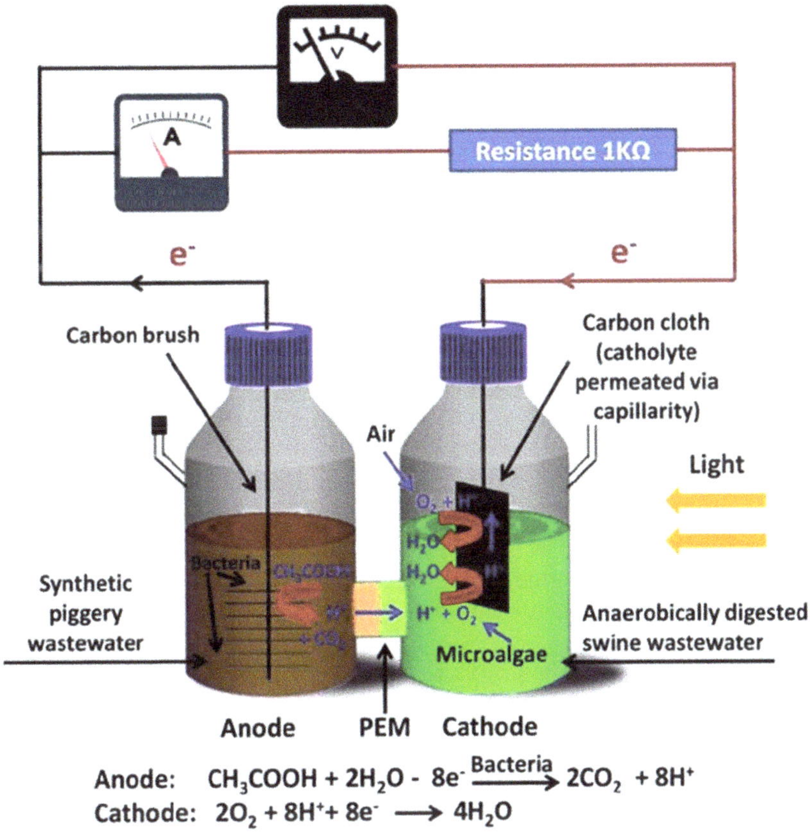

Figure 4.5. H-cell reactors. (Adapted with permission from [46]. MDPI Open Access.)

mutually as the anode and cathode in a typical arrangement. The MES investigations' poor space–time profits also suggest that such a reactor may not be appropriate for achieving commercial significance with MESs [45]. In other circumstances, however, profits may be restricted by the micro-organisms rather than the reactor system itself.

4.4.2 Concentric tubular reactors

Circular tubular cells have been employed as a laboratory-scale equivalent to H-cells in various studies. The anode chamber is enclosed inside the larger cathode compartment, divided by a cylinder-shaped glass tube with a layer window, which is commonly located at the bottom of the anode chamber. Xu *et al* [47] conducted an innovative study on this sort of reactor, comparing methane generation in a biogas-updating procedure in H-cells and a concentrical tubular MEC. They determined that the concentric cylindrical MEC performed much better; nevertheless, the evaluation is challenging to make due to the differences in sheath zone, substratum, and procedure type. Jourdin *et al* [48] showed fed-batch control of this reactor type in an acetate-producing MES. When associated with a H-cell, this reactor has the benefit of being easier to set up and taking up less space; also, the membrane area may be readily modified when utilizing a 'membrane bag' instead of a membrane win. By packaging the membrane around a permeable glass container and forming an electrode–membrane construction, the electrodes can all be near each other to reduce the system's internal resistance. In H-cells, the electrode SVR seems to be reduced. Sasaki *et al* [49], Carmona-Martinez *et al* [50], and Jeon *et al* [51] have all employed similar ideas on a wider scale. In a concentric cylindrical reactor, Sasaki *et al* [49] generated methane on a 2–4 l scale. Carmona-Martinez *et al* [50] additionally looked at the system's long-term performance and hydrogen manufacture. A 1 l MES generating acetic acid, propionic acid, and butyric acid was studied by Jeon *et al* [51]. The yields achieved while electrochemically creating hydrogen with a utilized voltage are comparable to a system growing with 2 atm of hydrogen above pressure, which is an interesting result of this investigation. This study demonstrates that the hydrogen produced by the hydrogen process at the cathode may be the cause of the bioelectrocatalysis increase over a conventional culture without electrochemistry. The discussion in this section demonstrates that this technology is further ascendable than the H-cell. Low-velocity regions and intensity slopes, which are dependent on the diffusivity of the substratum, may emerge in larger structures because mixing is impeded by the implanted chamber [51]. This problem has been addressed in studies by using an external loop to promote mixing; continuous running of the system has been recorded.

4.4.3 Two-chamber cube-type reactors

The cube-type or tubular MEC is another popular container. This reactor is essentially the same as the H-cell. The dual compartments, however, are not two containers, but two cylinder-shaped, flat, disk-designed compartments created by cutting cubic plastic materials, which are separated by a membrane, as shown in figure 4.6. As a result, compared to the H-cell, the sheath area comparative to volume can be enlarged or reduced by utilizing a tiny membrane window. In addition, because this sort of device does not always have a headspace for gas collection, a gas pipe (pipe or tube) on the upper side of the two compartments is required. The cube cells with higher headspace were shown to be more applicable in terms of sampling when compared to cube cells with very little headspace. Liu *et al* [52] examined the presentation of a cube MEC and an H-cell during hydrogen manufacture in a 2005 study. The cube system produced less hydrogen, presumably due to the creation of diffusion across a wider membrane surface to the anode chamber. Villano *et al* [45] used a comparable reactor design to make a 'fixed bed reactor' by adding graphite grains to the chambers. In such a strategy, medium recirculation is used to ensure mixing. This reactor has an advantage over the H-cell in that the distance between the cathode and anode is shorter, lowering the internal resistance. This system is further more adaptable and ascendable than an H-cell since the membrane zone, gas intake and exit, substrate inlet, and electrode space may all be changed. There have also been reports of continuous operating modes for

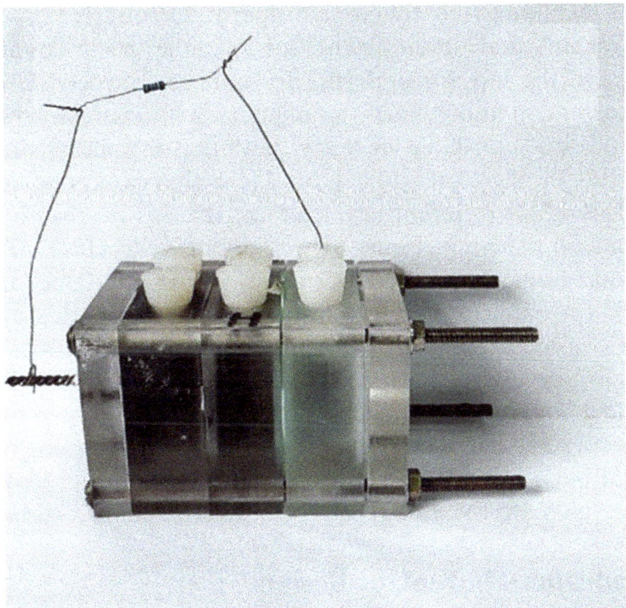

Figure 4.6. Two-chamber cube-type reactors of MFCs. (Reproduced with permission from [54]. MDPI Open Access.)

such cubic reactor designs [53]. By adding an additional layer to generate a three-chamber structure, this device was able to achieve immediate attention of the MES product acetate.

4.5 Significance evaluation of MET designs

In section 4.2 numerous reactor types are explained, and in the following they are rated on the basis of several performance and design factors. The selected projects are supplied with the appropriate literature and model assessments, since the application of individual reactors varies substantially. The anode–cathode distance and current assembly, electrolyte mixing SVR (which affects volumetric current densities and production yields), chemical short circuits, are all considered when evaluating the performance. In addition, the design's scalability, estimated costs, and complexity are rated because these factors influence its applicability and practical implementation. When determining the values of different rating limitations, it is important to consider the purpose of the reactor in a specific case, particularly when preforming experimental laboratory work with important research questions versus application-driven experiments. The anode–cathode gap, for example, is insignificant in three-electrode settings for investigating microbe–electrode connections, but it is critical in a scaled-up apparatus. It is challenging to make a definitive proposal on the best reactor designs due to the poor comparability of key performance indicators. Many of the reactor systems presented here fail to meet several essential conditions for microbial electrochemical technology to function properly. Obviously, this is not always necessary; in research and development, for example, basic designs that can be repeated simply in the lab are preferred to more elaborately designed systems. Some of the current methods, such as reactors with successive anode–cathode positions (cathode-on-top), have so many flaws that it appears prudent to phase them out [8]. After all, there are some natural laws that cannot be overturned. Due to the varying needs for different applications, there is no reactor configuration that outperforms all others in terms of well-planned structures. Pilot tests have indicated that some reactors, including the level plate and lamellar methods, are quite well suited for applications. Internal resistance and mixing concerns should be given more consideration while building reactors. The central message with regard to reactor design is that existing information should be given more consideration (e.g. from the electrochemistry field). Merging projects that hold to the key physical rules is crucial if microbial electrobiotechnologies such as MES are to gain credibility. Commercial reactors can be extremely beneficial in this situation as they increase the global approach to well-defined technology.

4.6 Recent advances in reactor design

The elementary components of reactors have been used in a wide range of designs in various bioelectrochemical systems (BESs). Reactors come in a variety of sizes and types, ranging from microfluidic reactors and testing instruments with contents as little as a few milliliters, to pilot-scale devices for manufacturing and public effluent action with capacities as large as 1 m^3. The current work in reactor projects at these diverse scales is examined in the sections that follow.

4.6.1 Miniaturized MFCs and screening tools

Because of their small electrode spacing (down to 100 mm), smaller MFCs have a lower internal resistance. Testing for electrochemically effective micro-organisms, optimum development requirements, and promising electrode fabrics, and acting as power sources for sub-100 mW products are some of the main applications of such devices [52]. Potential new applications are being looked into, such as using implanted devices to power medical implants such as pacemakers. Several microfluidic MFCs with various electrode materials have been constructed with capacities ranging from 1.25 ml to several microliters, allowing for a lab-on-a-chip method to existing manufacture with pure and mixed cultures [52]. The future problems of miniaturizing MFCs with increased power densities have been discussed elsewhere extensively. Different tiny MFCs have been examined in the laboratory, but their functional applicability remains a concern. These miniature-scale methods are useful for investigating how device architecture interacts with active biocatalysts. Fundamental issues with electron transmission at the natural electrode contact can also be investigated [55]. Because the best process results were obtained in very small-scale BESs, the outcomes can be utilized as a benchmark for larger structures. Half-cell experiments are frequently used in MFC research to evaluate bacteria and electrode materials for the first time. To date, several systems have been published.

For the optimization of BESs and the clarification of the essential aspects affecting functioning, more effective screening platforms are required. As a result similar low-cost systems are required. Recently, an assessment centered on tungsten fragments, which change color from white to blue when reduced, was created [56]. Electroactive organisms such as *Shewanella oneidensis* can reduce these particles, and electro-activity compares with the color shift, which can be calculated using a standard micro-titer plate (MTP) reader. The units can also be combined with Lauren broth (LB) agar to detect the development of electroactive bacteria through shade change [57, 58]. An alternative method involves integrating electrodes (A electrode = 2 cm^2) into an adapted MTP lid to create an electrochemical MTP (eMTP), which enables the broadcast and optimization of electroenzymatic processes [59]. The eMTP allows for parallelized electrochemical trials in conjunction with simultaneous optical data collection, and it can be utilized to design processes quickly. Miniaturized systems for evaluating BESs have also been created. A parallel 24-well MFC array (anodic chamber volume = 650 ml, A electrode = 38.5 mm^2) separated by a proton exchange membrane was used to screen for electroactive bacteria in ambient samples to select interesting candidates for further characterization investigations. Small-scale, sheath-free devices have also been constructed to assess the electroactivity of various bacteria and electrode materials [8]. The fundamental features and possible electron givers of electroactive bacteria, as well as enzymatic fuel cells (EFCs), have been screened for using modified pipette tips (anodic chamber volume = 500 ml) [23]. The cathode was created by pressing graphite felt into the minor half of a 1 ml pipette tip. The cathodic chamber was then coated with a prepared nafion sheath, and the anodic compartment was created by filling the upper section of the pipette slope with graphite felt. There have been a few fascinating instances of screening tools revealed thus far, but high throughput systems are still lacking.

4.6.2 Laboratory-scale bioelectrochemical systems

On the laboratory scale (several milliliters up to 1 l) a wide range of BES reactors have been described. Their reactor designs can be used to classify them: H-shaped reactors with an ion exchange sheath dividing the anodic and cathodic cavities that connects two bottles; single-compartment reactors, in which electrodes are exposed to air or additional gases that can contribute to the reaction [60]; the cube-shaped and cylindrical reactors, in which the cavities are divided by methods described previously; and the single-chamber reactors, in which conductors are unprotected by air or other gases that can contribute to the response [61]. In the case of microbial systems, the initial characterization tests are usually carried out in H-shaped reactors. However, because of the ohmic losses produced by a distal electrode arrangement, the performance in these reactors is generally low, and parallelization is problematic. In addition, this system is typically run in batch mode. Continuous systems, on the other hand, are desirable because they provide an adequate supply of substrate for the electroactive biocatalyst [62]. Because of the narrow electrode spacing, the lack of centrifuges, and appropriate counter reactions, numerous designs have been created and assessed to ensure low interior resistances. Tubular reactors have also been created in a number of configurations for MFCs and EFCs containing a wide range of electrode materials [34–36, 40]. For power generation from wastewater, dissimilar electrode pairings can be used, such as assembled graphite disc anodes paired with cathodes subjected to air. A long-term study of a 10 l serpentine (tubular)-type MFCS stack processing brewery wastewater was conducted using a comparable device [63]. The serpentine reactor including up to 40 tubular single MFCs with air cathodes and carbon felt anodes was able to harvest control over a 120 day period [34]. Another approach is to construct an anodic compartment near a catholyte-perfused inner tube. Immobilized enzymes and mediators have been employed in these systems for glucose-based EFCs. In cylindrical reactors for MFCs and electroenzymatic methods, fluidized and packed beds acting as electrodes can be employed. The use of a small flow-through structure with carbon cloth as the anode and cathode material increased long-term power generation (63 days). Low internal resistances were achieved by reducing electrode spacing and using a low-cost fabric separator. In comparison to feeding systems, the electro-generation of hydrogen peroxide as a co-substrate for enzymatic reactions may be helpful as a dose-up mechanism. Hydrogen peroxide is produced in small but significant volumes utilizing a separated system with a 3D graphite particle-packed bed cathode. The usage of a gas dispersal electrode as a cathode in a complete flow-across system is an alternative approach for electrochemical hydrogen peroxide synthesis [60]. When the apparent surface area of the gas distribution electrode-based system was compared to the immovable bed reactor system, the hydrogen peroxide productivity was higher. If enzymes rely on cofactors, industrial applications will necessitate the redevelopment or substitution of classical cofactors such as NADH or NADPH. In a scalable system, redox active mediators can be used for incidental regeneration. In a tubular reaction system with a packed bed, NADP+ or NAD+ is reinforced electrochemically, and alcohol dehydrogenase-catalyzed corrosions have been observed. Several types of reactors have been constructed, however, there has yet to be a comparison of their advantages and disadvantages. Due to variances in settings, lost characteristics such as the electroactive shallow region, or the wide variety in the living constituents utilized (different combined cultures, micro-organisms, or enzymes), it is often impossible to compare results [2, 64]. In addition, alternative styles of process should be examined because they all produce distinct results.

4.6.3 Technical scale bioelectrochemical systems

Only MFCs and MEC-type systems, to our knowledge, are close to being used for manufacturing applications. Companies such as Emefcy (http://www.emefcy.com) are distributing the first commercial systems for effluent remediation and consequent energy production [8]. Furthermore, benthic MFCs, which are designed for use as long-term power supplies in aquatic environments to power nautical devices such as anchored buoys, have been described as being close to industrial operation [65, 66].

Other BESs, such as those for enzymatic or microbial electrosynthesis, are still in the early stages of growth and have a long way to go before they can be used in industry. Finding low-cost electrode resources and creating condensed, separated reactor strategies are two common scale-up issues for MFCs [67–69]. Increasing the number of basic cell components can be used as an alternative to scaling up to boost productivity. Connecting them in series or parallel to stacked reactors can boost power production and current generation. Due to reduced substrate concentrations, the last cells in a serial assemblage will have lower productivities. Various reactor projects and procedure modes with volumes ranging from 1 to 1000 l have been studied elsewhere comprehensively. The majority of the systems studied thus far have been tubular molded and flat plate designs. To ensure proper component location, systems with several electrodes might be employed as an alternative. There is also a way to avoid ohmic failures and edge properties, which disrupt the electric field and lead to higher power densities. The most viable design approach appears to be stackable reactor systems, which achieve the volumes of focus by increasing the number of fixed units [70]. Different reactors can have well-defined performance characteristics, making total productivity estimation simple. To our knowledge, the main BES to date is a 1000 l MEC pilot plant that was used to create hydrogen and methane at the cathode using a bacterial anode [71]. The results of new studies have revealed that process performance at the practical scale is greatly inferior to that of small-scale strategies. This could be attributed to the immunization and enrichment techniques, as well as process control, natural feedstock variation, and long-term stability difficulties with the resources used for the filters and electrodes. Nonetheless, the statistics demonstrate the industrial potential of MFCs and MECs. The productivity of large-scale systems could be improved by re-engineering created on current results and fresh insights into the exact mechanisms. These re-engineering experiences will undoubtedly result in some common broad, relevant scale-up standards for BESs.

4.7 Conclusion and future considerations

In addition to reactor design, there are at least two other factors to consider when designing new MEC and MES reactors. In a two-chamber system, these are (i) the electrode material and shape, and (ii) the sheath fabric. The performance of the process is influenced greatly by the electrode. As a result, the electrode material must be conductive, support biofilm formation (if necessary), or at the very least be biocompatible, and have a suitable SVR. Finally, (low) cost and scalability are desirable features for electrode resources, according to Krieg *et al* [3]. Even though precious-metal electrodes have the best conductivity, their exorbitant cost prevents them from being used in larger systems. Alternatives have been identified, with some relying on material arrangements that affect the bacteria's surface adhesion. Membranes come in a variety of shapes and sizes, with cation and anion exchange membranes being the most common. Further resources, such as earthenware or permeable glass diaphragms/centrifuges, have been employed in place of membranes, but their selectivity and permeability are much less precise and different to ionic exchange films since they are essentially permeable bars. When selecting a material, these characteristics must be considered and compared. Different membrane materials have been compared in the past, and the results have been published elsewhere. Due to the limits of the designed reactors and the parts thereof, scaling up some reactors may also be impossible. It needs to be considered whether the scaling-up tactics are likely to increase bioelectrochemical reactor performance on a commercial level. Cusick *et al* [28] described using 24 electrode components in their microbial electrolysis cell for a research study. Scalability is determined by the requirements and should be complemented by the well-established science of electrochemistry. Electrodes are scaled up to the extreme (2 m^2) in this field to provide excellent current collection and mixing. This is something that should be replicated in the field of METs. To avoid difficulties such as ohmic loss, we consider it is inappropriate to seek system reduction. A good reactor design will lower costs, while a larger scale will save on costs and maintenance. The necessity for some expense is self-evident, but the combination of capital and functioning expenses must be inexpensive. METs, for example, will be more costly than anaerobic digesters due to the high material demand; nevertheless, depending on the desired outcome, added value will be provided. For example, practical and interesting metabolic pathways have been studied [72] to find products that may hypothetically gain from electrical aid in MES or electrofermentation, and it has been proved what is necessary for this. Yield and production excellence will be crucial in this case. If the system is used to treat waste, an additional condition must be met, treatment efficiency, which includes the removal of carbon dioxide, nitrogen, and phosphorus. The most difficult restriction to evaluate is individuality. When seeing the MFCs as a major example of MET, there are things to be learned for MES as well. Because anaerobic absorption is highly developed in this sector, MFCs that produce power from wastewater have yet to find applications. Furthermore, when a digester changes a load of 50 kg carbon dioxide into a theoretical current flow of 6979 A, this corresponds to a hypothetical current flow of 6979 A. MFCs are quite unlikely to achieve this. MFCs, on the other hand, have several unique characteristics: they work in an oxidative way, resulting in sulfide-free effluents; they work as a biofilm procedure; and they can operate at low temperatures, even on the seafloor. These distinguishing characteristics must guide application development and, as a result, system architecture. MESs also have the distinct feature of allowing the selective creation of multi-carbon compounds from CO_2 in a single step, which can be connected to power either directly or indirectly (e.g. H_2 generation). This can be combined with *in situ* withdrawal, which reduces the number of withdrawal stages required.

References

[1] Hongo M and Iwahara M 1979 Application of electro-energizing method to L-glutamic acid fermentation *Agric. Biol. Chem.* **43** 2075–81

[2] Schröder U, Harnisch F and Angenent L T 2015 Microbial electrochemistry and technology: terminology and classification *Energy Env. Sci.* **8** 513–9

[3] Krieg T, Sydow A, Schröder U, Schrader J and Holtmann D 2014 Reactor concepts for bioelectrochemical syntheses and energy conversion *Trends Biotechnol.* **32** 645–55
[4] Schievano A, Sciarria T P, Vanbroekhoven K, De Wever H, Puig S, Andersen S J, Rabaey K and Pant D 2016 Electro-fermentation—merging electrochemistry with fermentation in industrial applications *Trends Biotechnol.* **34** 866–78
[5] Idris M O, Guerrero-Barajas C, Kim H C, Yaaqop A A and Ibrahim 2022 Scalability of biomass-derived graphene derivative materials as viable anode electrode for a commercialized microbial fuel cell: a systematic review *Chin. J. Chem. Eng.* **2** 1–16
[6] Yaqoob A A, Fadzli F S, Ibrahim M N and Yaqoob A S 2022 Benthic microbial fuel cells: a sustainable approach for metal remediation and electricity generation from sapodilla waste *Int. J. Environ. Sci. Technol.* **10** 1–4
[7] Roy S, Schievano A and Pant D 2016 Electro-stimulated microbial factory for value added product synthesis *Bioresour. Technol.* **213** 129–39
[8] Krieg T, Madjarov J, Rosa L F, Enzmann F, Harnisch F, Holtmann D and Rabaey K 2018 Reactors for microbial electrobiotechnology *Bioelectrosynthesis* **167** 231–71
[9] Patil S A, Gildemyn S, Pant D, Zengler K, Logan B E and Rabaey K 2015 A logical data representation framework for electricity-driven bioproduction processes *Biotechnol. Adv.* **33** 736–44
[10] Simonte F, Sturm G, Gescher J and Sturm-Richter K 2017 Extracellular electron transfer and biosensors *Bioelectrosynthesis* **167** 15–38
[11] Yaqoob A A, Al_Zaqri N, Yaakop A S and Umar K 2022 Potato waste as an effective source of electron generation and bioremediation of pollutant through benthic microbial fuel cell *Sust. Energy Technol. Assess.* **53** 102–5
[12] Fruehauf H M, Enzmann F, Harnisch F, Ulber R and Holtmann D 2020 Microbial electrosynthesis—an inventory on technology readiness level and performance of different process variants *Biotechnol. J.* **15** 2000066
[13] Ter Heijne A, Geppert F, Sleutels T H, Batlle-Vilanova P, Liu D and Puig S 2017 Mixed culture biocathodes for production of hydrogen, methane, and carboxylates *Bioelectrosynthesis* **167** 203–29
[14] Torlaema T A, Ibrahim M N, Ahmad A, Guerrero-Barajas C, Alshammari M B, Oh S E and Hussain F 2022 Deregulation of hydroquinone coupled with energy generation through microbial fuel cells energized by organic waste *Process* **10** 1–20
[15] Yaqoob A A, Khatoon A, Mohd Setapar S H, Umar K, Parveen T, Mohamad Ibrahim M N, Ahmad A and Rafatullah M 2020 Outlook on the role of microbial fuel cells in remediation of environmental pollutants with electricity generation *Catalysts* **10** 819
[16] Sharma M, Bajracharya S, Gildemyn S, Patil S A, Alvarez-Gallego Y, Pant D, Rabaey K and Dominguez-Benetton X 2014 A critical revisit of the key parameters used to describe microbial electrochemical systems *Electrochim. Acta* **140** 191–208
[17] Zhang Y and Angelidaki I 2014 Microbial electrolysis cells turning to be versatile technology: recent advances and future challenges *Water Res.* **56** 11–25
[18] Geppert F, Liu D, van Eerten-Jansen M, Weidner E, Buisman C and Ter Heijne A 2016 Bioelectrochemical power-to-gas: state of the art and future perspectives *Trends Biotechnol.* **34** 879–94
[19] Lovley D R 2011 Powering microbes with electricity: direct electron transfer from electrodes to microbes *Env. Microbiol. Rep.* **3** 27–35

[20] Logan B E, Call D, Cheng S, Hamelers H V, Sleutels T H, Jeremiasse A W and Rozendal R A 2008 Microbial electrolysis cells for high yield hydrogen gas production from organic matter *Env. Sci. Technol.* **42** 8630–40

[21] Fadzli F S, Rashid M, Yaqoob A A and Ibrahim M N M 2021 Electricity generation and heavy metal remediation by utilizing yam (*Dioscorea alata*) waste in benthic microbial fuel cells (BMFCs) *Biochem. Eng. J.* **172** 108067

[22] Call D and Logan B E 2008 Hydrogen production in a single chamber microbial electrolysis cell lacking a membrane *Env. Sci. Technol.* **42** 3401–6

[23] Fu Q, Kuramochi Y, Fukushima N, Maeda H, Sato K and Kobayashi H 2015 Bioelectrochemical analyses of the development of a thermophilic biocathode catalyzing electromethanogenesis *Env. Sci. Technol.* **49** 1225–32

[24] Xafenias N and Mapelli V 2014 Performance and bacterial enrichment of bioelectrochemical systems during methane and acetate production *Int. J. Hydrogen Energy* **39** 21864–75

[25] Logan B E 2010 Scaling up microbial fuel cells and other bioelectrochemical systems *Appl. Microbiol. Biotechnol.* **85** 1665–71

[26] Rader G K and Logan B E 2010 Multi-electrode continuous flow microbial electrolysis cell for biogas production from acetate *Int. J. Hydrogen Energy* **35** 8848–54

[27] Cheng S, Xing D, Call D F and Logan B E 2009 Direct biological conversion of electrical current into methane by electromethanogenesis *Env. Sci. Technol.* **43** 3953–8

[28] Cusick R D, Bryan B, Parker D S, Merrill M D, Mehanna M, Kiely P D, Liu G and Logan B E 2011 Performance of a pilot-scale continuous flow microbial electrolysis cell fed winery wastewater *Appl. Microbiol. Biotechnol.* **89** 2053–63

[29] Bo T, Zhu X, Zhang L, Tao Y, He X, Li D and Yan Z 2014 A new upgraded biogas production process: coupling microbial electrolysis cell and anaerobic digestion in single-chamber, barrel-shape stainless steel reactor *Electrochem. Commun.* **45** 67–70

[30] Troshina O, Oshurkova V, Suzina N, Machulin A, Ariskina E, Vinokurova N, Kopitsyn D, Novikov A and Shcherbakova V 2015 *Sphaerochaeta associata* sp. nov., a spherical spirochaete isolated from cultures of *Methanosarcina mazei* JL01 *Int. J. Syst. Evol. Microbiol.* **65** 4315–22

[31] Call D F and Logan B E 2011 A method for high throughput bioelectrochemical research based on small scale microbial electrolysis cells *Biosens. Bioelectron.* **26** 4526–31

[32] Guo Z, Thangavel S, Wang L, He Z, Cai W, Wang A and Liu W 2017 Efficient methane production from beer wastewater in a membraneless microbial electrolysis cell with a stacked cathode: the effect of the cathode/anode ratio on bioenergy recovery *Energy Fuels* **31** 615–20

[33] Gajda I, Obata O, Greenman J and Ieropoulos I A 2020 Electroosmotically generated disinfectant from urine as a by-product of electricity in microbial fuel cell for the inactivation of pathogenic species *Sci. Rep.* **10** 1–10

[34] Guo K, Tang X, Du Z and Li H 2010 Hydrogen production from acetate in a cathode-on-top single-chamber microbial electrolysis cell with a mipor cathode *Biochem. Eng. J.* **51** 48–52

[35] Liu J, Zhang F, He W, Yang W, Feng Y and Logan B E 2014 A microbial fluidized electrode electrolysis cell (MFEEC) for enhanced hydrogen production *J. Power Sources* **271** 530–3

[36] Katuri K P, Werner C M, Jimenez-Sandoval R J, Chen W, Jeon S, Logan B E, Lai Z, Amy G L and Saikaly P E 2014 A novel anaerobic electrochemical membrane bioreactor (AnEMBR) with conductive hollow-fiber membrane for treatment of low-organic strength solutions *Env. Sci. Technol.* **48** 12833–41

[37] Li J and He Z 2016 Development of a dynamic mathematical model for membrane bioelectrochemical reactors with different configurations *Env. Sci. Pollut. Res.* **23** 3897–906

[38] Giddings C G, Nevin K P, Woodward T, Lovley D R and Butler C S 2015 Simplifying microbial electrosynthesis reactor design *Front. Microbiol.* **6** 468

[39] Dange P, Pandit S, Jadhav D, Shanmugam P, Gupta P K, Kumar S, Kumar M, Yang Y-H and Bhatia S K 2021 Recent developments in microbial electrolysis cell-based biohydrogen production utilizing wastewater as a feedstock *Sustainability* **13** 8796

[40] Cheng K Y, Ho G and Cord-Ruwisch R 2011 Novel methanogenic rotatable bioelectrochemical system operated with polarity inversion *Env. Sci. Technol.* **45** 796–802

[41] Yaqoob A A, Ibrahim M N M and Guerrero-Barajas C 2021 Modern trend of anodes in microbial fuel cells (MFCs): an overview *Env. Technol. Innov.* **23** 101579

[42] Yaqoob A A, Mohamad Ibrahim M N, Rafatullah M, Chua Y S, Ahmad A and Umar K 2020 Recent advances in anodes for microbial fuel cells: an overview *Materials* **13** 2078

[43] Yaqoob A A, Ibrahim M N M, Umar K, Parveen T, Ahmad A, Lokhat D and Setapar S H M 2021 A glimpse into the microbial fuel cells for wastewater treatment with energy generation *Desalin. Water Treat.* **214** 379–89

[44] Nevin K P, Hensley S A, Franks A E, Summers Z M, Ou J, Woodard T L, Snoeyenbos-West O L and Lovley D R 2011 Electrosynthesis of organic compounds from carbon dioxide is catalyzed by a diversity of acetogenic microorganisms *Appl. Environ. Microbiol.* **77** 2882–6

[45] Villano M, Monaco G, Aulenta F and Majone M 2011 Electrochemically assisted methane production in a biofilm reactor *J. Power Sources* **196** 9467–72

[46] Ling J, Xu Y, Lu C, Lai W, Xie G, Zheng L, Talawar M P, Du Q and Li G 2019 Enhancing stability of microalgae biocathode by a partially submerged carbon cloth electrode for bioenergy production from wastewater *Energies* **12** 3229

[47] Xu H, Wang K and Holmes D E 2014 Bioelectrochemical removal of carbon dioxide (CO_2): an innovative method for biogas upgrading *Bioresour. Technol.* **173** 392–8

[48] Jourdin L, Freguia S, Donose B C, Chen J, Wallace G G, Keller J and Flexer V 2014 A novel carbon nanotube modified scaffold as an efficient biocathode material for improved microbial electrosynthesis *J. Mater. Chem.* A **2** 13093–102

[49] Sasaki D, Sasaki K, Watanabe A, Morita M, Igarashi Y and Ohmura N 2013 Efficient production of methane from artificial garbage waste by a cylindrical bioelectrochemical reactor containing carbon fiber textiles *AMB Express* **3** 1–9

[50] Sivret E C, Wang B, Parcsi G and Stuetz R M 2016 Prioritisation of odorants emitted from sewers using odour activity values *Water Res.* **88** 308–21

[51] Jeon B Y, Jung I L and Park D H 2012 Conversion of carbon dioxide to metabolites by *Clostridium acetobutylicum* KCTC1037 cultivated with electrochemical reducing power *Adv. Microbiol.* **2** 332

[52] Liu H, Grot S and Logan B E 2005 Electrochemically assisted microbial production of hydrogen from acetate *Env. Sci. Technol.* **39** 4317–20

[53] Gildemyn S, Verbeeck K, Slabbinck R, Andersen S J, Prévoteau A and Rabaey K 2015 Integrated production, extraction, and concentration of acetic acid from CO_2 through microbial electrosynthesis *Env. Sci. Technol. Lett.* **2** 325–8

[54] Ai C, Yan Z, Hou S, Zheng X, Zeng Z, Amanze C, Dai Z, Chai L, Qiu G and Zeng W 2020 Effective treatment of acid mine drainage with microbial fuel cells: an emphasis on typical energy substrates *Minerals* **10** 443

[55] Emde R and Schink B 1990 Enhanced propionate formation by *Propionibacterium freudenreichii* subsp. *freudenreichii* in a three-electrode amperometric culture system *Appl. Environ. Microbiol.* **56** 2771–6

[56] Park D, Laivenieks M, Guettler M, Jain M and Zeikus J 1999 Microbial utilization of electrically reduced neutral red as the sole electron donor for growth and metabolite production *Appl. Environ. Microbiol.* **65** 2912–7

[57] Yaqoob A A, Ibrahim M N M, Yaakop A S and Ahmad A 2021 Application of microbial fuel cells energized by oil palm trunk sap (OPTS) to remove the toxic metal from synthetic wastewater with generation of electricity *Appl. Nanosci.* **11** 1949–61

[58] Chuo S C, Mohamed S F, Mohd Setapar S H, Ahmad A, Jawaid M, Wani W A, Yaqoob A A and Mohamad Ibrahim M N 2020 Insights into the current trends in the utilization of bacteria for microbially induced calcium carbonate precipitation *Materials* **13** 4993

[59] Rozendal R A, Hamelers H V, Euverink G J, Metz S J and Buisman C J 2006 Principle and perspectives of hydrogen production through biocatalyzed electrolysis *Int. J. Hydrogen Energy* **31** 1632–40

[60] Kipf E, Koch J, Geiger B, Erben J, Richter K, Gescher J, Zengerle R and Kerzenmacher S 2013 Systematic screening of carbon-based anode materials for microbial fuel cells with *Shewanella oneidensis* MR-1 *Bioresour. Technol.* **146** 386–92

[61] Wang A, Liu W, Cheng S, Xing D, Zhou J and Logan B E 2009 Source of methane and methods to control its formation in single chamber microbial electrolysis cells *Int. J. Hydrogen Energy* **34** 3653–8

[62] Fan Y, Hu H and Liu H 2007 Enhanced Coulombic efficiency and power density of air-cathode microbial fuel cells with an improved cell configuration *J. Power Sources* **171** 348–54

[63] Clauwaert P and Verstraete W 2009 Methanogenesis in membraneless microbial electrolysis cells *Appl. Microbiol. Biotechnol.* **82** 829–36

[64] Yaqoob A A, Ibrahim M N M, Yaakop A S, Umar K and Ahmad A 2021 Modified graphene oxide anode: a bioinspired waste material for bioremediation of Pb^{2+} with energy generation through microbial fuel cells *Chem. Eng. J.* **417** 128052

[65] Yaqoob A A, Mohamad Ibrahim M N, Umar K, Bhawani S A, Khan A, Asiri A M, Khan M R, Azam M and AlAmmari A M 2021 Cellulose derived graphene/polyaniline nano-composite anode for energy generation and bioremediation of toxic metals via benthic microbial fuel cells *Polymers* **13** 135

[66] Daud N N M, Ahmad A, Yaqoob A A and Ibrahim M N M 2021 Application of rotten rice as a substrate for bacterial species to generate energy and the removal of toxic metals from wastewater through microbial fuel cells *Env. Sci. Pollut. Res.* **28** 62816–27

[67] Yaqoob A A, Ibrahim M N M and Yaakop A S 2021 Application of oil palm lignocellulosic derived material as an efficient anode to boost the toxic metal remediation trend and energy generation through microbial fuel cells *J. Clean. Prod.* **314** 128062

[68] Yaqoob A A, Ibrahim M N M and Umar K 2021 Biomass-derived composite anode electrode: synthesis, characterizations, and application in microbial fuel cells (MFCs) *J. Env. Chem. Eng.* **9** 106111

[69] Yaqoob A A, Ibrahim M N M, Rodríguez-Couto S and Ahmad A 2021 Preparation, characterization, and application of modified carbonized lignin as an anode for sustainable microbial fuel cell *Process Safety Env. Protect.* **155** 49–60

[70] Yaqoob A A, Serrà A, Ibrahim M N M and Yaakop A S 2021 Self-assembled oil palm biomass-derived modified graphene oxide anode: an efficient medium for energy transportation and bioremediating Cd (II) via microbial fuel cells *Arabian J. Chem.* **14** 103121

[71] Baudler A, Schmidt I, Langner M, Greiner A and Schröder U 2015 Does it have to be carbon? Metal anodes in microbial fuel cells and related bioelectrochemical systems *Energy Env. Sci.* **8** 2048–55

[72] Kracke F and Krömer J O 2014 Identifying target processes for microbial electrosynthesis by elementary mode analysis *BMC Bioinf.* **15** 1–14

IOP Publishing

Microbial Fuel Cells

Emerging trends in electrochemical applications

Mohamad Nasir Mohamad Ibrahim, Asim Ali Yaqoob and Akil Ahmad

Chapter 5

Conventional electrode materials for electrochemical performance in microbial fuel cells

Mary L Monisha, Rajakumar Merlyn Sujatha and J Jayapriya

Microbial fuel cells (MFCs) have significant potential in sustainable energy production, treating waste, and reducing CO_2 emissions. Electrode materials play a crucial role in reducing the internal resistance of the system and hence the electrode cost will be the key deciding factor for the large-scale implementation of MFC technology. Carbon-based materials are cost-effective alternatives to traditional metal-based MFC electrode materials because they are very cheap and promote bacterial growth. Recent progress in the understanding of carbon-based anode and cathode materials used in MFCs for different applications is reviewed in this chapter, highlighting the characteristics and modifications that influence MFC performance.

5.1 Introduction

A fuel cell is an electrochemical device that converts the free energy generated during a redox reaction to electrical energy. A microbial fuel cell (MFC) generates electricity by diverting the electrons from the anaerobic oxidation of organic and inorganic compounds on the anode by using microbes as catalysts. Typically, in an MFC, the anode and cathode chambers are separated by a proton exchange membrane (figure 5.1).

The micro-organisms oxidize the substrate in the anodic chamber and transfer the electrons to the cathode through an external circuit, thus producing a flowing electric current. The protons produced during oxidation travel across the membrane separator or diffuse through the electrolyte to the cathode, where they form water by combining with oxygen and electrons. MFCs are basically of two types, namely mediated MFCs and mediator-less MFCs, depending on the mode of electron transfer from the

doi:10.1088/978-0-7503-4791-4ch5

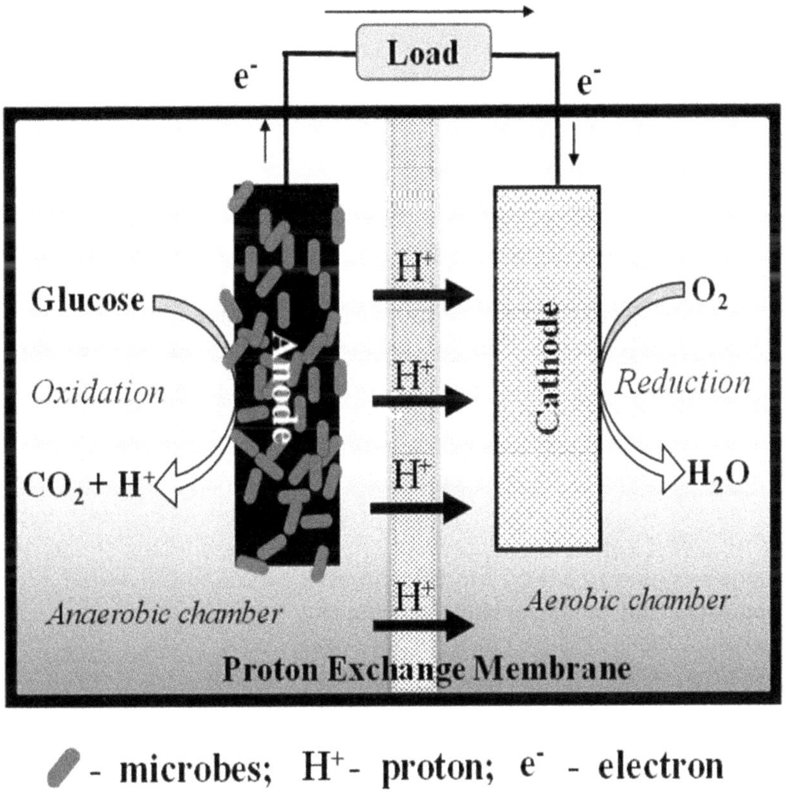

Figure 5.1. Schematic representation of a microbial fuel cell.

bacteria to the anode surface. In mediator MFCs, the bacteria are incapable of direct electron transfer since the proteins associated with electron transport are restricted within the cell membrane [1]. Hence, exogenous mediators, such as thionine, methyl blue, methyl viologen, and humic acid, are used to shuttle electrons between the microbial biocatalytic system and the electrode via redox coupling [2]. However, the disadvantages of exogenous mediators, include high cost, short lifetime, and toxicity to the micro-organisms, limit the general application of this approach. A possible solution to circumvent these issues is to use native electron shuttles (endogenous) produced by the micro-organism itself [3]. The topmost endogenous electron shuttles reported in MFCs are phenazines, such as pyocyanin and phenazine-1-carboxamide from *Pseudomonas aeruginosa* and quinone-based redox shuttles from *Shewanella oneidensis*. On the other hand, in mediator-less MFCs, the micro-organisms typically take advantage of electrochemically active components, such as membrane-associated cytochromes, conductive pili, and secreted redox-mediators, which facilitate electron transfer. The breakthrough was achieved when it was observed that some dissimilatory metal-reducing microbes could directly transfer electrons to the anode in the absence of a shuttle. Many micro-organisms, including metal-reducing bacteria such as *Geobacter metallireducens*, *Rhodoferax ferrireducens*, *Shewanella putrefaciens*, and

Aeromonas hydrophila make use of such a process [4–6]. With the cost of the mediator eliminated, mediator-less MFCs can be considered a sustainable biotechnological solution that can meet future energy needs; however, the current version of the technology suffers from several bottlenecks, which reduce the power output and thereby limit its commercial development/applications. Most factors that impede MFC performance stem from the disparity in the laboratory- and commercial-scale MFC operation in terms of microbes, electrode materials, proton-conducting material, catalysts, electrode spacing, and system configuration. Electrode composition is the top determinant of the power generation efficiency of MFCs. The commercial high-efficiency electrodes commonly applied in conventional fuel cells, such as metal-coated carbon–polymer composites and noble-metal catalysts, are not biocompatible and are prone to catalyst poisoning in the presence of industrial effluents. Hence, it is essential to design and explore cheap, non-corrosive electrode materials (anode) that have easy processing and are compatible with the microbes and their metabolism in terms of redox behavior for the successful expansion of MFC operation. Similarly, cathode materials with high redox potential are required to facilitate high oxygen reduction. In this chapter, recent progress in the understanding of the anode and cathode materials used in MFCs for different applications are reviewed, highlighting the characteristics and modifications that influence MFC performance.

5.2 Electrodes

Generally, MFCs suffer from high internal resistance and exhibit low power output because of the low electron transfer rate from the bacterial membranes to the anode/cathode. In contrast, the redox compounds secreted by electrochemically active bacteria are key to indirect electron transfer to the electrodes. Therefore, MFC electrodes that can easily accept these electrons are necessary to boost the performance. Hence, electrode cost is the fundamental determinant of the large-scale implementation of MFC technology. Recent advances in electrode materials and their modification have helped lab-scale MFC operation step closer to practical applications.

5.2.1 Anode requirements for MFCs

Currently, efforts to improve the performance of mediator-less MFCs by efficiently wiring the micro-organisms to the anode are in progress to achieve better electron transfer. The conventional high-efficiency electrodes (e.g. metal-coated composites, carbon–polymer composites, and noble-metal catalysts) used in traditional fuel cells cannot be applied in MFCs because of their poor biocompatibility and high susceptibility to catalyst poisoning when used with industrial waste/aquatic sediments. Thus, MFC-specific anodic materials with a conductive surface and appropriate redox behavior to support microbial metabolism are needed; in addition, they must also be economical, easily processable, biocompatible, and scalable. In this view, several substances varying from non-corrosive stainless steel to various forms/configurations of versatile carbons have been tested as MFC anodes [7]. Carbon-based materials are the cost-effective alternatives to traditional metal-based MFC anode materials because they are very cheap and promote bacterial growth.

Carbon electrodes as anodes

Carbon electrodes have the distinctive benefits of a wide potential window, low residual current, low cost, excellent biocompatibility, and outstanding electrical conductivity, in addition to long-term stability, corrosion resistance, chemical stability, and low-density thermal expansion [8]. Many homogeneous carbon forms, including glassy carbon, graphite, vitreous carbon, screen-printed carbon, fullerenes, and carbon nanotubes/brushes/fibers, as well as heterogeneous carbons, such as carbon paste electrodes and carbon-fiber-reinforced plate, have been evaluated in MFCs (table 5.1). Despite this, MFCs have continued to produce low power outputs because of the high activation overpotential resulting from the slow electrode reactions. Each carbon electrode has specific characteristics, as shown in table 5.2. Hence, it is difficult to find a material that meets all the requirements of a suitable MFC anode (pristine) material to improve the power density, such as mechanical strength, good flexibility, conducive to biofilm adhesion, and facilitation of proper substrate diffusion to reach the bacterial cells.

5.2.2 Cathode requirements for MFCs

Carbon-based materials with a high redox potential to accept electrons are usually preferred as the cathode as they greatly affect the power capacity of MFCs. In general, the terminal electron acceptor in the cathodic reduction reaction in MFCs is oxygen since it has a high standard potential ($O_2 + 2H_2O + 4e^- \rightarrow 4 OH^-$, $E'^{\circ} = 0.815$ V, pH = 7). In acidic solutions, the cathodic oxygen reduction reaction (ORR) may occur via two distinct processes: (i) a four-electron pathway: $O_2 + 4H^+ + 4e^- \rightarrow 2H_2O$ ($E'^{\circ} = 0.816$ V) or (ii) a two-electron pathway: $O_2 + 2H^+ + 2e^- \rightarrow H_2O_2$ ($E'^{\circ} = 0.295$ V) [42]. Unfortunately, although the two-electron pathway is thermodynamically advantageous, it produces hydrogen peroxide, which is toxic to the microbes and electrodes, leading to poor MFC performance. Therefore, promoting the four-electron reaction by employing a noble-metal cathode catalyst, such as Pt, is crucial. Although platinum is the best-known oxygen reduction catalyst, its high cost and sensitivity to pH prohibit its application in large-scale MFCs.

Carbon electrodes as cathodes

Different carbon materials have been tested so far as cathodes in MFCs, as listed in table 5.1. Their greatest disadvantage is the slow kinetics of the oxygen reduction reaction at their surface, which requires catalysts or artificial electron mediators for open circuit potential (OCP) enhancement. The current generation of MFCs is heavily reliant on the reduction of ferricyanide $[Fe(CN)_6]^{3-}$, a widely used artificial electron mediator, instead of oxygen because of the considerably lower oxidation rate of $[Fe(CN)_6]^{4-}$ by oxygen than its reduction rate. Moon *et al* (2006) designed an MFC that generated a power density of 150 mW m^{-2} with graphite felt cathode containing platinum (Pt) which was three times that obtained with a graphite

Table 5.1. Comparison of power density generated in an MFC using different conventional carbon electrodes.

Anode	Cathode	MFC configuration	Biocatalyst	Substrate	Power density	References
Graphite plates	Graphite plates	Two chamber	Electrochemically active bacteria	Glucose	15.2 W m^{-3}	Rahimnejad *et al* (2011) [9]
Carbon mesh	Activated carbon (AC)	Cubic shaped chamber	Domestic wastewater	Wastewater	240 ± 22 mW m^{-2}	Hays *et al* (2011) [10]
Graphite felt	Graphite felt	Two chamber	Actinobacteria, B-Proteobacteria,	Glucose	339 mW m^{-3}	Tao *et al* (2011) [11]
Graphite plates	Air	Two chamber	Mixed culture	Glucose	283 mW m^{-2}	Rahimnejad *et al* (2011) [12]
Non-wetproof carbon paper	Carbon paper	Two chamber	Cellulose degrading bacteria	Cellulose	188 mW m^{-2}	Hassan *et al* (2012) [13]
Graphite felt	Pt coated sheet	Two chamber	Wastewater of complex organic substrates	Alcohol distillery	1000 mA m^{-2}	Ishii *et al* (2012) [14]
Graphite plate	Graphite plate	Single chamber	*Phragmites australis*	Sludge-based	9.4 mW m^{-2}	Zhao *et al* (2013) [15]
Carbon brush	Carbon cloth	Two chamber	Anaerobic consortia	sodium acetate	6.58 W m^{-2}	Choi *et al* (2013) [16]
Graphite plate	Graphite plate	Single chamber	Sediment zone	Aquaculture pond water	106.7 ± 1.1 (μW m^{-2})	Sajana *et al* (2013) [17]
Graphite felt	Air-cathode	Single chamber	Animal carcass wastewater	Glucose	2.19 W m^{-3}	Li *et al* (2013) [18]
Graphite	Graphite	Single chamber	*Phragmites australis*	Glucose	43 mW m^{-2}	Villaseñor *et al* (2013) [19]
Graphite fiber	Graphite rod	Two chamber	Digester sludge	Landfill leachate	41 mA m^{-2}	Mahmoud *et al* (2014) [20]
Graphite fiber	Graphite felt	Tubular two chamber	Anaerobic sludge	Wastewater	2.6 mW m^{-2}	Huang *et al* (2014) [21]
Graphite felt	Carbon cloth	Two chamber	*Pseudomonas aeruginosa*	Anaerobic mixed consortia	93 mW m^{-3}	Chang *et al* (2014) [22]

(*Continued*)

Table 5.1. (*Continued*)

Anode	Cathode	MFC configuration	Biocatalyst	Substrate	Power density	References
Carbon cloth	Carbon cloth	Two chamber	*S. putrefaciens* CN32	Sodium acetate	679.7 mW m^{-2}	Qiao *et al* (2015) [23]
Graphite fiber	Platinum	Single chamber	White wine wastewater	Sodium acetate	111 mW m^{-2}	Sciarria *et al* (2015) [24]
Carbon cloth	Carbon cloth	Two chamber	*B. subtilis*	Dichlorophenol	9.5 W m^{-3}	Hassan *et al* (2016) [25]
Plain carbon cloth	Pt catalyst	Two chamber	Anaerobic bacteria	Synthetic wastewater	72 mW m^{-2}	Almatouq *et al* (2016) [26]
Graphite plate	Graphite plate	Two chamber	Dairy industry wastewater	Dairy Industry wastewater	621.13 mW m^{-2}	Radha *et al* (2016) [27]
Graphite plate	Graphite plate	Two chamber	Anaerobic sludge	Glucose	58.726 mW m^{-2}	Zinadini *et al* (2017) [28]
Carbon brush	Carbon cloth	Two chamber	Mixed consortium	Sodium acetate	293.7 mW m^{-2}	Li *et al* (2017) [29]
Carbon felt	Air-cathode	Single chamber	*Thauera*, *Trichococcus*, Rhodocyclaceae	Synthetic wastewaster	4.38 mW m^{-2}	Liu *et al* (2017) [30]
Carbon cloth	Carbon cloth	Single chamber	Mixed consortium	Sodium acetate	40 mW m^{-2}	Marzorati *et al* (2018) [31]
Carbon felt	Carbon felt	Single chamber	*Nitrosomonas*, *Alishewanella*, *Arcobacter*, *Thauera*, and *Rheinheimera*	Acetate	104 mW m^{-2}	Yang *et al* (2018) [32]
Carbon felt	Air-cathode	Single chamber	*Geobacter*, *Pseudomonas*, *Rhodococcus*	Sodium acetate	19.8 mW m^{-2}	Wang *et al* (2019) [33]
Graphite plate	Graphite plate	Two chamber	Anaerobic sludge	Sodium acetate	844.0 mW m^{-2}	Liu *et al* (2019) [34]
Carbon cloth	Carbon cloth	Two chamber	*Nitrosomonas* SM1A02, *Thauera*, *Stenotrophomonas*, *Flavobacterium*, *Marinobacter*, and *Thioalkalispira*	Mustard tuber wastewater	1.45 W m^{-3}	Zhang *et al* (2019) [35]

Graphite felt	Graphite felt	Two chamber	*Thauera* and *Pseudomona*, *Klebsiella* and *Alkaliphilus*	Municipal wastewater	3.02 mA cm^{-2}	Vijay *et al* (2019) [36]
Graphite felt	Carbon fiber brush	Two chamber	Anaerobic microbes	Artificial municipal wastewater	253.84 mW m^{-2}	Ye *et al* (2020) [37]
Carbon cloth	Carbon cloth	Single chamber	*Shewanella oneidensis* and *Clostridium butyricum*, *Daphnia similis*	Dairy wastewater	0.48 W m^{-3}	Paiva *et al* (2020) [38]
Carbon cloth	Carbon cloth	Single chamber	*Shewanella oneidensis* and *Clostridium butyricum*	Dairy wastewater	3.5 W m^{-3}	Marassi *et al* (2020) [39]
Carbon fiber sponge and cloth	Carbon	Single chamber	Mixed consortium	Acetate	17.4 A m^2	Xu *et al* (2021) [40]
Carbon fiber felt	Carbon	Single chamber	Anaerobic sludge	Aquaculture wastewater	369 mW m^{-2}	Pugazhendi *et al* (2021) [41]

Table 5.2. Characteristics of different carbon materials tested as electrodes in MFCs.

Carbon electrode (anode)	Characteristics of carbon electrode materials
Graphite brush	Large electrode spacing in the electrode architecture can lead to high ohmic resistance.
Graphite rod	Good mechanical strength; however, the smooth surface adversely affects biofilm accumulation on the electrode.
Reticulated glassy carbon	Reticulated (vitreous) glassy carbon shows high corrosion resistance and electrical conductivity. An open-pore foam material with poor mechanical strength.
Carbon mesh (CM)	Open weaving gives a more open structure than carbon cloth which can aid in the reduction of biofouling.
Carbon cloth (CC)	Weaved using thin fibers predominantly made of carbon atoms, CC is highly porous but very expensive (~\$1000 m^{-2}).
Carbon fiber electrodes	Good flexibility and high conductivity; due to its relatively high thickness, biofilm adhesion on the surface hinders substrate diffusion.
Graphite felt (GF)	High surface area compared with graphite rods; however, it possesses lower electrochemical activity.

cathode [43]. Despite its outstanding ability to boost the MFC performance, ferricyanide is not considered a viable mediator because of its low re-oxidation rate which necessitates the continuous replacement of the catholyte and incurs a high cost. Therefore, to accomplish economically sustainable large-scale applications, the Pt catalyst must be substituted by low-cost metal compounds, such as transition metal oxides or macrocyclic organic compounds with Fe/Co/Ni metal centers [44].

5.3 Electrode modification

Despite the differences in the physicochemical properties, electrochemical activity, and structural features of carbon materials, they can be adapted easily by surface or bulk matrix modifications that can alter the electrochemical behavior to suit MFC applications. Cost-effective and easy-to-process MFC-specific anodic materials that have a conductive surface, suitable redox behavior for microbial metabolism, biocompatibility, and the prospect for scale-up are in high demand. Similarly, precious-metal catalysts (e.g. platinum) and expensive cathode materials are not economically viable and must be avoided during MFC scale-up. However, fabricating air cathodes with long-term stability at different pH values and low-cost metal catalysts for ORR is a challenging task. The immobilization of metals, metal oxides, or other active compounds on carriers, such as carbon materials and conductive polymers, is a popular approach to improve MFC performance. Various chemical modification methods, including ammonia or acid treatment, sintering, soaking methods, chemical vapor deposition, and carburization, are known to have beneficial effects on the electrode surface properties.

5.3.1 Carbon-conducting polymer based composite electrodes

The structure of conducting polymers (CPs) contains electron-deformation pairs or even electron–phonon cloud pairs in interconnected chains (i.e. alternate single and double bonds). The double bonds in these molecules constantly move to maintain the charge balance in the adjacent atoms. This movement eventually confers conductivity to the CPs. The most advantageous CPs used in MFCs are polypyrrole (PPy), polyaniline (PANI), polythiophene (PTh), and poly (3,4-ethylenedioxythiophene) (PEDOT). Due to their excellent electrical conductivity, diverse polymeric structure, chemical stability, and non-biodegradability, conducting polymer electrodes are suitable for scale-up. Therefore, substantial work on bulk and surface modifications in insulating and semiconducting/conducting polymers has been done in the recent past.

Carbon-conducting polymer based composite anode

Carbon and polymer based composite electrodes are commonly used in microbial fuel cells after bulk modification or surface modification. In recent years, three-dimensional electrodes have been in high demand because they offer increased biofilm adhesion, thereby boosting the volumetric power density. Polyurethane foam (PUF) is macroporous and exhibits good mechanical strength and low toxicity. Polyurethane (PU) foam/graphite/polypyrrole (PU/Gr/PPy) composites have been tested as anodes in microbial fuel cells with municipal wastewater under anaerobic conditions [45]. In another case, a low-cost MFC operated with PU/Gr/PPy-$FeCl_3$ as the anode exhibited 2.13 times higher efficiency than that using pure PU foam. Jayapriya and co-workers (2012) [3] showed that using graphite electrodes with 24% (w/w) epoxy resin as the binder in the presence of different metal salts reduced the high overpotential of the MFC. Similarly, this group has also explored unsaturated polyester resins (UPRs) as binders in graphite electrodes with metal salts as dopants. Among the different combinations of electrodes tested in MFCs, the highest power density was achieved with Mn and Zn doped graphite polyester composite electrodes as anode materials (1200.77 ± 59 μW m^{-2}; 1187.95 ± 0.045 μW m^{-2}) and graphite block as the cathode, along with a simultaneous increase in the decolorization of azo dyes (97% ± 1.4%) [46]. Jayapriya and co-workers prepared a surface-modified carbon cloth-based composite material with two redox species, namely nickel cobaltite ($NiCo_2O_4$) and a conductive polymer (PANI), through a hydrothermal method and electropolymerization, respectively, to effectively improve the electrochemical properties. The MFC with the polyaniline-coated carbon electrode as the anode produced nine-fold higher power density than that obtained with bare CC electrodes, whereas the double-modified $NiCo_2O_4$/PANI/CC (cathode)–carbon cloth (anode) pair exhibited the best performance with the highest power density (12.19 ± 0.59 mW m^{-2}) [44].

Carbon-conducting polymer based composite cathodes

Different carbon-conducting polymer composites have been tested as cathodes in MFCs, as shown in table 5.3. Polyaniline (PANI)-modified graphene nanosheet (GNS) cathodes were fabricated through *in situ* aniline polymerization in a solution

Table 5.3. Comparison of the power density generated in an MFC using different carbon-conducting polymer based composite cathodes.

Cathode	Anode	Reactor configuration	Biocatalyst	Substrate	Buffer in catholyte	Power density	References
Graphite felt–PANI	Graphite	Sediment MFC	Sewage sludge	Sodium acetate	Phosphate buffer	99 mW m^{-2}	Ren *et al* (2013) [47]
Graphite block–Ni-GPECE	Graphite block	Two chamber	*Pseudomonas aeruginosa*	Glucose	Phosphate buffer	1575 ± 223.26 $\mu W\ m^{-2}$	Narayanasamy *et al* (2018) [46]
Carbon cloth–Ppy–MnO_2	Carbon cloth	Single chamber	*Shewanella putrefaciens*	Glucose	Phosphate buffer	0.63 W m^{-2}	Karthick *et al* (2020) [49]
Carbon cloth–Ppy	Carbon cloth	Single chamber	Activated sludge	Glucose	Phosphate buffer	401.8 mW m^{-2}	Yuan *et al* (2010) [50]
Carbon paper–PPy–AC	Plain carbon paper	Cubic shaped chamber	Palm oil mill effluent	Glucose	Phosphate buffer	72.1 mW m^{-2}	Esmaeili *et al* (2014) [51]
Graphite felt–PANI–β-MnO_2	Graphite felt	Two chamber	Anaerobic digester sludge	Glucose	Phosphate buffer	204 mW m^{-2}	Zhou *et al* (2018) [52]
Carbon cloth–PPy–NP	Graphite felt	Single chamber	*Shewanella putrefaciens*	Glucose	Phosphate buffer	~190 ± 4 mW m^{-2}	Sumisha *et al* (2020) [48]

containing homogeneously dispersed GNS to improve the electricity generation capacity [47]. Saranya *et al* (2021) synthesized bulk-modified graphite polyester composite electrodes (GPECE) doped with metal salts (MS-GPECE) by casting [46]. The highest power density (1575 ± 223.26 μW m^{-2}) was obtained with Ni-GPECE as the cathode. In another study, a polypyrrole nanoparticle (Ppy-NP)-modified carbon cloth (CC) cathode was employed in a single-chamber membraneless microbial fuel cell (SCMFC) for bioenergy production and iron removal using *Shewanella putrefaciens* as the microbial catalyst. The maximum power density and coulombic efficiency obtained with PPy NPs were found to be ~190 ± 4 mW m^{-2} [48]. However, it has been suggested that the lower catalytic activity of PPy results in fewer electron-transfers than polyaniline.

5.3.2 Metal/metal oxide based electrodes

Metals/metal oxides are non-carbon materials that have been tested as electrodes in MFCs due to their higher conductivity and wide redox potential. However, it is suggested to not use them directly as electrodes on the larger scale as they are highly susceptible to corrosion [53]. Adding a thin layer of metal/metal oxide coating on carbon materials significantly enhances biofilm formation and anode surface wettability, thereby decreasing the charge transfer resistance. In recent years, some researchers have revealed that the use of metal and metal oxide nanostructures, as well as their alloys, decreases the charge transfer resistance at the anode/cathode.

Metal/metal oxide based anodes

Dumas *et al* (2008) designed stainless steel (SS) based anodes and obtained current densities in the range of 2.4–3.1 A m^{-2} [54]. However, the use of stainless steel mesh (SSM) or a composite SS anode and modifying the surface with heat treatment, natural gas flame treatment, carbon coating, polyaniline coating (PANI), neutral red grafting (NR), and surface hydrophilization (SDBS) have been reported to enhance the adhesion of biofilm and thus increase the current output [55]. TiO_2 possesses low conductivity, and hence it is often used in hybrid anodes for MFCs. For example, a TiO_2 thin film-modified stainless steel mesh (SSM/TiO_2-H) was prepared by an electrochemical method, and these nanostructured graphene/TiO_2 hybrid anodes were tested in MFCs [56]. A carbonized microwire-modified titanium mesh (CM/TiM) electrode was prepared by the *in situ* colonization of filamentous fungi and a carbonization method, which could convert the attached mycelia into carbon microwires. Because of the biocompatibility and interlaced 3D structure of the carbon microwires, the energy production increased by more than two-fold on the CM/TiM anode compared with the TiM control [57].

Metal/metal oxide based cathodes

Although precious-metal catalysts, such as platinum, palladium, and gold, present high catalytic activity in oxygen reduction, their high cost, limited availability, and high sensitivity toward biological and chemical fouling hinder their application in large-scale MFCs. Thus, various non-noble metal-based materials, including manganese oxides, iron phthalocyanine (FePc), cobalt and iron-copper phthalocyanine

(CoPc and FeCuPc), cobalt tetramethoxyphenylporphyrin (CoTMPP), and iron-cobalt tetramethoxyphenylporphyrin (FeCoTMPP), have been investigated as cathode catalysts in MFCs for simultaneous cost reduction and improvement of ORR kinetics. It is known that silver/silver oxide possesses antimicrobial properties, which hinder biofilm adhesion on anodes with silver metallic coatings, thereby reducing the power density. However, Pu *et al* (2014) enhanced the ORR electro-catalytic activity by electrodepositing silver on an activated carbon (AC) air-cathode for a duration of 50 s and generated 69% higher maximum power density (1080 ± 60 mW m^{-2}) than that of the bare AC cathode [58]. Metal oxides, such as lead oxide, cobalt oxide, titanium dioxide, vanadium oxide, and manganese oxide, have been used as ORR catalysts and have been shown to boost MFC performance as they augment the surface area and decrease the overpotential (table 5.4). Bimetallic catalysts include the iron phthalocyanine (FePc)–MnOx and Fe–Mn–AAPyr composite catalysts show even better catalytic activity than their monometallic counterparts, and they can be synthesized easily by the hydrothermal method [59]. Liu and Vipulanandan (2016) improved MFC performance by using rod-shaped (Fe, Ni, and Fe/Ni) metallic nanoparticles produced by the precipitation/co-precipitation method, demonstrating enhanced power production up to 66.4 mW m^{-3} [60]. They also found that the rod-shaped nanoparticles presented larger specific surface areas than did the spheres and cubes, thus resulting in high catalytic activity towards ORR. In addition to the electrodeposition technique, Veeramani *et al* (2020) employed successive ionic layer adsorption and reaction (SILAR) to coat cobalt oxide on carbon cloth and used it as a cathode in MFC [61]. In addition to these, since the cost-effective and eco-friendly layered double hydroxides (LDHs) possess high electrochemical activity and chemical stability, they are considered suitable ORR catalysts [62]. Further, the high surface-area-to-volume ratios of LDHs lead to augmented electron transfer and electrode–electrolyte contact areas. Some LDHs, including CoMn-LDH, CoNiMn-LDH, NiAl-LDH, and CoFe-LDH, have been explored as catalysts to improve the ORR activity of carbon-based materials. A remarkable improvement in the power density of graphite plates modified with CoNiAl-LDH was reported by using the electrophoretic deposition (EPD) method [63]. This LDH-modified cathode produced a power density of 87.91 mW m^{-2} in a two-chambered MFC. Various techniques used for coating metals/metal oxides on carbon electrodes are listed in table 5.5 along with their advantages and limitations.

5.3.3 Graphene and its modification

Graphene (Gr) is an allotrope of carbon with a two-dimensional honeycomb lattice nanostructure, which consists of a single layer of carbon atoms that are bonded together in a repeating pattern of hexagons. Graphene is more biocompatible than CNTs and is also synthesized at room temperature at a lower cost than that required for multi-walled CNTs. Therefore, it is a promising anode and cathode candidate both in the pristine form and when modified with polymer or metal catalysts. It can be synthesized by either the top-down or bottom-up approaches, as shown in figure 5.2.

Table 5.4. Power generation performance of MFCs operated with different metal/metal oxide based cathodes.

Cathode	Anode	MFC type	Biocatalyst	Substrate	Buffer in catholyte	Power density	Reference
Carbon black–Mn_2O_3	Carbon felt	Two chamber	Anaerobic mesophilic sludge	Glucose	Air-cathode	31.7 W m^{-3}	Martin *et al* (2011) [64]
Carbon brush–Ni–C	Carbon paper	Two chamber	Palm oil effluent	Glucose	Phosphate buffer	94.4 mW m^{-2}	Ghasemi *et al* (2013) [65]
Carbon cloth–$AuCl^{-4}$	Carbon brush	Two chamber	Anaerobic sludge	Artificial wastewater	Tetrachloroaurate	6.58 W m^{-2}	Choi *et al* (2013) [66]
Carbon black–CuPc	Carbon paper	Two cubic shaped chambered	Palm oil mill effluent	Glucose	Phosphate buffer	118.2 mW m^{-2}	Mostafa *et al* (2013) [65]
Stainless steel mesh	Carbon felt	Single chamber	Domestic wastewater	Wastewater	Phosphoric acid buffer	1080 ± 60 mW m^{-2}	Pu *et al* (2014) [58]
Stainless steel mesh–N-type Cu_2O–Ac	Carbon felt	Single chamber	Domestic wastewater	Wastewater sodium acetate	Phosphate buffer	1390 ± 76 mW m^{-2}	Zhang *et al* (2015) [67]
Carbon cloth–V_2O_5–MF (microflowers)	SS mesh	Single chamber	Anaerobic septic tank sludge	Fish market wastewater	Air-cathode	6.06 W m^{-3}	Noori *et al* (2016) [68]
SS mesh	Carbon felt	Single chamber	Domestic wastewater	Sodium acetate	Phosphate buffer	2000 ± 59 mW m^{-2}	Liangtao *et al* (2017) [69]
SS mesh	Carbon felt	Single chamber	Domestic wastewater	Acetate solution	Phosphate buffer	2252 ± 46 mW m^{-2}	Song *et al* (2018) [70]
Carbon cloth	Graphite brush	Single chamber	Domestic wastewater	Sodium acetate	Phosphate buffer	180 ± 4.65 mW m^{-2}	Majidiv *et al* (2018) [71]
SnO_2–PANI carbon felt	Carbon felt	Single chamber	Anaerobic sludge	Acetate	Air-cathode	65 mW m^{-2}	Tiwari *et al* (2019) [72]
Carbon cloth–cobalt oxide	Carbon cloth	Two chamber	Dairy wastewater	Dairy wastewater	Potassium ferricyanide	170 mW m^{-2}	Veeramani *et al* (2020) [61]
CoAiA–LDH–graphite plates	Graphite plates	Two chamber	Anaerobic activated sludge	Glucose	Potassium dihydrogen phosphate K_2HPO_4	87.91 mW m^{-2}	Khajeh *et al* (2020) [63]

Table 5.5. Advantages and limitations of the techniques used for coating metals/metal oxides on carbon electrodes.

Technique	Remarks
Electrodeposition. The use of electric current to reduce the cations of desired metal(s) in an electrolyte to form a thin conductive metal coating on the electrode surface	**Advantages** • A low-temperature process. • Easily controls the growth of nanoparticles (size and film morphology). • Self-purification as well as extrinsic and intrinsic doping are possible. • Results in improved interfacial bonding between the coating material and the substrate before heat treatment or sintering. • The uniformity and morphology of the coating depend on the polymerization pattern or the polymer used. • A continuous and compact coating with preferred orientation is possible. **Limitations** Use of reactive precursor compounds and toxic solvents.
Electrophoretic deposition. Colloidal particles, such as polymers and metal/metal oxides, suspended in a liquid medium migrate under the influence of an electric field and deposit onto an electrode.	**Advantages** • High deposition rate. • Simple process. • No binder required.
Successive ionic layer adsorption and reaction (SILAR). SILAR is based on the immersion of the substrate into separate cationic and anionic precursor solutions, followed by a reaction to form a thin film.	**Advantages** • Simple, less expensive, and less time-consuming. • Applicable for the deposition of large-area thin films. • The chemicals used in SILAR are recyclable.
Hydrothermal method. Reduction of graphite oxide to reduced graphene oxide in a sealed vessel at both high temperature and pressure.	**Advantages** • Particularly suitable for the growth of large, good-quality crystals while maintaining control over their composition. • Nanoparticles of desired size and shape can be prepared. • Well-crystallized powder can be formed. • Produces nanocrystals with high crystallinity. **Limitations** • The method needs expensive autoclaves, and observing the crystal is impossible as it grows in a steel tube. • Involves difficult-to-control processes. • Reliability and reproducibility issues.

Figure 5.2. Graphene synthesis techniques.

Graphene-based anodes

Mesoporous anodes originally do not support bacterial growth inside their hierarchical porous structure. However, mesoporous graphene electrodes modified with metals or polymers could promote bacterial attachment, thereby increasing the extracellular electron transfer in MFCs (table 5.6). Chen *et al* (2014) fabricated a flexible macroporous 3D graphene sponge anode using an ice template during a freeze-drying process to tailor the pore structure, and it produced a maximum power density of 427.0 W m $^{-3}$, which was higher than that obtained using a carbon felt anode [73]. Reduced graphene oxide is currently used most often for electrode modification since it possesses a longer shelf-life without agglomeration, more stability in organic solvents, and higher conductivity than graphene oxide and graphite. Pareek *et al* (2019) employed a modified Hummers' method and L-cysteine as the reducing agent to synthesize chemically reduced graphene oxide (rGO) [74]. The carbon cloth modified with the porous CGO structure exhibited improved electrical capacitance and charge transfer due the abundance of electrocatalytic sites. Sayed *et al* (2021) [75] developed a graphene-based anode (RG0-CB) by modifying a carbon brush anode with a layer of deoxygenated graphene oxide via electrophoretic deposition, which increased the power density by more than ten times (from 33 mW m^{-2} for CB to 381 mW m^{-2}).

Table 5.6. Comparison of power density generated in an MFC using different graphene-based anodes.

Anode	Cathode	Reactor configuration	Biocatalyst	Substrate	Power density	References
Crumpled graphene	Carbon cloth	Two chamber	Anaerobic sludge	Sodium acetate	3.6 W m^{-3}	Xiao *et al* (2012) [76]
Graphene–polyaniline	Pt mesh	Two chamber	Anaerobic sludge	Sodium acetate	1390 mW m^{-2}	Hou *et al* (2013) [77]
Porous graphene–nickel	Titanium plate	Two chamber	*Shewanella putrefaciens*	Acetate	3903 mW m^{-2}	Qiao *et al* (2014) [78]
Graphene-modified anode	Activated carbon and CNT-modified stainless steel	Two chamber	Mixed consortium	Acetate	2142 mW m^{-2}	Hou *et al* (2014) [79]
Graphene sponge	Carbon paper	Two chamber	Anaerobic sludge	Acetate	427 W m^{-3}	Chen *et al* (2014) [73]
Graphene microsheets	Graphite rods	Single chamber	*Escherichia coli*	Glucose	2.85 W m^{-2}	Najafabadi *et al* (2016) [24]
Graphene-based aerogel	Air-cathode	Single chamber	*Rhodopseudomonas palustris*	Glucose	3.51 W m^{-2}	Call *et al* (2017) [77]
Carbon cloth–rGO	SS mesh	Two chamber	Anaerobic sludge	Glucose	6 mW m^{-2}	Pareek *et al* (2019) [74]
Carbon black–rGO	Air-cathode	Single chamber	Anaerobic sludge	Sewage wastewater	33.076 mW m^{-2} to 381 mW m^{-2}	Sayed *et al* (2021) [82]

Graphene-based cathodes

Graphene oxide (GO) is an amphiphilic material containing single 2D sp^2-hybridized carbon layers, along with hydrophilic regions consisting of carboxylic, hydroxyl, and epoxy groups; hence it can be incorporated easily with functional materials, such as polymers, metal nanoparticles, and metal–organic frameworks (MOFs). Many studies have reported the MFC performance of various cathode catalysts, such as MnO_2–GO, Fe–Ag–NG and V_2O_5–GO, Ni–Co–MGO, and conducting polymers (table 5.7). Meng *et al* (2017) deposited the GO–MgO catalyst on carbon cloth and obtained a chemical oxygen demand removal efficiency of 79.5% and a coulombic efficiency of 31.6% [83]. The power density of the MFC catalyzed by GO–MgO was 755.63 mW m^{-2}, which was equivalent to 86.78% of that of MFCs catalyzed by the Pt/C (870.75 mW m^{-2}). Moreover, nitrogen-doped graphene nanosheets are efficient electrocatalysts of O_2 reduction and follow the four-electron pathway to generate H_2O. Wen *et al* (2016) presented a facile route involving carbonization and an activation step to fabricate porous nitrogen-doped carbon nanosheets (PNCN) on graphene derived from the GO-PANI hybrid [84]. As a result, a higher maximum power density of 1159.34 mW m^{-2} was obtained in the MFC with the PNCN catalyst, than that with the Pt/C catalyst (858.49 mW m^{-2}). Gautam *et al* (2016) compared the performance of different cathodes, such as α-MnO_2 nanorods (MN), α-MnO_2 nanorods supported on reduced graphene oxide (MN–rGO), and α-MnO_2 nanorods supported on nitrogen-doped reduced graphene oxide (MN–NrGO), through a facile one-step hydrothermal method [83]. It was found that the MN–NrGO electrocatalyst-based MFCs produced the maximum power density even at very low electrocatalyst loadings (0.5 mg cm^{-2}). Normally, thermal exfoliation of GO at 200 °C is employed to reduce GO (rGO), which is obtained as flat wrinkled sheets. Luo *et al* (2015) applied an aerosol-assisted capillary compression process to convert the flat sheets into crumpled paper ball-like structures by [85]. The crumpled graphene cathode electrodes elevated the maximum power density from 2.5 W m^{-3} to 3.3 W m^{-3}, which is an 11-fold increase compared with that of the unmodified carbon cloth electrode (0.3 W m^{-3}). Metal doping in GO is usually mediated by covalent or non-covalent linkages. For example, the Ni–Co–GO nanocomposite reported by Papiya *et al* (2020) consisted of a graphene oxide (GO) support that was covalently modified with g-aminopropyl tri-ethoxysilane (APTES) for improved performance. The Ni–Co–MGO hybrid nanocomposite showed a maximum power density of 1003.18 mW m^{-2} [86].

5.4 Conclusion

Although MFCs can be scaled up to suit wastewater treatment, hydrogen production, desalination, and bioremediation, some difficulties associated with the cost of materials and the biocompatibility of anodes and ORR cathode catalysts still persist. Recent findings on the bulk and surface modification of carbon electrodes with conductive polymers and metal oxides/sulfates seem very promising, viable, and scalable for the MFC architecture, as discussed above. Some polymeric materials, such as polyaniline, polypyrrole, epoxy resin, and polyester, show great potential as

Table 5.7. Comparison of power density generated in an MFC using different graphene-based cathodes.

Cathode	Anode	Reactor configuration	Biocatalyst	Substrate	Buffer in catholyte	Power density	References
NG modified carbon cloth	Carbon cloth	Two chamber	Anaerobic sludge	Acetate	Phosphate buffer	776 ± 12 mW m^{-2}	Liu *et al* (2013) [87]
Carbon cloth–MnO_2 nanorods N-doped rGO	Carbon cloth	Single chamber	Anaerobic sludge	Glucose	Phosphate buffer	85.45 mW m^{-2}	Gautam *et al* (2016) [83]
Bio-rGO–graphite felt	Graphite felt	Two chamber	Anaerobic sludge	Glucose	Phosphate buffer	163.8 mW m^{-2}	Song *et al* (2016) [88]
Carbon cloth–GO–MgO	Carbon felt	Single chamber	Municipal sludge	Sodium acetate	Phosphate buffer	755.63 mW m^{-2}	Meng *et al* (2017) [84]
Stainless steel–V_2O_5–rGO	SS wire mesh	Single chamber	Fish market waste sludge	Fish market wastewater	Phosphate buffer	533 ± 37 mW m^{-2}	Noori *et al* (2017) [89]
Carbon cloth–GO-Zn–Co	Carbon cloth	Single chamber	Mixed bacteria	Acetate	Phosphate buffer	773 mW m^{-2}	Yang *et al* (2018) [90]
Stainless steel–Fe-Ag-NG	Carbon felt	Single chamber	Domestic wastewater	Sodium acetate	Phosphate buffer	1956.45 mW m^{-2}	Lv *et al* (2019) [91]
Carbon cloth–CGO-GO–SS mesh	Carbon cloth	Two chamber	Mixed consortium	Synthetic wastewater/ glucose	Saline buffer	6 mW m^{-2}	Pareek *et al* (2019) [74]
Carbon cloth–SGO–TiO_2–PANI	Carbon cloth	Single chamber	Mixed bacterial culture	Mixed bacterial culture	Air-cathode	904.18 mW m^{-2}	Papiya *et al* (2020) [92]

they can perform the dual role of binding the metal ions/oxide with the carbon materials and effectively increasing the electron transfer rate across the anode–biofilm interface, thus decreasing overpotential losses in MFCs. On the other hand, the use of metal–nitrogen–carbon catalysts in cathodes can significantly increase MFC performance in terms of power output with simultaneous wastewater treatment. Thus, inexpensive non-Pt-group metals may be viable cathode catalyst options during the scale-up and commercialization of MFCs.

References

[1] Logan B E, Hamelers B, Rozendal R, Schröder U, Keller J, Freguia S, Aelterman P, Verstraete W and Rabaey K 2006 Microbial fuel cells: methodology and technology *Env. Sci. Technol.* **40** 5181–92

[2] Park D H, Kim S K, Shin I H and Jeong Y J 2000 Electricity production in biofuel cell using modified graphite electrode with Neutral Red *Biotechnol. Lett.* **22** 1301–4

[3] Jayapriya J and Ramamurthy V 2012 Use of non-native phenazines to improve the performance of *Pseudomonas aeruginosa* MTCC 2474 catalysed fuel cells *Bioresour. Technol.* **124** 23–8

[4] Anh C, Je S, Thu N, Lee J, Seop I, Hong B, Yi H and Chun J 2003 A novel electrochemically active and Fe (III)-reducing bacterium phylogenetically related to *Aeromonas hydrophila*, isolated from a microbial fuel cell *FEMS Microbiol. Lett.* **223** 129–34

[5] Jayapriya J and Ramamurthy V 2015 Challenges to and opportunities in microbial fuel cells *Bioenergy Opportunities and Challenges* (New York: Apple Academic) pp 88–124

[6] Jayapriya J and Ramamurthy V 2014 The role of electrode material in capturing power generated in *Pseudomonas* catalysed fuel cells *Can. J. Chem. Eng.* **92** 610–4

[7] Bond D R and Lovley D R 2003 Electricity production by *Geobacter sulfurreducens* attached to electrodes *Appl. Env. Microbiol.* **69** 1548–55

[8] Chaudhuri S K and Lovley D R 2003 Electricity generation by direct oxidation of glucose in mediatorless microbial fuel cells *Nature Biotechnol.* **21** 1229–32

[9] Rahimnejad M, Najafpour G and Ghoreyshi A A 2009 *Effect of Mass Transfer on Performance of Microbial Fuel Cell* (Rijeka: Intech) pp 233–50

[10] Hays S, Zhang F and Logan B E 2011 Performance of two different types of anodes in membrane electrode assembly microbial fuel cells for power generation from domestic wastewater *J. Power Sources* **196** 8293–300

[11] Tao H, Liang M, Li W, Zhang L, Ni J and Wu W 2011 Removal of copper from aqueous solution by electrodeposition in cathode chamber of microbial fuel cell *J. Hazard. Mater.* **189** 186–92

[12] Rahimnejad M, Ghoreyshi A A, Najafpour G and Jafary T 2011 Power generation from organic substrate in batch and continuous flow microbial fuel cell operations *Appl. Energy* **88** 3999–4004

[13] Hassan S H A, Seong Y and Oh S 2012 Power generation from cellulose using mixed and pure cultures of cellulose-degrading bacteria in a microbial fuel cell *Enzyme Microbial Technol.* **51** 269–73

[14] Ishii S, Suzuki S, Norden-Krichmar T M, Nealson K H, Sekiguchi Y, Gorby Y A and Bretschger O 2012 Functionally stable and phylogenetically diverse microbial enrichments from microbial fuel cells during wastewater treatment *PLoS One* **7** e30495

[15] Zhao Y, Collum S, Phelan M, Goodbody T, Doherty L and Hu Y 2013 Preliminary investigation of constructed wetland incorporating microbial fuel cell: batch and continuous flow trials *Chem. Eng. J.* **229** 364–70

[16] Choi C and Hu N 2013 The modeling of gold recovery from tetrachloroaurate wastewater using a microbial fuel cell *Bioresour. Technol* **133** 589–98

[17] Sajana T K, Ghangrekar M M and Mitra A 2013 Application of sediment microbial fuel cell for *in situ* reclamation of aquaculture pond water quality *Aquacult. Eng.* **57** 101–7

[18] Li X, Zhu N, Wang Y, Li P, Wu P and Wu J 2013 Animal carcass wastewater treatment and bioelectricity generation in up-flow tubular microbial fuel cells: effects of HRT and non-precious metallic catalyst *Bioresour. Technol.* **128** 454–60

[19] Villaseñor J, Capilla P, Rodrigo M A, Cañizares P and Fernández F J 2013 Operation of a horizontal subsurface flow constructed wetland—microbial fuel cell treating wastewater under different organic loading rates *Water Res.* **47** 6731–8

[20] Mahmoud M, Parameswaran P, Torres C I and Rittmann B E 2014 Fermentation pre-treatment of landfill leachate for enhanced electron recovery in a microbial electrolysis cell *Bioresour. Technol.* **151** 151–8

[21] Huang L, Shi Y and Wang N 2014 Anaerobic/aerobic conditions and biostimulation for enhanced chlorophenols degradation in biocathode microbial fuel cells *Biodegradation* **25** 615–32

[22] Chang S H, Wu C H, Chang D K and Lin C W 2014 Effects of mediator producer and dissolved oxygen on electricity generation in a baffled stacking microbial fuel cell treating high strength molasses wastewater *Int. J. Hydrogen Energy* **39** 11722–30

[23] Qiao Y, Wen G-Y, Chang D K and Lin C W 2013 Anaerobic/aerobic conditions and biostimulation for enhanced chlorophenols degradation in biocathode microbial fuel cells *Biodegradation* **25** 615–32

[24] Najafabadi A T, Ng N and Gyenge E 2016 Electrochemically exfoliated graphene anodes with enhanced biocurrent production in single-chamber air-breathing microbial fuel cells *Biosens. Bioelectron.* **81** 103–10

[25] Hassan H, Jin B, Dai S, Ma T and Saint C 2016 Chemical impact of catholytes on *Bacillus subtilis*-catalysed microbial fuel cell performance for degrading 2,4-dichlorophenol *Chem. Eng. J.* **301** 103–14

[26] Almatouq A and Babatunde A O 2016 Concurrent phosphorus recovery and energy generation in mediator-less dual chamber microbial fuel cells: mechanisms and influencing factors *Int. J. Env. Res. Public Health* **13** 375

[27] Radha M and Kanmani S 2017 Evaluation of paper recycling industry wastewater treatment and simultaneous bioelectricity generation in a microbial fuel cell *Desalin. Water Treat.* **62** 161–7

[28] Zinadini S, Zinatizadeh A A, Rahimi M, Vatanpour V, Rahimi Z, Zinatizadeh A A, Rahimi M, Vatanpour V and Rahimi Z 2017 High power generation and COD removal in a microbial fuel cell operated by a novel sulfonated PES/PES blend proton exchange membrane *Energy* **125** 427–38

[29] Li Y, Styczynski J, Huang Y, Xu Z, McCutcheon J and Li B 2017 Energy-positive wastewater treatment and desalination in an integrated microbial desalination cell (MDC)-microbial electrolysis cell (MEC) *J. Power Sources* **356** 529–38

[30] Liu J, Wang X, Wang Z, Lu Y, Li X and Ren Y 2017 Integrating microbial fuel cells with anaerobic acidification and forward osmosis membrane for enhancing bio-electricity and water recoveryfrom low-strength wastewater *Water Res.* **110** 74–82

[31] Marzorati S, Schievano A, Colombo A and Lucchini G 2018 Ligno-cellulosic materials as air-water separators in low-tech microbial fuel cells for nutrients recovery *J. Cleaner Prod.* **170** 1167–76

[32] Yang N, Zhan G, Wu T, Zhang Y, Jiang Q, Li D and Xiang Y 2018 Effect of air-exposed biocathode on the performance of a *Thauera*-dominated membraneless single-chamber microbial fuel cell (SCMFC) *J. Env. Sci.* **66** 216–24

[33] Wang X *et al* 2019 Synergic degradation of 2,4,6-trichlorophenol in microbial fuel cells with intimately coupled photocatalytic-electrogenic anode *Water Res.* **156** 125–35

[34] Liu Y, Song P, Gai R, Yan C, Jiao Y, Yin D, Cai L and Zhang L 2019 Recovering platinum from wastewater by charring biofilm of microbial fuel cells (MFCs) *J. Saudi Chem. Soc.* **23** 338–45

[35] Zhang L, Fu G and Zhang Z 2019 Simultaneous nutrient and carbon removal and electricity generation in self-buffered biocathode microbial fuel cell for high-salinity mustard tuber wastewater treatment *Bioresour. Technol.* **272** 105–13

[36] Vijay A, Chhabra M and Vincent T 2019 Microbial community modulates electrochemical performance and denitrification rate in a biocathodic autotrophic and heterotrophic denitrifying microbial fuel cell *Bioresour. Technol.* **272** 217–25

[37] Ye Y, Ngo H H, Guo W, Chang S W, Nguyen D D, Zhang X, Zhang S, Luo G and Liue Y 2020 Impacts of hydraulic retention time on a continuous flow mode dual-chamber microbial fuel cell for recovering nutrients from municipal wastewater *Sci. Total Environ.* **734** 139220

[38] Marassi R J, Quieroz L G, Silva D C V R, dos Santos F S, Silva G C and De Paiva T C B 2020 Long-term performance and acute toxicity assessment of scaled-up air–cathode microbial fuel cell fed by dairy wastewater *Bioproc. Biosyst. Eng.* **43** 1561–71

[39] Marassi R J, Quieroz L G, Silva D C V R, da Silva F T, Silva G C and De Paiva T C B 2020 Performance and toxicity assessment of an up-flow tubular microbial fuel cell during long-term operation with high-strength dairy wastewater 259 *J. Cleaner Prod.* **259** 120882

[40] Xu T, Song J, Lin W, Fu B, Guo X, Huang X, Wu H and Zhang X 2021 A freestanding carbon submicro fiber sponge as high-efficient bioelectrochemical anode for wastewater energy recovery and treatment *Appl. Energy* **281** 1–13

[41] Pugazhendi A, Ghazi G and Jamal M T 2021 Bioenergy production and treatment of aquaculture wastewater using saline anode microbial fuel cell under saline condition *Env. Technol. Innov.* **21** 1–9

[42] Narayanasamy S and Jayaprakash J 2020 Application of carbon-polymer based composite electrodes for microbial fuel cells *Rev. Env. Sci. Biotechnol.* **19** 595–620

[43] Moon H, Chang I S and Kim B H 2006 Continuous electricity production from artificial wastewater using a mediator-less microbial fuel cell *Bioresour. Technol.* **97** 621–7

[44] Narayanasamy S and Jayaprakash J 2021 Carbon cloth/nickel cobaltite ($NiCo_2O_4$)/polyaniline (PANI) composite electrodes: preparation, characterization, and application in microbial fuel cells *Fuel* **301** 121016

[45] Rodríguez-Rey M *et al* 2016 A comprehensive database of quality-rated fossil ages for Sahul's Quaternary vertebrates *Sci. Data.* **3** 1–7

[46] Narayanasamy S and Jayaprakash J 2018 Improved performance of *Pseudomonas aeruginosa* catalyzed MFCs with graphite/polyester composite electrodes doped with metal ions for azo dye degradation *Chem. Eng. J.* **343** 258–69

[47] Ren Y, Pan D, Li X, Fu F, Zhao Y and Wang X 2013 Effect of polyaniline-graphene nanosheets modified cathode on the performance of sediment microbial fuel cell *J. Chem. Technol. Biotechnol.* **88** 1946–50

[48] Sumisha A and Haribabu K 2020 Nanostructured polypyrrole as cathode catalyst for Fe (III) removal in single chamber microbial fuel cell *Biotechnol. Bioprocess Eng.* **25** 78–85
[49] Karthick S and Haribabu K 2020 Bioelectricity generation in a microbial fuel cell using polypyrrole–molybdenum oxide composite as an effective cathode catalyst *Fuel* **275** 117994
[50] Yuan Y, Zhou S and Zhuang L 2010 Polypyrrole/carbon black composite as a novel oxygen reduction catalyst for microbial fuel cells *J. Power Sources* **195** 3490–3
[51] Esmaeili C, Ghasemi M, Heng L Y, Hassan S H A, Abdi M M, Daud W R W, Ilbeygi H and Ismail A F 2014 Synthesis and application of polypyrrole/carrageenan nano-bio composite as a cathode catalyst in microbial fuel cells *Carbohydr. Polym.* **114** 253–9
[52] Zhou X, Xu Y, Mei X, Du N, Jv R, Hu Z and Chen S 2018 Polyaniline/β-MnO_2 nanocomposites as cathode electrocatalyst for oxygen reduction reaction in microbial fuel cells *Chemosphere* **198** 482–91
[53] Sonawane J M, Yadav A, Prakash C and Adeloju S B 2016 Recent advances in the development and utilization of modern anode materials for high performance microbial fuel cells *Biosens. Bioelectron.* **90** 558–76
[54] Dumas C and Bergel A 2008 Biofilm on stainless steel cathodes *Electrochim. Acta* **53** 2494–500
[55] Liang Y, Feng H, Shen D, Li N and Guo K 2017 Enhancement of anodic biofilm formation and current output in microbial fuel cells by composite modification of stainless steel electrodes *J. Power Sources* **342** 98–104
[56] Ying X, Shen D, Wang M, Feng H, Gu Y and Chen W 2017 Titanium dioxide thin film-modified stainless steel mesh for enhanced current-generation in microbial fuel cells *Chem. Eng. J.* **333** 260–7
[57] Ying X, Feng H, Shen D, Wang M, Xu Y, Chen T and Zhu Y 2019 Science of the total environment sustainable synthesis of novel carbon microwires for the modification of a Ti mesh anode in bioelectrochemical systems *Sci. Total Env.* **669** 294–302
[58] Pu L, Li K, Chen Z, Zhang P, Zhang X and Fu Z 2014 Silver electrodeposition on the activated carbon air cathode for performance improvement in microbial fuel cells *J. Power Sources* **268** 476–81
[59] Yang Y J, Duan M, Yan C, Zhao D, Jiang C, Duan X and Song X 2019 Facile synthesis of CoFe-LDH/MWCNT/rGO nanocomposite as efficient bifunctional electrocatalysts for oxygen evolution and reduction *J. Electroanal. Chem.* **856** 113697
[60] Liu J and Vipulanandan C 2017 Effects of Fe, Ni, and Fe/Ni metallic nanoparticles on power production and biosurfactant production from used vegetable oil in the anode chamber of a microbial fuel cell *Waste Manag.* **66** 169–77
[61] Veeramani V, Rajangam K and Nagendran J 2020 Performance of cobalt oxide/carbon cloth composite electrode in energy generation from dairy wastewater using microbial fuel cells *Sustain. Envron. Res.* **4** 0–7
[62] Guo X, Jia J, Dong H, Wang Q, Xu T, Fu B, Ran R, Liang P, Huang X and Zhang X 2019 Hydrothermal synthesis of Fe–Mn bimetallic nanocatalysts as high-efficiency cathode catalysts for microbial fuel cells *J. Power Sources* **414** 444–52
[63] Khajeh R T, Aber S and Zarei M 2020 Comparison of $NiCo_2O_4$, CoNiAl-LDH, and CoNiAl-LDH@$NiCo_2O_4$ performances as ORR catalysts in MFC cathode *Renew. Energy* **154** 1263–71
[64] Martin E, Tartakovsky B and Savadogo O 2011 Cathode materials evaluation in microbial fuel cells: a comparison of carbon, Mn_2O_3, Fe_2O_3 and platinum materials *Electrochim. Acta* **58** 58–66

[65] Ghasemi M, Ramli W, Daud W, Rahimnejad M and Rezayi M 2013 Copper-phthalocyanine and nickel nanoparticles as novel cathode catalysts in microbial fuel cells *Int. J. Hydrogen Energy* **38** 9533–40

[66] Choi C and Hu N 2013 The modeling of gold recovery from tetrachloroaurate wastewater using a microbial fuel cell *Bioresour. Technol.* **133** 589–98

[67] Zhang X, Li K, Yan P, Liu Z and Pu L 2015 N-type Cu_2O doped activated carbon as catalyst for improving power generation of air cathode microbial fuel cells *Bioresour. Technol.* **187** 299–304

[68] Noori T, Ghangrekar M M and Mukherjee C K 2016 V_2O_5 microflower decorated cathode for enhancing power generation in air-cathode microbial fuel cell treating fish market wastewater *Int. J. Hydrogen Energy* **41** 3638–45

[69] Pu L, Liu D, Li K and Wang J 2017 Carbon-supported binary transition metal chalcogenide used as cathode catalyst for oxygen reduction in microbial fuel cell *Int. J. Hydrogen Energy* **42** 14253–63

[70] Zhang S, Su W, Wang X, Li K and Li Y 2018 Bimetallic metal-organic frameworks derived cobalt nanoparticles embedded in nitrogen-doped carbon nanotube nanopolyhedra as advanced electrocatalyst for high-performance of activated carbon air-cathode microbial fuel cell *Biosens. Bioelectron.* **127** 181–7

[71] Majidi M R, Farahani F S, Hosseini M, Ahadzadeh I and CR P T 2018 Low-cost nanowired α-MnO2/C as an ORR catalyst in air-cathode microbial fuel cell *Bioelectrochemistry* **125** 38–45

[72] Tiwari A K, Jain S, Mungray A A and Mungray A K 2019 SnO_2:PANI modified cathode for performance enhancement of air-cathode microbial fuel cell *Biochem. Pharmacol.* **8** 103590

[73] Chen W, Huang Y-X, Li D-B, Yub H-Q and Yan L 2014 Preparation of a macroporous flexible three dimensional graphene sponge using an ice-template as the anode material for microbial fuel *RSC Adv.* **4** 21619–24

[74] Pareek A, Shanthi Sravan J and Venkata Mohan S 2019 Exploring chemically reduced graphene oxide electrode for power generation in microbial fuel cell *Mater. Sci. Energy Technol.* **2** 600–6

[75] Sayed E T, Wilberforce T, Elsaid K, Rabaia M K H, Abdelkareem M A, Chae K J and Olabi A G 2021 A critical review on environmental impacts of renewable energy systems and mitigation strategies: wind, hydro, biomass and geothermal *Sci. Total Env.* **766** 144505

[76] Pareek A, Sravan J S and Mohan S V 2019 Exploring chemically reduced graphene oxide electrode for power generation in microbial fuel cell *Mater. Sci. Energy Technol.* **2** 600–6

[77] Call T P, Carey T, Bombelli P, Lea-smith D J, Hooper P, Howe J and Torrisi F 2017 Cathodes for single chamber microbial fuel cells *J. Mater. Chem. A* **1** 23872–86

[78] Najafabadi A T and Ng N 2016 Electrochemically exfoliated graphene anodes with enhanced biocurrent production in single-chamber air-breathing microbial fuel cells *Biosensors Bioelectron.* **81** 103–10

[79] Hou J, Liu Z, Yang S and Zhou Y 2014 Three-dimensional macroporous anodes based on stainless steel fiber felt for high-performance microbial fuel cells *J. Power Sources* **258** 204–9

[80] Hou J, Liu Z and Zhang P 2013 A new method for fabrication of graphene/polyaniline nanocomplex modified microbial fuel cell anodes *J. Power Sources* **224** 139–44

[81] Xiao L, Damien J, Luo J, Dong H, Huang J and He Z 2012 Crumpled graphene particles for microbial fuel cell electrodes *J. Power Sources* **208** 187–92

[82] Sayed E T, Alawadhi H, Olabi A G, Jamal A, Almahdhi M S, Khalid J and Abdelkareem M A 2020 Electrophoretic deposition of graphene oxide on carbon brush as bioanode for microbial fuel cell operated with real wastewater *Int. J. Hydrogen Energy* **46** 5975–83

[83] Gautam R K, Bhattacharjee H, Venkata Mohan S and Verma A 2016 Nitrogen doped graphene supported α-MnO_2 nanorods for efficient ORR in a microbial fuel cell *RSC Adv.* **6** 110091–101

[84] Li M, Zhou S and Xu M 2017 Graphene oxide supported magnesium oxide as an efficient cathode catalyst for power generation and wastewater treatment in single chamber microbial fuel cells *Chem. Eng. J.* **328** 106–16

[85] Liu Y, Jin X, Dionysiou D D, Liu H, Huang Y and A C S C 2015 Homogeneous deposition-assisted synthesis of iron-nitrogen composites on graphene as highly efficient non-precious metal electrocatalysts for microbial fuel cell power generation *J. Power Sources* **278** 773–87

[86] Papiya F, Das S, Pattanayak P and Kundu P P 2019 The fabrication of silane modified graphene oxide supported Ni–Co bimetallic electrocatalysts: a catalytic system for superior oxygen reduction in microbial fuel cells *Int. J. Hydrogen Energy* **44** 25874–93

[87] Liu Y, Liu H, Wang C, Hou S X and Yang N 2013 Sustainable energy recovery in wastewater treatment by microbial fuel cells: stable power generation with nitrogen-doped graphene cathode *Env. Sci. Technol.* **47** 13889–95

[88] Song T-S, Jin Y, Bao J, Kang D and Xie J 2016 Graphene/biofilm composites for enhancement of hexavalent chromium reduction and electricity production in a biocathode microbial fuel cell *J. Hazard. Mater.* **317** 73–80

[89] Noori M T, Mukherjee C K and Ghangrekar M M 2017 Enhancing performance of microbial fuel cell by using graphene supported V_2O_5-nanorod catalytic cathode *Electrochim. Acta* **228** 513–21

[90] Yang W, Chata G, Zhang Y, Peng Y, Lu J E, Wang N, Mercado R, Li J and Chen S 2019 Graphene oxide-supported zinc cobalt oxides as effective cathode catalysts for microbial fuel cell: high catalytic activity and inhibition of biofilm formation *Nano Energy* **57** 811–9

[91] Lv C, Liang B, Zhong M, Li K and Qi Y 2019 Activated carbon-supported multi-doped graphene as high-efficient catalyst to modify air cathode in microbial fuel cells *Electrochim. Acta* **304** 360–9

[92] Papiya F, Pattanayak P, Kumar V, Das S and Kundu P P 2020 Sulfonated graphene oxide and titanium dioxide coated with nanostructured polyaniline nanocomposites as an efficient cathode catalyst in microbial fuel cells *Mater. Sci. Eng.* C **108** 110498

Chapter 6

The impact of biomass-derived electrodes on the electrochemical performance of microbial fuel cells (MFCs)

Najwa Najihah Mohamad Daud, Asim Ali Yaqoob, Mohamad Nasir Mohamad Ibrahim and Khalid Umar

Microbial fuel cells (MFCs) are novel bioelectrochemical techniques based on the metabolic activities of inoculation bacteria that serve as biocatalysts for natural conversion of waste materials into energy. The electrode is one of the most important MFC components in terms of energy generation and pollutant removal. As a result, it is in direct contact with the bacteria and is in charge of the pace at which electrons are transferred from the electroactive bacteria to the electrode surface. The electrode material and cost are now growing concerns in MFC research. Natural biomass-derived materials have been identified as one of the best options for addressing the current problems of materials for anodes in MFCs. Naturally derived anodes have several drawbacks that can be overcome using metal oxide composites, as explained in this chapter. This chapter outlines various natural biomass materials and for electrode fabrication in MFCs to reduce cost and improve performance.

6.1 Introduction

The health of humans and the health of the ecosystem are intricately linked. Today, the production, manufacturing, distribution, and utilization of fuels and other energy sources are ever-increasing throughout the world, which in turn all have a significant impact on human health and ecosystem imbalance [1]. Therefore, the exploitation of alternative sources of energy that are both economically and environmentally beneficial is gaining increasing attention. Microbial fuel cell (MFC) technology has shown promise in addressing these challenges. A series of electrochemical reactions mediated by micro-organisms converts the chemical energy held in wastewater organics directly into electrical power [2, 3]. Energy

doi:10.1088/978-0-7503-4791-4ch6

extraction from wastewater is regarded as the most promising future method for the generation of renewable energy and wastewater treatment [4]. There are two chambers in MFCs, the anode and cathode, separated by a proton exchange membrane (PEM). It necessitates continual anode electron release and cathode electron consumption [5]. The anode bacteria oxidize organic substrates to produce electrons and protons through respiration [6, 7]. The electrons are then sent from the anode to the cathode through an external circuit. In contrast, protons move directly to the cathode side and react with oxygen to form water. There are numerous benefits of MFCs over existing technologies in power production and waste treatment, including a lower creation of activated sludge, no need for aeration energy, simple operation under normal conditions, and environmental friendliness [8–10]. A schematic of MFCs is shown in figure 6.1(A), along with a list of their advantages.

Even though MFCs have significant strengths, many researchers also agree that they have some drawbacks. MFCs are currently impractical for commercial and/or industrial use because of their inability to produce adequate energy, inefficient wastewater treatment, and enormous cost. The anode electrode's material and design are two of the most significant hurdles in MFC design and operation. Metals/metal oxides, conducting polymers, and materials derived from carbon have all been investigated for use as the anode electrode in MFCs [11–13]. However, the standard materials derived from carbon cannot generate adequate electricity or efficiently treat wastewater [10]. The previously mentioned standard materials have numerous drawbacks, including low chemical stability, a limited surface area, poor mechanical properties, non-biocompatibility, corrosion, a large number of pores, and high cost [13].

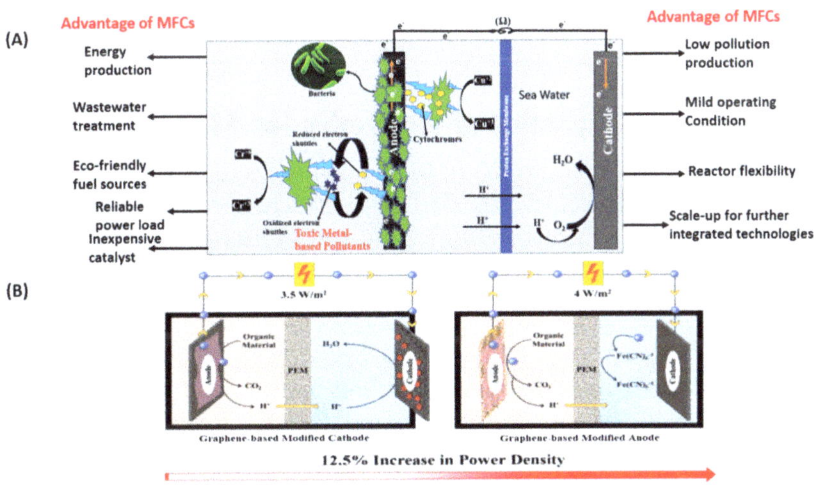

Figure 6.1. (A) Diagram of MFCs along with their advantages. (B) The advantages of a graphene-based modified anode over a graphene-based modified cathode. (Reproduced with permission from [13]. Copyright 2020 Elsevier.)

As a result, a better quality material is needed to overcome these limitations and build a stable anode. Cai *et al* recently examined a variety of natural wastes that could be employed as materials for anode electrodes in MFCs [14]. These anode electrodes made from natural waste outperformed those constructed with standard materials due to their increased area and precise porosity. Their electrochemical efficiency, however, remained unsatisfactory.

Furthermore, natural waste can be used to make graphene and its derivatives [15, 16]. Modifying carbon-based materials that are very conductive, such as graphene and its derivatives (e.g. with metal oxides), may improve the performance of MFCs. A study of the literature on graphene and its derivative based electrodes for MFCs found that an anode electrode based on modified graphene generated more energy than the cathode electrode in power generation [17]. As a result, the anode electrode based on modified graphene improved the degradation of the substrate by fostering the optimal development of micro-organisms on the anode electrode's surface, significantly increasing the movement of electrons through a system (figure 6.1(B)). According to the literature, there has been no study into using graphene anodes made from waste in MFCs for the objectives of energy generation and pollution reduction. The use of waste-derived graphene in the anode electrode lowers the cost of MFC operation while simultaneously improving the electrochemical efficiency. Furthermore, graphene modification with metal oxides such as ZnO, AgO, CuO, TiO_2, and others significantly lowers the expenditure and prevents corrosion [18].

The following review outlines the various types of electrode materials and some novel modifying approaches for improving electrochemical performance. In addition, the critical parameters for anode electrode materials are discussed in this chapter, which have been rarely documented in other literature. Furthermore, several natural materials and modified methodologies are studied thoroughly to solve the current concerns regarding electrodes in MFCs, including the costs and durability.

6.2 Required characteristics of electrodes for MFCs

The preferred material for electrodes in MFCs continues to be a complicated subject for researchers aiming to achieve the desired electrochemical performance, electron transfer, and bacterial adherence [19]. Figure 6.2(A) lists and illustrates several of the most crucial features that an optimal electrode must have.

6.2.1 Conductivity

The fundamental attribute of electrode materials is their conductivity. Conductivity is a critical attribute of an electrode material since it transmits the electrons generated by bacteria to pass from an anode to the cathode through an external circuit. As a result, an anode electrode facilitates and accelerates electron flow by lowering the resistivity of a bulk solution employing highly conductive chemicals [20–22]. In addition, the contact resistance between the substrates and electrodes needs to be minimal to maximize the electron transfer [22]. Before fabricating an electrode, the conductivity of materials is often investigated.

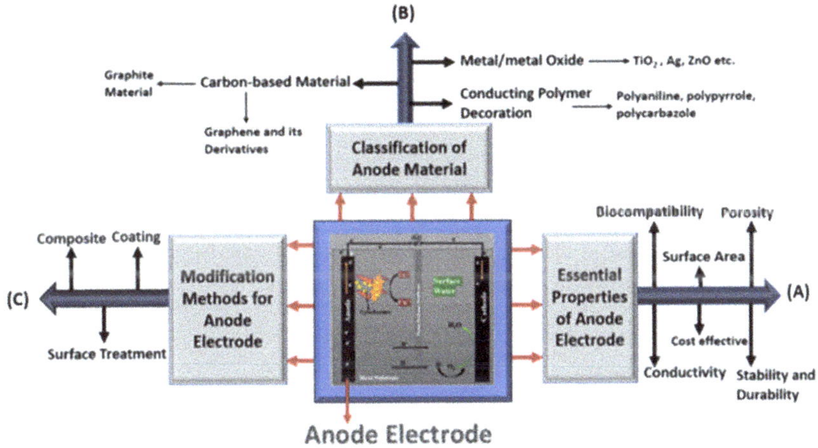

Figure 6.2. Diagram of an anode. (A) Required characteristics of an anode. (B) Classification of anode material. (C) Modification procedures of the anode. (Reproduced with permission from [7]. Copyright 2020 Elsevier.)

6.2.2 Area and porosity of the electrode surface

The area of the electrode surface has a significant effect on the energy output in MFCs [23]. Since the resistance of the electrode corresponds to the voltage drop across the circuit of the fuel cell, the most suitable approach is to enlarge its surface. Furthermore, a high surface area provides more bacterial growth sites, improving the kinetics of fuel cell electrode efficiency. Bacterial species including Geobacter, *Pseudomonas*, and *Escherichia coli* were rendered immobile on the electrode surfaces, resulting in successful electron transmission [12]. Anode electrode surfaces are subjected to biological processes; hence, the surface area significantly affects the MFC performance [24]. Graphene and its derivatives have a surface area that is more extensive than standard carbon materials, as shown in table 6.1 [24–26].

6.2.3 Biocompatibility

A biologically compatible electrode material is critical in MFC operation, where it comes into direct interference with bacteria and cellular respiration. Certain elements, including gold, silver, and copper, are unsuitable as electrodes in MFCs because of their corrosive nature [5, 26, 27]. The toxic effect of such compounds will hinder the development of bacteria in MFCs, resulting in reduced power production.

6.2.4 Stability and long-term durability

In general, prolonged contact of standard electrodes with the substrate and inoculation micro-organisms results in swelling due to mechanical and chemical instability, resulting in a profound effect on the electrode's physical stability. Corrosion, thermal fluctuation, and low mechanical strength are the factors that contribute to the swelling [28]. Additionally, a rough electrode surface provides more active sites for bacterial adhesion [26]. The cost and availability of electrode

Table 6.1. Materials often utilized as MFC anodes.

Electrodes	Anode size		Anode surface area	Energy efficiency	Reference
Material for anode	Material for cathode				
Carbon paper	Carbon paper	2.5 cm × 4.5 cm	22.5 cm^2	600 mW m^{-2}	[11]
Carbon felt	Carbon fiber felt	2.5 × 2.5 cm	2.5 cm^2	784 mW m^{-2}	[30]
Carbon cloth	Carbon cloth	2 cm × 2 cm	4 cm^2	679.7 mW m^{-2}	[31]
Carbon brush		Length of anode: 2.5 cm Diameter of anode: 2.5 cm Length of cathode: 1.7 cm Diameter of cathode: 1.3 cm	16 cm^2	4.25 mW m^{-2}	[32]
Graphene oxide–CNTs	Carbon cloth	–	–	434 mW m^{-2}	[33]
Graphene–Ppy	Carbon cloth	–	136 g m^{-2}	145 mW m^{-2}	[34]
Graphene powder–polytetrafluoroethylene on carbon cloth	Carbon cloth	4 cm × 4 cm		0.329 mW m^{-2}	[35]
Graphene coating on carbon cloth	Carbon cloth	Cathode: 4 cm × 4 cm Anode: 1 cm × 2 cm	4 cm^2	52.5 mW m^{-2}	[36]
N-doped graphene nanosheets (NGNS) on carbon cloth	Carbon cloth	1 cm × 1.5 cm	597 m^2 g^{-1}	1008 mW m^{-2}	[37]
Zero-dimension nitrogen-doped carbon dots modification with carbon paper	Carbon paper	2.5 cm × 2.5 cm		0.32 mW m^{-2}	[38]
Nickel foam–CNTs–PANI	Carbon cloth	–	1 cm^2 of anode surface area	113 W m^{-2}	[39]
Polypyrrole–graphene oxide	Carbon paper	Cathode: 5 cm × 5 cm Anode: 2 cm × 1 cm		102 mW m^{-2}	[40]
Graphene oxide	Carbon felt	3.0 cm × 2.0 cm × 0.5 cm	–	1326 mW m^{-2}	[40]

(*Continued*)

Table 6.1. (*Continued*)

Electrodes	Anode size		Anode surface area	Energy efficiency	Reference
Material for anode	Material for cathode				
Graphene oxide modification with carbon paper	Carbon paper	5 cm × 3 cm	–	368 mW m^{-2}	[41]
Graphene–Au composite	Carbon paper	–	6 cm^2	508 mW m^{-2}	[42]
rGO–Ppy	Carbon paper	1 cm × 1.5 cm		1068 mW m^{-2}	[43]
rGO–SnO_2–carbon cloth composite	Pt rode	3 cm × 2 cm	6 cm^2	1624 mW m^{-2}	[44]
Polyaniline PANI networks on graphene nanorods (GNRs) coated on carbon paper (CP–GNRs–PANI)	Carbon paper	2 cm × 2 cm	4 cm^2	856 mW m^{-2}	[45]
rGO–carbon cloth–PANI	Carbon felt	1.8 cm × 1.8 cm	–	1390 mW m^{-2}	[40]
rGO sheets–carbon cloth	Carbon cloth	–	4.5 cm^2	2.5 W m^{-2}	[46]
Graphene nanosheet coating on carbon paper	Carbon cloth	–	–	610.0 mW m^{-2}	[47]
Graphene	Carbon cloth	–	4 cm^2	2850 mW m^{-2}	[48]
3D-graphene	Carbon cloth	Anode: 30 mm × 5 mm (diameter × thickness)	9.41 m^2	1516 ± 87 mW m^{-2}	[49]
Graphite brush	Carbon cloth	3 cm × 2 cm	8 cm^2	1280 mW m^{-2}	[50]
TiO_2 and rGO composite	Carbon fiber/brush	1 cm × 1 cm	Anode projected surface area 1 cm^2	3169 mW m^{-2}	[13]
Glassy carbon	Carbon cloth	Length of anode and cathode: 1.7 cm Diameter of anode: 1.8 cm Diameter of cathode: 3.0 cm	7 cm^2	1905 mW m^{-2}	[51]
Activated carbon cloth	Graphite foil	–	1.5 cm^2 in projected area	0.51 mW m^{-2}	[52]
Ti/TiO_2	Pt meshes	2 cm × 4 cm	–	2317 W m^{-2}	[53]
Titanium rod	Graphite felt	20 mm	20 ± 1 cm^2	–	[54]
Titanium		2 cm × 2 cm	–	–	[55]

materials are critical elements in the selection process since they significantly affect the overall price of MFCs. For example, gold, platinum, and silver are costly and difficult to obtain. In MFCs, electrode materials made of metal composites and carbon-based materials might be intriguing alternatives to expensive metals [29].

6.3 Biomass-derived composite electrodes

6.3.1 Carbon anodes derived from biomass

The most frequently employed precursors for the manufacture of carbon anodes are plants. Carbon in green plants is derived through photosynthesis by converting carbon dioxide into glucose. However, the heteroatom content is intimately tied to the enzymes involved in photosynthesis [56]. In many cases, direct pyrolysis of their green predecessors at elevated temperatures is used to produce porous carbons. Plants function as self-doped heteroatom based carbon materials, while water evaporates during high-temperature pyrolysis to form a porous carbon structure. For example, three-dimensional microporous carbon foams (NPSCFs) were synthesized by Zhang *et al* by direct pyrolysis of commercial bread [57]. Large pores, a high specific surface area (295.07 $m^2 g^{-1}$), a homogeneous distribution of N, P, and S dopants, and excellent electrical conductivity were all characteristics of these foams. The maximum power density and current density of an MFC coupled with an NPSCF anode were 3134 mW m^{-2} and 7.56 A m^{-2}, respectively, which were nearly 2.5 times those of a carbon cloth anode, which were 1218 mW m^{-2} and 2.87 A m^{-2}.

As detailed in table 6.2, various natural resources have been utilized to fabricate anodes for MFCs. The surface properties and porosity are the primary parameters affecting the functionality of the device and important factors to consider while preparing an anode. For example, Chen *et al* constructed a hierarchically organized anode utilizing chestnut shells as a predecessor because of the spherical shell's brush-like structure with thick and stiff thorns, which provides a surface that is accessible for the biofilm [58]. The anode is designed to obtain a maximum power density of 759 ± 38 mW m^{-2} and Coulombic efficiency of 76%, equivalent to that of a standard graphite brush anode. Other research discovered that the chestnut shell-derived anode gave a maximum power density of 23.6 W m^{-3}, 2.3 times that (10.4 W m^{-3}) of a carbon cloth anode after chemical activation [59]. To improve the microbial loading while maintaining the anode's volume and weight, Zhu *et al* used porous structure fibers as the material for the anode, which allowed electron transport on both the exterior and inner surfaces [60]. Indeed, the anode's weight-based power density, which is 104.1 mW g^{-1}, was about 20 times that of a standard solid carbon fiber electrode with 5.5 mW g^{-1}.

Along with the pyrolysis route, covering natural precursors with activated carbon nanoparticles is an energy-efficient method of electrode synthesis. For example, Xie *et al* produced a CNT carbon electrode by soaking it in aqueous carbon nanotube textile (CNT) ink and drying it [61]. Compared to a standard carbon cloth electrode, the electrode produced by this process had a maximum current density 157% higher and a maximum power density 68% higher. Zhou *et al* used a natural loofah sponge soaked in Chinese ink to create a three-dimensional electrode [62]. The resulting

Table 6.2. List of biomass-derived carbon anodes in MFCs reported in the recent literature.

Biomass-derived anode sources	Synthesis methods	Type of MFC	Organic substrate	Source of inoculum	Power density	References
Pomelo peel	Carbonization		Acetate	Domestic wastewater	4.0 mA cm^{-2}	[67]
Bamboo	Simple carbonization	Double chamber	Acetate	Pre-acclimated bacteria	1652 ± 18 mW m^{-2}	[68]
Neem wood	Carbonization	Double chamber	Glucose	Pre-acclimated bacteria	256 ± 25 mW m^{-2}	[69]
Kapok fibers	Carbonization	Single chamber	Acetate	Activated sludge	1738.1 mW m^{-2}	[70]
King mushroom	Carbonization	–	Acetate	Anaerobic digester sludge	2.09 mW cm^{-2}	[71]
Wild mushroom	Carbonization	–	Acetate	Anaerobic digester sludge	3.02 mW cm^{-2}	[71]
Corn stem	Carbonization	–	Acetate	Anaerobic digester sludge	3.12 mW cm^{-2}	[71]
Chestnut shell	Carbonization and activation	Single chamber	Acetate	Pre-acclimated bacteria	759 ± 38 mW m^{-2}	[72]
Loofah sponges	Polymerization and carbonization	Single chamber	Acetate	Pre-acclimated bacteria	1090 ± 72 mW m^{-2}	[73]
Silk cocoon	Simple carbonization	Single chamber	Acetate	Activated anaerobic sludge	5.0 mW m^{-2}	[74]
Municipal sludge	Carbonization and polymerization	Double chamber	Lactate	*Shewanella oneidensis* MR-1	568.5 mW m^{-2}	[75]
Sugarcane	Carbonization	Double chamber	Acetate	Pre-acclimated bacteria	59.94 ± 2.81 W m^{-2}	[76]
Loofah sponges	Chinese black coating	Single chamber	Acetate	Pre-acclimated bacteria	0.82 mW cm^{-2}	[77]
Sewage sludge	Carbonization	Double chamber	Acetate	Pre-acclimated bacteria	3.2 mW m^{-2}	[78]

Coconut shell	Carbonization	Single chamber	Acetate	Pre-acclimated bacteria	1069 ± 15 W m^{-2}	[79]
Cake	Carbonization and polymerization	Single chamber	Acetate	Pre-acclimated bacteria	1307 mW m^{-2}	[80]
Cocklebur fruit	Carbonization	Single chamber	Acetate	Pre-acclimated bacteria	572.57 ± 24.90 μW	[81]
Cotton textile	Carbonization and polymerization	Single chamber	Acetate	Activated sludge	931 ± 61 mW m^{-2}	[82]
Chestnut shell	Carbonization and activation	Single chamber	Acetate	Municipal waste	23.6 W m^{-2}	[83]
Sewage sludge	Methane chemical vapor	Double chamber	Acetate	Pre-acclimated bacteria	2228 W m^{-2}	[84]
Bread	Simple carbonization	Double chamber	Acetate	Pre-acclimated bacteria	134 mW m^{-2}	[85]
Pinecone	Simple carbonization	Double chamber	Acetate	Pre-acclimated bacteria	10.88 W m^{-3}	[86]

sponge was remarkably flexible and had an open porosity structure. Additionally, the electrode demonstrated good stability during a 70 day trial period and gave a maximum current density of 16.3 0.5 mA cm^{-3} higher than a 3D electrode created using the carbonization procedure. In a comparative study, it has been shown that the soaking method can be used to fabricate electrodes. Zhang *et al* obtained a maximum power density of 61.7 ± 0.6 W m^{-3} with MFCs based on a carbon-backed loofah sponge electrode, which generally has a low cost [63]. Direct pyrolysis of biomass precursors at elevated temperatures is the most common way to make anodes. This process turns biomass into porous carbon by incorporating heteroatoms into it. The high energy input required for pyrolysis, on the other hand, invariably raises the cost of electrode preparation. Encapsulating natural precursors with conductive carbon nanoparticles, on the other hand, is an innovative, low-cost, and easy electrode production process. Plants, animal waste, and wastewater sludge have been used as precursors in the production of electrodes. Silk cocoons are common natural polymer composites with 15% nitrogen content and naturally random layering, with pores of a silk fiber in a three-dimensional nonwoven structure. Due to these fascinating characteristics, silk cocoons are an excellent option for electrode manufacture. For example, Lu *et al* carbonized silk cocoons in a single process to develop a cantilevered and versatile electrode with a high nitrogen content and hierarchical pores [64]. The developed flexible carbon fiber anode demonstrated excellent mechanical properties and 3.1 times the maximum gravimetric power density of a carbon cloth anode in MFCs. On the other hand, a by-product of the treatment of effluents, sewage sludge has multiple functional groups on its surface and a high carbon concentration, making it an ideal catalyst for chemical reactions [65, 66]. In fact, sewage sludge has been utilized to make anodes. Yuan *et al* carbonized the sewage sludge directly at 900 °C in order to manufacture monolith anodes with a maximum power density of 486 ± 18 mW m^{-2}, which was higher than the power density of a standard graphite plate anode (410 ± 10 mW m^{-2}) [66]. Despite the abundance of functional groups on the anode surface, the pores of sewage sludge-derived carbon anodes are generally in the micrometer range; hence, obstruction of pores can significantly restrict microbial colonization and volume usage.

6.3.2 Carbon cathodes derived from biomass

While platinum as well as its alloys are generally known to be the most active ORR catalysts, their high cost and low resistance to chemicals make them unsuitable for commercialization in MFCs. Natural biomass has the potential to be a good precursor for producing heteroatom-doped carbon materials as viable substitutes. Plant-derived biomass is a common precursor for ORR catalysts because it is high in nitrogen and phosphorus, and is primarily used as a precursor for heteroatoms and carbon. The creation of cathodes made from plant materials can minimize the necessity of intricate reaction techniques, expensive specialty precursors and equipment, and various dangerous chemicals due to the natural inclusion of heteroatoms and the inexpensive cost of raw precursors. N-doped hierarchically porous carbon (HPC) has been fabricated using watermelon skins as an N-rich and very stable precursor and has been shown to have a redox peak comparable to that of Pt/C at current densities of 0.19 mA cm^{-2} but with lower charge transfer (20.63 Ω) than Pt/C (37.56 Ω) [87]. The ORR catalyst developed by Fan *et al* by directly pyrolyzing *Chlorella pyrenoidosa* at 900 °C was expected to yield a maximum power density of 2068 ± 30 mW m^{-2} of 13% greater than the maximum power density generated by the MFCs using 20% Pt/C [88]. Similar to creating anodes, biomass from wastewater and animals has been employed to develop a carbon-based cathode catalyst, owing to the high nitrogen content. Simple pyrolysis is used to introduce N into the carbon framework's porous structure. A sustainable electrocatalyst made from spider silk by Yuan *et al* was shown to have an 1800 ± 82 mW m^{-2} power density at a 2 mg cm^{-2} loading, which was much higher than the power density of PtC in MFCs [89]. Additionally, the catalyst in its as-prepared state maintained a consistent production of the maximum power density over a period of up to 90 days. In a separate work, chitin was hydrothermally treated and pyrolyzed in a two-step technique to synthesize an N-doped carbon sheet. In alkaline media, the derived catalyst showed an analysis potential of 0.15 V and a maximum power density of 705 ± 5 mW m^{-2} comparable to that of Pt/C (0.13 V, 727 ± 5 mW m^{-2}) in MFCs [89].

Waste recycling techniques that are both sustainable and ecologically benign include the use of cathode catalysts made from sewage sludge, as well as animal-derived biomass. Deng *et al* synthesized sustainable biocarbon from livestock waste-water that exhibited a high ORR activity and 0.07 V half-wave potential comparable to 0.04 V Pt/C in a phosphate buffer solution following KOH activation [90]. They discovered that the carbon catalyst in its as-prepared state had a maximum power density of 1273 3 mW m^{-2}, which was comparable to 1294 mW m^{-2} Pt/C and shown good stability after three months of operation (989 ± 4 mW m^{-2} for the as-prepared catalyst against 462 ± 5 mW m^{-2} for Pt/C after three months). Biocarbon generated from trees, plants, animals, and wastewater precursors all performed satisfactorily and remained stable in MFCs in these investigations.

Notably, it has been discovered that the majority of carbon extracted from biomass is produced via pyrolysis. Pyrolysis is a thermochemical degradation process that may be used to treat any organic (carbon-based) matter that occurs in the absence of water, oxygen, or other reagents subjected to elevated temperatures in an inert gas atmosphere [29]. For example, Li *et al* synthesized a biocarbon by sun-drying corn cobs for five days and then grinding them into a powder [91]. A controlled furnace was used to heat the powder between 250 °C and 750 °C for 2 h in a N_2 environment. In MFCs, the developed catalysts had a power density of 458.85 mW m^{-2}. In addition to pyrolysis, a hydrothermal process can be used to prepare the catalyst. The hydrothermal process in an aqueous medium at moderate temperatures, usually 180 °C, converts biomass to carbon. For example, Zhu *et al* established a straightforward approach for synthesizing carbon nanodots as electro-catalysts from raw soy milk and a hydrothermal technique at 180 °C for 3 h [92]. Carbon nanoparticles are formed during the hydrothermal process due to the carbonization of tiny organic molecules. The comparatively moderate temperatures utilized in the hydrothermal technique, on the other hand, frequently result in a restricted degree of graphitization of the catalysts, resulting in low electrical conductivity. Hydrothermal treatment and pyrolysis can be used together to alleviate this problem. For example, Zhou *et al* generated carbon nanoparticle-coated porous carbon from a natural plant moss (*Weisiopsis anomala*) by hydrothermal treatment at 180 °C for 24 h followed by carbonization at 900 °C in N_2 atmosphere [93]. They discovered that the hydrothermal procedure resulted in the creation of carbon nanodots that aggregated on porous carbon, enhancing the active locations' particular surface area and accessibility. The as-produced catalysts demonstrated an $E_{1/2}$ of +0.832 V against reversible hydrogen electrode in 0.1 M KOH and a power density of 703 mW m^{-2} in MFCs, which were significantly greater than the +0.762 V and 494 mW m^{-2} for the sample prepared only by direct carbonization.

6.4 Economic feasibility of biomass electrodes

Anodes are essential components of MFCs because they offer the surface area required for microbial growth to generate electrons and oxidization during the electrochemical process. Even though MFCs are capable of being employed on a broad scale, their operation is hindered by their high cost and the fact that they have non-renewable anode electrodes. As a result, numerous studies have calculated that the electrode material cost might account for around 20%–50% of the entire cost of MFCs [94, 95]. Considering their high catalytic properties, the most often used anode electrode materials in MFCs are graphite and granular activated carbon (GAC) [96]. According to Huggins *et al* the estimated costs for the materials range from 500 to 2500 US dollars per ton, which is substantially lower than the cost of carbon paper and carbon cloth (100 000–500 000 US\$ m^{-2}). However, the cost of GCs and GACs remains prohibitive. Additionally, their manufacturing techniques are unsustainable, resulting in environmental problems and the greenhouse effect [97]. Within the context of designing an anode that is environmentally friendly and cost-effective, biomass waste, such as biochar, appears to be a viable option. Therefore, several studies have shown that biomass waste, including sewage sludge [98, 99], kenaf (*Hibiscus cannabinus*) stems [100] and fibers [101], palm kern shell [102], and sewage sludge combined with fly ash [66], can be used to create anode electrodes for MFCs. Table 6.3 compares the cost and performance of electrodes produced from biomass

Table 6.3. Comparison of commercial and biomass-derived anode electrode materials. (Table reproduced with permission from [7].)

Nature of material	Anode material	Surface area (m^2)	Material cost (US\$ ton^{-1})	Maximum power density W m^{-2}	Maximum power density mW m^{-2}	Cost of power (US\$ W^{-2})
Biomass-derived	BCP	0.52	51–381	5.78 ± 18	532 ± 18	17.27
Biomass-derived	BCC	0.32	51–381	4.97 ± 20	457 ± 20	35.79
Commercial	GAC	3.68	800–2500	7.32 ± 10	674 ± 10	402.80
Commercial	GGs	0.003	500–800	6.15 ± 5	566 ± 5	392.62

BCP: biochar from compressed milling waste; BCC: biochar from forestry waste; GAC: granular activated carbon; GCs: graphite granules.

for MFCs. As demonstrated, biochar materials are just as effective as commercial materials but at a significantly lower price. In addition, biochar is made of biowaste. Comparing MFC electrodes manufactured from biomass waste with those made from commercial materials is a topic that has received very little attention in the scientific literature. Hence, additional research is required to determine the industrial and technological potential of biomass-derived anode electrodes for MFCs.

6.5 Challenges and future perspectives of electrodes in MFCs

The development of electrode materials for MFCs has taken place over the course of several decades in order to improve their overall performance. Although significant enhancements have been made, the present MFCs paradigm is still not suited for large-scale use at this time. As an important field of study for future MFCs applications, electrode development has a number of challenges to overcome. The current drawbacks and issues in the development of electrodes include the electrode material's cost, as the electrode is a critical component of MFCs, as well as other factors. As a result of the high electrode material costs, the MFC strategy as a whole is not feasible on a commercial scale. Numerous resources have been investigated over the last decade to reduce costs, but standard resources are incapable of producing enough energy or removing enough pollutants to meet current global demands. Similarly, another issue is the anode electrode's design and configuration, as the anode's design and scaling are critical factors in providing a defined surface area for microbial attachment and growth [3]. Therefore, choosing the right anode material is an important step in creating a high-performance MFCs. The development of healthy bacteria is hindered, and electron flow is less efficiently supported when the design and/or configuration is inadequate. Long-term stability is also a significant challenge for contemporary research, as all standard resources have been unable to sustain the functioning of MFCs over an extended period of time. The anode's durability is highly dubious, as it cannot maintain stability over extended periods due to its mechanical and chemical instability [26].

Natural biomass waste materials are a possible source of the most promising outcomes for constructing cost-effective, high-quality electrodes, and therefore future research into building electrodes for MFCs should focus on these characteristics. As summarized in table 6.2, several anodes made from biomass waste were investigated and significant results were obtained. However, biomass anodes with three-dimensional macropores, such as loofahs, pomelo peels, and kenaf, have demonstrated better power generation.

Additionally, electrodes generated from natural resources are widely accessible at a low cost, which meets the fundamental condition of an ideal electrode: readily available at a fair cost [14]. Numerous materials, such as metal-based materials, can achieve high performance with modest adjustments, but their availability and cost are inappropriate. It is also important to note that natural biomass waste provides a rational surface area with adjustable pores, which is critical because pore size has a significant impact on performance; larger pores in anode materials disrupt extracellular electron transfer mechanisms, resulting in lower performance. Carbonization enables the material's pore size to be adjusted to meet specific requirements, and material generated from natural biomass can be further modified by adding other materials to increase its conductivity and electrocatalytic performance. At the moment, a moderately developed 3D printing process is being used to maintain the structure of the anode. The right design and scale can be accomplished through the use of 3D printing technology, which increases the surface area and allows for precise tuning of the pore size. Thus, it is advantageous to enhance bacterial attachment and electrocatalytic sites without impairing substrate diffusion. Additionally, modification methodologies must be implemented to address the long-term stability of current anodes, as multidisciplinary research may yield unexpected results. As illustrated in table 6.4, natural biomass supplies, metals/metal oxides, conducting polymers, carbon nanotubes, and commercial graphene have both benefits and drawbacks. Thus, the concept of a composite material is a superior technique for mitigating material drawbacks while enhancing their favorable attributes. Natural waste-derived materials can be further processed to produce graphene-based materials, also considered modern materials for MFC efficiency. The alteration of natural biomass-derived graphene derivatives with metal oxides might provide surprising outcomes, as seen with GO/ZnO, GO/TiO_2, GO/CuO, and GO/AgO, among others. Graphene derivatives generated from natural biomass exhibited typically poor mechanical strength and high electrical resistance, whereas metal oxides exhibited good mechanical strength and low electrical resistance. As a result, combining these materials may result in substantial outcomes. Now, much work is required to implement MFCs on a wide scale.

Table 6.4. A comparison of different anode materials. (Reproduced with permission from [7].)

Material used for anode	Benefits	Drawbacks
Biomass resources in nature	• Accessibility • Biocompatible • Inexpensive • Convenient and simple preparation procedure • Unique 3D microporous scaffolding for organic substrate diffusion within the anode material	• High resistance to electric current • Mechanical instability issues
Conventional carbon-based materials	• Biocompatible • Accessibility	• Poor electrical conductivity • Mechanical stability issues • Insufficient durability • Moderate cost

CNTs	• High modulus of elasticity • High tensile strength • High electrical resistance • Enhanced electrocatalytic performance • Large surface area	• Biotoxicity • Expensive
Metal/metal oxides	• Outstanding catalytic performance • Low electrical resistance • Good conductivity • Good mechanical stability • Stable at high temperatures	• Highly prone to corrosion, causing toxicity towards bacteria • Chemical instability • Decrease in long-term stability
Conducting polymers	• Environmentally friendly material • High power density • Huge surface area • Affordable cost • High degree of scalability and chemical stability	• Potential danger of polymer shedding • Difficult to synthesize • Moderate conductivity
Graphene derivatives	• High electrical conductivity • Good in chemical stability • Huge surface area	• Moderately expensive
Composite material with natural biomass-derived GO (GO–metal oxides)	• High electrical conductivity • High-strength, long-term stability • Good mechanical stability • Good chemical stability • Stable at high temperatures • Large surface area • Tunable pore size of material • Biocompatible • Inexpensive • Accessible • Simple preparation	• Time-consuming procedure

6.6 Conclusions

MFCs are a novel, cutting-edge energy technology that may be used for both energy conversion and wastewater treatment. MFC electrodes made from biomass-derived carbons are a major and intriguing technique. We outline current advancements in the manufacturing of biocarbon-based electrodes for MFC applications in this paper. Porosity in an anode is a critical property because it allows bacteria to colonize its internal surface and allows for bulk transport of substrates and ions through the internal surface of an anode. Moreover, a vast particular surface area and a considerable amount of heteroatom doping are necessary to improve the quantity and availability of the active area, and the cathode's design process must allow for electron transmission as well as bulk transport of ions and intermediates reactions. MFCs electrodes have been made using biomass from sewage sludge, plants, and animals. Biomass-based anodes such as chestnut shells, loofah sponges, pomelo peels, and silk cocoons, for example, have a broad pore size ranging from a few micrometers to several hundred micrometers, resulting in excellent anode performance and utilization rates. The majority of alternative biomass-derived carbons have pores ranging in size from nanometers to a few micrometers, and do not meet the criteria for microbial colonization, resulting in a poor rate of anode interior utilization. Because the anode's porosity is nearly entirely determined by the raw materials used, it is critical to investigate biomass precursors with pore diameters ranging from tens to many hundreds of micrometers for the design of the anode architecture.

Furthermore, heteroatom doping is necessary for cathode manufacturing to achieve a suitable ORR catalytic efficiency. Plant-based biomass has a low nitrogen level of just 0.515 wt percent in roots and stems and 35% wt percent in leaves, as well as small specific surface area; moreover, some of the nitrogen is lost during the process of pyrolysis. Thus, additional heteroatom doping or activation is frequently required to enhance ORR activity. It is necessary to achieve an optimal balance between higher preparation costs and enhanced power output by heteroatom doping or activation in MFCs since the power density of MFCs is at least 23 orders of magnitude lower than the power density of chemical fuel cells. In terms of electrode performance, biomass-derived electrodes have unique qualities that allow for optimal electrochemical performance, including increased electrode surface area for bacterial loading and increased ORR active sites, a high number of pores for mass transfer of ions and oxygen, low electrode resistance and superior electrical conductivity that allow for quick electron transfer and inexpensive preparation costs. However, the stability and lifespan of electrodes generated from biomass have not been examined extensively in actual wastewater since the bulk of investigations have been performed in simulated wastewater and longevity experiments are seldom reported. Additionally, the reproducibility of biomass-derived carbon materials is essential for commercialization as a result abundance biomass can alter the electrode composition, porosity, mechanical properties, and geometrical structure. As a result, these areas of study are crucial to advancing MFC technology in a variety of applications, including sewage treatment, sustainable energy recovery, and biosensors.

References

[1] Yaqoob A A, Parveen T, Umar K and Mohamad Ibrahim M N 2020 Role of nanomaterials in the treatment of wastewater: a review *Water* **12** 495

[2] Yaqoob A A, Khatoon A, Mohd Setapar S H, Umar K, Parveen T, Mohamad Ibrahim M N, Ahmad A and Rafatullah M 2020 Outlook on the role of microbial fuel cells in remediation of environmental pollutants with electricity generation *Catalysts* **10** 819

[3] Asim A Y, Mohamad N, Khalid U, Tabassum P, Akil A, Lokhat D and Siti H 2021 A glimpse into the microbial fuel cells for wastewater treatment with energy generation *Desalin. Water Treat.* **214** 379–89

[4] Bajracharya S, Sharma M, Mohanakrishna G, Benneton X D, Strik D P, Sarma P M and Pant D 2016 An overview on emerging bioelectrochemical systems (BESs): technology for sustainable electricity, waste remediation, resource recovery, chemical production and beyond *Renew. Energy* **98** 153–70

[5] Yaqoob A A, Mohamad Ibrahim M N, Rafatullah M, Chua Y S, Ahmad A and Umar K 2020 Recent advances in anodes for microbial fuel cells: an overview *Materials* **13** 2078

[6] Logan B E and Rabaey K 2012 Conversion of wastes into bioelectricity and chemicals by using microbial electrochemical technologies *Science* **337** 686–90

[7] Yaqoob A A, Ibrahim M N M and Guerrero-Barajas C 2021 Modern trend of anodes in microbial fuel cells (MFCs): an overview *Env. Technol. Innov.* **23** 101579

[8] Yaqoob A A, Ibrahim M N M and Umar K 2021 Biomass-derived composite anode electrode: synthesis, characterizations, and application in microbial fuel cells (MFCs) *J. Env. Chem. Eng.* **9** 106111

[9] Fadzli F S, Rashid M, Yaqoob A A and Ibrahim M N M 2021 Electricity generation and heavy metal remediation by utilizing yam (*Dioscorea alata*) waste in benthic microbial fuel cells (BMFCs) *Biochem. Eng. J.* **172** 108067

[10] Yaqoob A A, Ibrahim M N M, Yaakop A S, Umar K and Ahmad A 2021 Modified graphene oxide anode: a bioinspired waste material for bioremediation of Pb^{2+} with energy generation through microbial fuel cells *Chem. Eng. J.* **417** 128052

[11] Wei J, Liang P and Huang X 2011 Recent progress in electrodes for microbial fuel cells *Bioresour. Technol.* **102** 9335–44

[12] Sarathi V S and Nahm K S 2013 Recent advances and challenges in the anode architecture and their modifications for the applications of microbial fuel cells *Biosens. Bioelectron.* **43** 461–75

[13] Yaqoob A A, Ibrahim M N M and Rodríguez-Couto S 2020 Development and modification of materials to build cost-effective anodes for microbial fuel cells (MFCs): an overview *Biochem. Eng. J.* **164** 10–37

[14] Cai T, Meng L, Chen G, Xi Y, Jiang N, Song J, Zheng S, Liu Y, Zhen G and Huang M 2020 Application of advanced anodes in microbial fuel cells for power generation: a review *Chemosphere* **248** 125985

[15] Tomai T, Nakayasu Y, Okamura Y, Ishiguro S, Tamura N, Katahira S and Honma I 2020 Bottom-up synthesis of graphene via hydrothermal cathodic reduction *Carbon* **158** 131–6

[16] Wu Y, Wang S and Komvopoulos K 2020 A review of graphene synthesis by indirect and direct deposition methods *J. Mater. Res.* **35** 76–89

[17] ElMekawy A, Hegab H M, Losic D, Saint C P and Pant D 2017 Applications of graphene in microbial fuel cells: the gap between promise and reality *Renew. Sustain. Energy Rev.* **72** 1389–403

[18] Nitisoravut R, Thanh C N and Regmi R 2017 Microbial fuel cells: advances in electrode modifications for improvement of system performance *Int. J. Green Energy* **14** 712–23

[19] Sun Y, Wei J, Liang P and Huang X 2011 Electricity generation and microbial community changes in microbial fuel cells packed with different anodic materials *Bioresour. Technol.* **102** 10886–91

[20] Li S, Cheng C and Thomas A 2017 Carbon-based microbial-fuel-cell electrodes: from conductive supports to active catalysts *Adv. Mater.* **29** 1602547

[21] Gnana Kumar G, Kirubaharan C J, Udhayakumar S, Karthikeyan C and Nahm K S 2014 Conductive polymer/graphene supported platinum nanoparticles as anode catalysts for the extended power generation of microbial fuel cells *Ind. Eng. Chem. Res.* **53** 16883–93

[22] Yaqoob A A, Ibrahim M N, Yaakop A S and Rafatullah M 2022 Utilization of biomass-derived electrodes: a journey toward the high performance of microbial fuel cells *Appl. Water Sci.* **12** 1–20

[23] Yaqoob A A, Ibrahim M N M and Yaakop A S 2021 Application of oil palm lignocellulosic derived material as an efficient anode to boost the toxic metal remediation trend and energy generation through microbial fuel cells *J. Cleaner Prod.* **314** 12–24

[24] Long D, Peng J, Liu H, Feng Z, Chen L, Chen X and Lu M 2020 Graphene oxide discarded solution for high surface area photocatalyst *Sol. Energy Mater. Sol. Cells* **209** 110446

[25] Zhang S, Wang H, Liu J and Bao C 2020 Measuring the specific surface area of monolayer graphene oxide in water *Mater. Lett.* **261** 127098

[26] Hindatu Y, Annuar M and Gumel A 2017 Mini-review: anode modification for improved performance of microbial fuel cell *Renew. Sustain. Energy Rev.* **73** 236–48

[27] Yaqoob A A, Umar K and Ibrahim M N M 2020 Silver nanoparticles: various methods of synthesis, size affecting factors and their potential applications—a review *Appl. Nanosci.* **10** 1369–78

[28] Sauerteig D, Hanselmann N, Arzberger A, Reinshagen H, Ivanov S and Bund A 2018 Electrochemical–mechanical coupled modeling and parameterization of swelling and ionic transport in lithium-ion batteries *J. Power Sources* **378** 235–47

[29] Kang Y L, Pichiah S and Ibrahim S 2017 Facile reconstruction of microbial fuel cell (MFC) anode with enhanced exoelectrogens selection for intensified electricity generation *Int. J. Hydrogen Energy* **42** 1661–71

[30] Lv Z, Chen Y, Wei H, Li F, Hu Y, Wei C and Feng C 2013 One-step electrosynthesis of polypyrrole/graphene oxide composites for microbial fuel cell application *Electrochim. Acta* **111** 366–73

[31] Qiao Y, Wen G-Y, Wu X-S and Zou L 2015 L-cysteine tailored porous graphene aerogel for enhanced power generation in microbial fuel cells *RSC Adv.* **5** 58921–7

[32] Xiao L, Damien J, Luo J, Jang H D, Huang J and He Z 2012 Crumpled graphene particles for microbial fuel cell electrodes *J. Power Sources* **208** 187–92

[33] Hassan S H, Kim Y S and Oh S-E 2012 Power generation from cellulose using mixed and pure cultures of cellulose-degrading bacteria in a microbial fuel cell *Enzyme Microb. Technol.* **51** 269–73

[34] Yuan Y, Zhou S, Zhao B, Zhuang L and Wang Y 2012 Microbially-reduced graphene scaffolds to facilitate extracellular electron transfer in microbial fuel cells *Bioresour. Technol.* **116** 453–8

[35] Choi C and Cui Y 2012 Recovery of silver from wastewater coupled with power generation using a microbial fuel cell *Bioresour. Technol.* **107** 522–5

[36] Liu J, Qiao Y, Guo C X, Lim S, Song H and Li C M 2012 Graphene/carbon cloth anode for high-performance mediatorless microbial fuel cells *Bioresour. Technol.* **114** 275–80

[37] Yong Y-C, Dong X-C, Chan-Park M B, Song H and Chen P 2012 Macroporous and monolithic anode based on polyaniline hybridized three-dimensional graphene for high-performance microbial fuel cells *ACS Nano* **6** 2394–400

[38] Nourbakhsh F, Mohsennia M and Pazouki M 2017 Nickel oxide/carbon nanotube/polyaniline nanocomposite as bifunctional anode catalyst for high-performance *Shewanella*-based dual-chamber microbial fuel cell *Bioprocess. Biosyst. Eng.* **40** 1669–77

[39] Dominguez-Benetton X, Srikanth S, Satyawali Y, Vanbroekhoven K and Pant D 2013 Enzymatic electrosynthesis: an overview on the progress in enzyme-electrodes for the production of electricity, fuels and chemicals *J. Microb. Biochem. Technol.* **6** 7
[40] Hou J, Liu Z and Zhang P 2013 A new method for fabrication of graphene/polyaniline nanocomplex modified microbial fuel cell anodes *J. Power Sources* **224** 139–44
[41] Guo W, Cui Y, Song H and Sun J 2014 Layer-by-layer construction of graphene-based microbial fuel cell for improved power generation and methyl orange removal *Bioprocess. Biosyst. Eng.* **37** 1749–58
[42] Kumar G G, Hashmi S, Karthikeyan C, GhavamiNejad A, Vatankhah-Varnoosfaderani M and Stadler F J 2014 Graphene oxide/carbon nanotube composite hydrogels—versatile materials for microbial fuel cell applications *Macromol. Rapid Commun.* **35** 1861–5
[43] Gnana Kumar G, Kirubaharan C J, Udhayakumar S, Ramachandran K, Karthikeyan C, Renganathan R and Nahm K S 2014 Synthesis, structural, and morphological characterizations of reduced graphene oxide-supported polypyrrole anode catalysts for improved microbial fuel cell performances *ACS Sust. Chem. Eng.* **2** 2283–90
[44] Mehdinia A, Ziaei E and Jabbari A 2014 Facile microwave-assisted synthesized reduced graphene oxide/tin oxide nanocomposite and using as anode material of microbial fuel cell to improve power generation *Int. J. Hydrogen Energy* **39** 10724–30
[45] Chen J, Deng F, Hu Y, Sun J and Yang Y 2015 Antibacterial activity of graphene-modified anode on *Shewanella oneidensis* MR-1 biofilm in microbial fuel cell *J. Power Sources* **290** 80–6
[46] Zou L, Qiao Y, Wu X-S, Ma C-X, Li X and Li C M 2015 Synergistic effect of titanium dioxide nanocrystal/reduced graphene oxide hybrid on enhancement of microbial electrocatalysis *J. Power Sources* **276** 208–14
[47] Ma M, You S, Gong X, Dai Y, Zou J and Fu H 2015 Silver/iron oxide/graphitic carbon composites as bacteriostatic catalysts for enhancing oxygen reduction in microbial fuel cells *J. Power Sources* **283** 74–83
[48] Najafabadi A T, Ng N and Gyenge E 2016 Electrochemically exfoliated graphene anodes with enhanced biocurrent production in single-chamber air-breathing microbial fuel cells *Biosens. Bioelectron.* **81** 103–10
[49] Osgood H, Devaguptapu S V, Xu H, Cho J and Wu G 2016 Transition metal (Fe, Co, Ni, and Mn) oxides for oxygen reduction and evolution bifunctional catalysts in alkaline media *Nano Today* **11** 601–25
[50] Yang X, Ma X, Wang K, Wu D, Lei Z and Feng C 2016 Eighteen-month assessment of 3D graphene oxide aerogel-modified 3D graphite fiber brush electrode as a high-performance microbial fuel cell anode *Electrochim. Acta* **210** 846–53
[51] Pareek A, Sravan J S and Mohan S V 2019 Fabrication of three-dimensional graphene anode for augmenting performance in microbial fuel cells *Carbon Resour. Conver.* **2** 134–40
[52] Zhao F, Rahunen N, Varcoe J R, Chandra A, Avignone-Rossa C, Thumser A E and Slade R C 2008 Activated carbon cloth as anode for sulfate removal in a microbial fuel cell *Env. Sci. Technol.* **42** 4971–6
[53] Michaelidou U, Ter Heijne A, Euverink G J W, Hamelers H V, Stams A J and Geelhoed J S 2011 Microbial communities and electrochemical performance of titanium-based anodic electrodes in a microbial fuel cell *Appl. Env. Microbiol.* **77** 1069–75
[54] Guan Y-F, Zhang F, Huang B-C and Yu H-Q 2019 Enhancing electricity generation of microbial fuel cell for wastewater treatment using nitrogen-doped carbon dots-supported carbon paper anode *J. Clean. Prod.* **229** 412–9

[55] Slate A J, Whitehead K A, Brownson D A and Banks C E 2019 Microbial fuel cells: an overview of current technology *Renew. Sustain. Energy Rev.* **101** 60–81

[56] Antolini E 2016 Nitrogen-doped carbons by sustainable N-and C-containing natural resources as nonprecious catalysts and catalyst supports for low temperature fuel cells *Renew. Sustain. Energy Rev.* **58** 34–51

[57] Zhang L, He W, Yang J, Sun J, Li H, Han B, Zhao S, Shi Y, Feng Y and Tang Z 2018 Bread-derived 3D macroporous carbon foams as high performance free-standing anode in microbial fuel cells *Biosens. Bioelectron.* **122** 217–23

[58] Chen S, Tang J, Jing X, Liu Y, Yuan Y and Zhou S 2016 A hierarchically structured urchin-like anode derived from chestnut shells for microbial energy harvesting *Electrochim. Acta* **212** 883–9

[59] Chen Q, Pu W, Hou H, Hu J, Liu B, Li J, Cheng K, Huang L, Yuan X and Yang C 2018 Activated microporous–mesoporous carbon derived from chestnut shell as a sustainable anode material for high performance microbial fuel cells *Bioresour. Technol.* **249** 567–73

[60] Zhu H, Wang H, Li Y, Bao W, Fang Z, Preston C, Vaaland O, Ren Z and Hu L 2014 Lightweight, conductive hollow fibers from nature as sustainable electrode materials for microbial energy harvesting *Nano Energy* **10** 268–76

[61] Xie X, Hu L, Pasta M, Wells G F, Kong D, Criddle C S and Cui Y 2011 Three-dimensional carbon nanotube−textile anode for high-performance microbial fuel cells *Nano Lett.* **11** 291–6

[62] Zhou L, Sun L, Fu P, Yang C and Yuan Y 2017 Carbon nanoparticles of Chinese ink-wrapped natural loofah sponge: a low-cost three-dimensional electrode for high-performance microbial energy harvesting *J. Mater. Chem.* A **5** 14741–7

[63] Zheng J, Cheng C, Zhang J and Wu X 2016 Appropriate mechanical strength of carbon black-decorated loofah sponge as anode material in microbial fuel cells *Int. J. Hydrogen Energy* **41** 23156–63

[64] Lu M, Qian Y, Yang C, Huang X, Li H, Xie X, Huang L and Huang W 2017 Nitrogen-enriched pseudographitic anode derived from silk cocoon with tunable flexibility for microbial fuel cells *Nano Energy* **32** 382–8

[65] Feng H, Jia Y, Shen D, Zhou Y, Chen T, Chen W, Ge Z, Zheng S and Wang M 2018 The effect of chemical vapor deposition temperature on the performance of binder-free sewage sludge-derived anodes in microbial fuel cells *Sci. Total Env.* **635** 45–52

[66] Jia Y, Feng H, Shen D, Zhou Y, Chen T, Wang M, Chen W, Ge Z, Huang L and Zheng S 2018 High-performance microbial fuel cell anodes obtained from sewage sludge mixed with fly ash *J. Hazard. Mater.* **354** 27–32

[67] Mecheri B, Gokhale R, Santoro C, Costa de Oliveira M A, D'Epifanio A, Licoccia S, Serov A, Artyushkova K and Atanassov P 2018 Oxygen reduction reaction electrocatalysts derived from iron salt and benzimidazole and aminobenzimidazole precursors and their application in microbial fuel cell cathodes *ACS Appl. Energy Mater.* **1** 5755–65

[68] Sindhuja M, Harinipriya S, Bala A C and Ray A K 2018 Environmentally available biowastes as substrate in microbial fuel cell for efficient chromium reduction *J. Hazard. Mater.* **355** 197–205

[69] Song R B, Wu Y, Lin Z Q, Xie J, Tan C H, Loo J S C, Cao B, Zhang J R, Zhu J J and Zhang Q 2017 Living and conducting: coating individual bacterial cells with *in situ* formed polypyrrole *Angew. Chem.* **129** 10652–6

[70] You S, Gong X, Wang W, Qi D, Wang X, Chen X and Ren N 2016 Enhanced cathodic oxygen reduction and power production of microbial fuel cell based on noble-metal-free electrocatalyst derived from metal-organic frameworks *Adv. Energy Mater.* **6** 1501497
[71] Wang R, Yan M, Li H, Zhang L, Peng B, Sun J, Liu D and Liu S 2018 FeS_2 nanoparticles decorated graphene as microbial-fuel-cell anode achieving high power density *Adv. Mater.* **30** 1800618
[72] Zhou X, Chen X, Li H, Xiong J, Li X and Li W 2016 Surface oxygen-rich titanium as anode for high performance microbial fuel cell *Electrochim. Acta* **209** 582–90
[73] Islam M A, Ethiraj B, Cheng C K, Yousuf A and Khan M M R 2018 An insight of synergy between *Pseudomonas aeruginosa* and *Klebsiella variicola* in a microbial fuel cell *ACS Sust. Chem. Eng.* **6** 4130–7
[74] Wang Z, Mahadevan G D, Wu Y and Zhao F 2017 Progress of air-breathing cathode in microbial fuel cells *J. Power Sources* **356** 245–55
[75] Li J, Hu L, Zhang L, Ye D-d, Zhu X and Liao Q 2017 Uneven biofilm and current distribution in three-dimensional macroporous anodes of bio-electrochemical systems composed of graphite electrode arrays *Bioresour. Technol.* **228** 25–30
[76] Song R B, Zhu W, Fu J, Chen Y, Liu L, Zhang J R, Lin Y and Zhu J J 2020 Electrode materials engineering in electrocatalytic CO_2 reduction: energy input and conversion efficiency *Adv. Mater.* **32** 1903796
[77] Baudler A, Schmidt I, Langner M, Greiner A and Schröder U 2015 Does it have to be carbon? Metal anodes in microbial fuel cells and related bioelectrochemical systems *Energy Env. Sci.* **8** 2048–55
[78] Zhao C-e, Gai P, Song R, Chen Y, Zhang J and Zhu J-J 2017 Nanostructured material-based biofuel cells: recent advances and future prospects *Chem. Soc. Rev.* **46** 1545–64
[79] Pocaznoi D, Calmet A, Etcheverry L, Erable B and Bergel A 2012 Stainless steel is a promising electrode material for anodes of microbial fuel cells *Energy Env. Sci.* **5** 9645–52
[80] Bian B, Shi D, Cai X, Hu M, Guo Q, Zhang C, Wang Q, Sun A X and Yang J 2018 3D printed porous carbon anode for enhanced power generation in microbial fuel cell *Nano Energy* **44** 174–80
[81] Yang W, Peng Y, Zhang Y, Lu J E, Li J and Chen S 2018 Air cathode catalysts of microbial fuel cell by nitrogen-doped carbon aerogels *ACS Sust. Chem. Eng.* **7** 3917–24
[82] Oliot M, Galier S, de Balmann H R and Bergel A 2016 Ion transport in microbial fuel cells: key roles, theory and critical review *Appl. Energy* **183** 1682–704
[83] Chen Y-M, Wang C-T, Yang Y-C and Chen W-J 2013 Application of aluminum-alloy mesh composite carbon cloth for the design of anode/cathode electrodes in *Escherichia coli* microbial fuel cell *Int. J. Hydrogen Energy* **38** 11131–7
[84] Zhang J, Li L, Zheng J, Yang P, Wu X, Cheng C, Li J, Tian Y and Wang F 2019 Improved organic pollutants removal and simultaneous electricity production via integrating Fenton process and dual rotating disk photocatalytic fuel cell system using bamboo charcoal cathode *Chem. Eng. J.* **361** 1198–206
[85] Liu P, Liang P, Jiang Y, Hao W, Miao B, Wang D and Huang X 2018 Stimulated electron transfer inside electroactive biofilm by magnetite for increased performance microbial fuel cell *Appl. Energy* **216** 382–8
[86] Zhao H and Kong C-H 2018 Enhanced removal of p-nitrophenol in a microbial fuel cell after long-term operation and the catabolic versatility of its microbial community *Chem. Eng. J.* **339** 424–31

[87] Zhong K, Li M, Yang Y, Zhang H, Zhang B, Tang J, Yan J, Su M and Yang Z 2019 Nitrogen-doped biochar derived from watermelon rind as oxygen reduction catalyst in air cathode microbial fuel cells *Appl. Energy* **242** 516–25

[88] Fan Z, Li J, Zhou Y, Fu Q, Yang W, Zhu X and Liao Q 2017 A green, cheap, high-performance carbonaceous catalyst derived from *Chlorella pyrenoidosa* for oxygen reduction reaction in microbial fuel cells *Int. J. Hydrogen Energy* **42** 27657–65

[89] Yuan H, Deng L, Cai X, Zhou S, Chen Y and Yuan Y 2015 Nitrogen-doped carbon sheets derived from chitin as non-metal bifunctional electrocatalysts for oxygen reduction and evolution *RSC Adv.* **5** 56121–9

[90] Deng L, Yuan H, Cai X, Ruan Y, Zhou S, Chen Y and Yuan Y 2016 Honeycomb-like hierarchical carbon derived from livestock sewage sludge as oxygen reduction reaction catalysts in microbial fuel cells *Int. J. Hydrogen Energy* **41** 22328–36

[91] Li M, Zhang H, Xiao T, Wang S, Zhang B, Chen D, Su M and Tang J 2018 Low-cost biochar derived from corncob as oxygen reduction catalyst in air cathode microbial fuel cells *Electrochim. Acta* **283** 780–8

[92] Zhu C, Zhai J and Dong S 2012 Bifunctional fluorescent carbon nanodots: green synthesis via soy milk and application as metal-free electrocatalysts for oxygen reduction *Chem. Commun.* **48** 9367–9

[93] Zhou L, Fu P, Wen D, Yuan Y and Zhou S 2016 Self-constructed carbon nanoparticles-coated porous biocarbon from plant moss as advanced oxygen reduction catalysts *Appl. Catal.* B **181** 635–43

[94] Rozendal R A, Leone E, Keller J and Rabaey K 2009 Efficient hydrogen peroxide generation from organic matter in a bioelectrochemical system *Electrochem. Commun.* **11** 1752–5

[95] Rabaey K, Butzer S, Brown S, Keller J and Rozendal R A 2010 High current generation coupled to caustic production using a lamellar bioelectrochemical system *Env. Sci. Technol.* **44** 4315–21

[96] He L, Du P, Chen Y, Lu H, Cheng X, Chang B and Wang Z 2017 Advances in microbial fuel cells for wastewater treatment *Renew. Sustain. Energy Rev.* **71** 388–403

[97] Chen S, Liu Q, He G, Zhou Y, Hanif M, Peng X, Wang S and Hou H 2012 Reticulated carbon foam derived from a sponge-like natural product as a high-performance anode in microbial fuel cells *J. Mater. Chem.* **22** 18609–13

[98] Li M, Li Y-W, Yu X-L, Guo J-J, Xiang L, Liu B-L, Zhao H-M, Xu M-Y, Feng N-X and Yu P-F 2020 Improved bio-electricity production in bio-electrochemical reactor for wastewater treatment using biomass carbon derived from sludge supported carbon felt anode *Sci. Total Env.* **726** 138573

[99] Yuan Y, Yuan T, Wang D, Tang J and Zhou S 2013 Sewage sludge biochar as an efficient catalyst for oxygen reduction reaction in an microbial fuel cell *Bioresour. Technol.* **144** 115–20

[100] Chen S, He G, Hu X, Xie M, Wang S, Zeng D, Hou H and Schröder U 2012 A three-dimensionally ordered macroporous carbon derived from a natural resource as anode for microbial bioelectrochemical systems *ChemSusChem* **5** 1059–63

[101] Chen S, He G, Liu Q, Harnisch F, Zhou Y, Chen Y, Hanif M, Wang S, Peng X and Hou H 2012 Layered corrugated electrode macrostructures boost microbial bioelectrocatalysis *Energy Env. Sci.* **5** 9769–72

[102] Offei F, Thygesen A, Mensah M, Tabbicca K, Fernando D, Petrushina I and Daniel G 2016 A viable electrode material for use in microbial fuel cells for tropical regions *Energies* **9** 35

Chapter 7

The role of the proton exchange membrane (PEM) in microbial fuel cell performance

Tanzina Akter, Modhusudon Shaha, Mohammad Al Mamun, Md Abu Sayem Khan and Abu Hashem

A microbial fuel cell (MFC) is a bioelectrochemical system that aims to produce electricity by transforming the biochemical energy in organic molecules into electrical energy. MFCs also have the ability to manufacture hydrogen for fuel cells, purify saline water, and offer long-term energy to isolated locations. To keep the current flowing in an MFC, a proton must be carried for every electron produced, making the proton exchange membrane (PEM) one of the most crucial elements in the cell. It physically divides the anode and cathode while allowing protons to travel through and reach the cathode. As a result, the PEM plays a vital role in improving MFC operation. PEMs are most typically employed to separate the anode and cathode compartments in two-chamber MFC systems. PEMs have superior conductivity for protons and little internal resistance, which reduces power production. Other metal ions in the medium can replace the proton-attachment positions of the negatively charged sulphonate moieties in the PEM, which is the cause of membrane fouling. To ensure long-term energy production, the fouled PEM has to be retrieved or changed. The high cost of PEMs is a significant impediment to MFC scalability and potential practical deployment. In this chapter, we look at the different aspects of PEM application in MFCs as well as future recommendations for better performing PEMs for MFCs.

7.1 Introduction

A microbial fuel cell (MFC) is a bioelectrochemical system that aims to produce electricity by transforming the biochemical energy in organic molecules into electrical energy [1] utilizing electrons derived from biological reactions catalyzed by bacteria [2–4]. MFCs are a relatively recent approach for simultaneously generating power and decomposing organic waste or biowaste. In MFCs the

doi:10.1088/978-0-7503-4791-4ch7

utilization of micro-organisms to create energy is a continuous process since the microbes proliferate and generate power continuously if there are adequate food stuffs to nourish them [4]. MFCs also have the ability to manufacture hydrogen for fuel cells, purify saline sea water, and offer long-term energy to isolated locations [5].

Because of their unique potential to harvest bio-power from waste and wastewaters, MFCs have received a lot of interest in the research community during the last two decades [3, 6, 7]. In MFCs microbes in the anodic chamber (negative end) produce electrons through substrate oxidation, which are then transported to the cathode chamber (positive end) via a conductive material. In the cathode electrons combine with oxygen, while protons diffuse through a proton exchange membrane (PEM). In MFCs the anode must release electrons at a steady rate while the cathode consumes them at a constant rate. The gap within the anode potential and the substrate redox potential determines how much metabolic energy bacteria can obtain. The best MFC design is still being researched, and alternative electrode materials as well as further selective PEMs are currently being developed to improve its functioning [8].

The use of micro-organisms to generate power from waste and renewable biomass has attracted a lot of interest. An MFC can be divided into three parts: (a) an anode compartment in which fuel is oxidized by micro-organisms, yielding electrons and protons; (b) a cathode compartment in which protons and electrons are consumed and water is produced by mixing with oxygen; and (c) a PEM that sends protons from the anode compartment to an external electric circuit, where they assemble electrons transferred from the anode compartment to the cathode compartment [5].

The type of membrane, catalyst, substrate, configuration, temperature, and other parameters all have a role in MFC power generation. One of the most crucial components of an MFC is the membrane, which has a considerable impact on its performance. A PEM is the most commonly utilized membrane in MFCs because it has a reasonably high conductivity to cations and a low inner resistance in contrast to other cationic membranes, both of which are important in MFC performance [1]. The membrane must allow protons to flow from the anode to the cathode while blocking the flow of other substances such as oxygen, substrates, and minerals. If oxygen reaches the anode section, the activity in the anode becomes aerobic, lowering the energy density and effluent treatment efficacy. Furthermore, medium penetration to the cathode reduces power output and has an impact on the micro-organisms' efficiency in removing chemical oxygen demand (COD) [9, 10].

To keep the current flowing in an MFC, a proton must be carried for each electron produced, making the PEM one of the most crucial elements in the MFC. It physically divides the anode and cathode while allowing protons to travel through and reach the cathode. Nafions are the most commonly used PEMs in MFCs. The rate of oxygen transfer from the cathode to the anode, the transfer and build-up of cations rather than protons, biofouling, and substrate loss are all issues that come with using Nafion and other PEMs in MFCs [1]. As a result, the PEM plays an important role in improving MFC performance. In this chapter, we look at the antiquity of PEM use in MFC development, the different types of PEMs used in MFCs, the operating principles of PEM, the impact of PEM functionality on MFCs, the problems and causes of dysfunctionality, and the methods for improving PEM performance, as well as future recommendations for better performing PEMs along with existing perspectives, recent trends, challenges, and the future scope.

7.2 Chronology of PEM development

In the last century, the number of studies related to the development of novel PEMs for use in PEM fuel cell (PEMFC) has witnessed exponential growth [11]. In order to decrease the concentration of CO_2 in the atmosphere along with the development of environmentally friendly fuels, alternatives to conventional fossil fuels have acted as prime reasons for the research and development of novel polymers. The DuPont Corporation, USA, was the pioneer in producing polytetrafluoroethylene (PTFE) in 1938, with the trademark Teflon™ [12]. Due to the strongly hydrophobic nature of PTFE, a second modification known as the sulphonation process (the addition of a terminal sulphonic acid group (–SO_3H) to the skeleton of PTFE) was done to make it an ion conducting membrane. This chemical modification produced the Nafion® membrane which was a sodium salt with a balanced SO_3^- group and a Na^+ ion. The addition of a sulfonated group to the Nafion membrane made it hydrophilic, with the ability to absorb large amounts of water. Nonetheless, even then, the Nafion membrane did not achieve its desired σ. In order to acquire this property, an activation process consisting of a heat treatment using concentrated H_2SO_4 at approximately 80 °C was used to discard the Na^+ in the form of sodium sulphate, producing the membrane in its acid form (-SO_3H) [12]. In a hydrated state, the SO_3^- groups of the Nafion membrane attract H^+ ions, forming hydronium ions (H_3O^+) with water, allowing the transfer of H^+ ions across the membrane [12, 13]. Nafion increased the specific conductivity of the membrane by two times as well as increasing the lifetime by four times (104–105 h), making Nafion the standard for use in MFCs today. Min *et al* compared the performance of a Nafion and a salt bridge in a desalination (DS)-MFC and reported that the output of the MFC with the PEM and the salt bridge was close to 3.9 mW m^{-2} and 2.2 mW m^{-2}, respectively, representing an order of magnitude higher output of the PEM than the salt bridge. The lower internal resistance was considered to be the reason behind the higher power output in the case of Nafion [14].

Another cation exchange membrane (CEM), Ultrex CMI-7000, produced by Membranes Inc, USA, has been used as a PEM in MFCs. [15–18]. CMI-7000 consists of a cross-linked structure of divinylbenzene and a gel polystyrene having sulphoric acid groups, resulting in a strong acid polymer membrane. Although CMI-7000 possesses higher ohmic resistance than Nafion, other characteristics such as the proton conductivity and mechanical durability of this membrane are comparable to Nafion [18].

Hyflon, a less expensive CEM, is composed of perfluoropolymer and has better chemical stability and higher conductivity than Nafion, even though it has higher internal resistance. Zirfon is another cost-effective membrane compared to Nafion. This membrane is prepared by combining 85 wt% hydrophilic ZrO_2 powder with 15 wt% polysulfone, resulting in a macroporous organo-mineral material [19].

Zirfon is used not only for its lower cost but also for its lower resistance in comparison to Nafion.

Although CEMs are widely used as membrane separators, they still have major drawbacks, such as high initial cost, formation of biofilm, and pH splitting, which are major limitations of their use in long-term operations. Anion exchange membranes (AEMs) are found to have better proton conductivity than CEMs owing to the presence of phosphate or carbonate. In addition to a better proton transfer rate, phosphate anions keep the pH more balanced [20]. However, for the treatment of wastewater, the use of a phosphate buffer is not cost effective and feasible. Another type of membrane known as a bipolar membrane (BPM; a combination of two membranes, an AEM and CEM) allows the transfer of both protons (H^+) and hydroxide ions (OH^-) [21]. While this membrane allows better transportation of both cations and anions, the major concern of pH splitting is still present. In addition to this, a lot of research using ceramics, nylon fibres, degradable shopping bags, and glass fibres is ongoing to develop membranes that are inexpensive and have characteristics and performance similar to Nafion [22–28]. Ceramics are gaining popularity as a future MFC material due to their porosity, chemical inertness, and high strength. The first study, which used ceramic (porcelain membrane) as a PEM with graphite electrodes in 2003, showed the ability of protons to move through the porous membrane and made the first steps of ceramic material use in MFC [29]. Since then, a lot of research has been carried out utilizing different types of materials such as earthenware, terracotta, and clayware to evaluate their use as PEMs [30].

7.3 Types of PEMs used in MFCs

A PEM is a semipermeable membrane consisting primarily of ionomers that is meant to transmit protons while also acting as an electronic insulator and chemical barrier, for example, to oxygen and other substrates. PEMs can be made from either pure polymer membranes or composite membranes where additional substances are inserted into a polymer matrix. A representative scheme of a PEM in a double-chambered MFC is depicted in figure 7.1. Fundamentally, PEMs can be divided into two groups: ion-exchange membranes (IEMs) and porous membranes.

7.3.1 Ion-exchange membranes

7.3.1.1 Cation exchange membranes

CEMs allow the transfer of positive ions or cations through the membranes. In their hydrated state, their chemical structures consist of negative charge anions such as SO_3^-, COO^-, $PO3^-$, $HPO3^-$, and $C_6H_4O^-$ etc. CEMs can be classified on the basis of materials used for their preparation. Different types of materials are used, for example, perfluorinated ionomer polymers, non-fluorinated hydrocarbon membranes, and acid–base blends [32].

(a) *Perfluorinated ionomers polymers.* The perfluorinated sulphonic acid polymers (PFSA) are widely used in MFCs as membranes which consist of a backbone of a polytetrafluoroethylene (PTFE) and side chains of vinyl ethers (for example, $-O-CF_2-CF-O-CF_2-CF_2-$), and these side chains

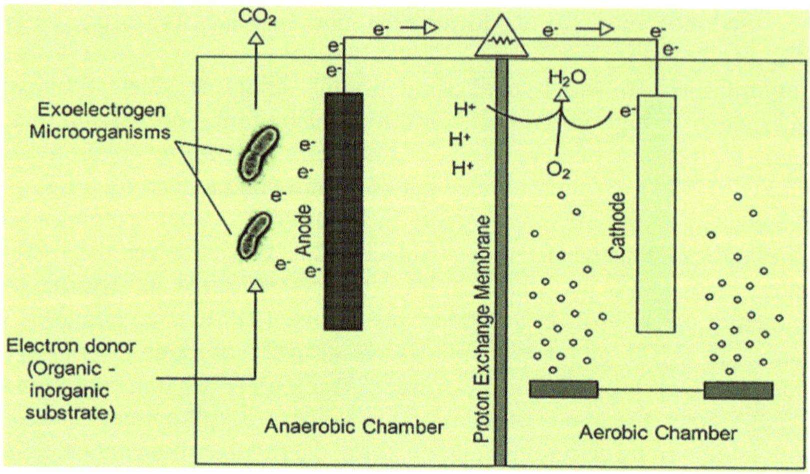

Figure 7.1. Schematic representation of a PEM in a double-chambered MFC. (Reproduced with permission from [31]. Creative Commons Attribution (CC BY) license.)

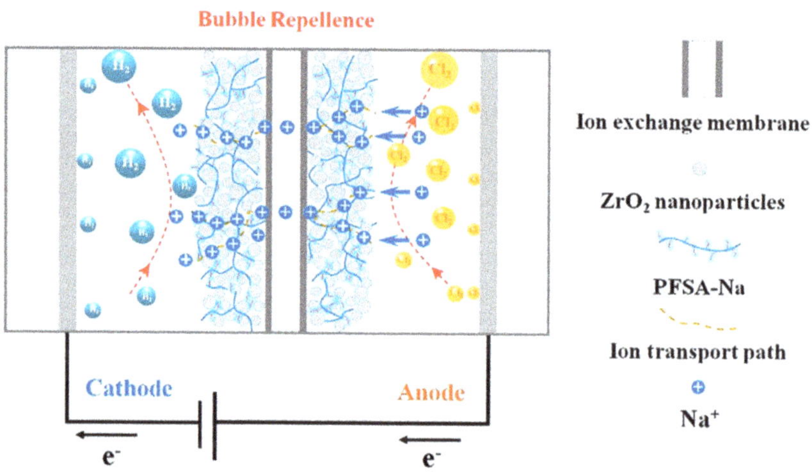

Figure 7.2. An under-brine superaerophobic perfluorinated ion-exchange membrane with ZrO_2 coating. (Reproduced with permission from [33]. Copyright 2021 Elsevier.)

contain sulfonic acid groups at the end of a cluster region (figure 7.2). The ion conducting properties are attributed due to the presence of sulfonic acid group clusters which are hydrophilic and in a hydrated state, the cation such as proton (H^+) can pass freely under a potential gradient. For excellent performance, PFSA membranes should have the following characteristics:

- High ionic conductivity in a hydrated state.
- High mechanical stability.
- Better chemical stability against O_2, H_2, strong oxidizing and reducing acids, and strong bases.

- No electronic conduction.
- Stability in long-term operation.
- Maximum reduced oxidant and fuel crossover.
- Low substrate crossover.
- Cost effectiveness.

The PFSA membrane family includes a wide variety of products from different companies under the trade names of Nafion™ (DuPont), Hyflon™ (Solvay-Solexis), Aquivion™ (Solvay), and Flemion™ (Asahi Glass). Among them, Nafion™ (DuPont) is the most popular and widely used membrane.

Nafion-117 (NF-117) is the most commonly used separator in different kinds of MFCs [22, 34, 35]. There are a number of reasons for its widespread use, such as: (a) a high σ PC value > 9.5 mS cm^{-1}, which depends on the hydration level, and is affected by the availability of sulfonic acid groups in the polymeric structure, the pressure, and the temperature; (b) high selectivity for H^+; (c) chemical and thermal stability; and (d) a low OR value, which indicates a high density of current [12, 22, 36].

However, the NF-117 membrane as a PEM has some challenges which are given below:

- *Highly expensive.* The NF-117 membrane accounts for more than 50% of the total cost of the MFC system, which acts as the main barrier for large scale MFCs.
- *Sensitivity to other cations.* Depending on the chemical nature of the membrane as well as the composition of the anolyte, apart from H^+, other cations such as Na^+, K^+, NH_4^+, Ca^{2+}, and Mg^{2+} can be transferred across the membrane. The transport of the other cations occurs, on the one hand, when the concentration is 10^5 times greater than the concentration of H^+ resulting from oxidation at the anode causing the acidification of the anolyte. On the other hand, an accumulation process of the other cations may also result until the concentration does not exceed 10^5.
- *Permeability of oxygen.* In order to carry out metabolic activity at the anode, the absence of oxygen is required for the anaerobic bacteria. A substantial amount of oxygen can pass from the cathode to the anode through the membrane, resulting in a reduction of anaerobic respiration of the bacteria, which directly affects the performance of the MFC in terms of electrical power.
- *Crossover and loss of substrate.* Substrates crossing the membrane result in a reduction in fuel which is converted into electricity and the substrate availability for the bacteria. It also causes the acidification of the anode due to the inhibition of proton transfer, which affects the growth of bacteria, leading to a decrease in electricity production.
- *Temperature sensitivity.* The presence of the sulfonic acid groups (HSO_3^-) which are hydrophilic in nature, attracts large amounts of water molecules, leading to hydration of the membrane. The level of hydration plays an important role in the transfer of protons through the membrane. High temperatures cause dehydration of the membrane, making the membrane unsuitable for working at elevated temperatures.

- *Biofouling.* Biofouling is a phenomenon that occurs when substrates, microorganisms, and their metabolic products adhere to the membrane and act as a barrier for proton conductivity, leading to a reduction in electricity production. Biofouling due to the result of biofilm formation in long-term operation is considered as a negative factor [35, 37].

(b) *Non-fluorinated hydrocarbon membranes.* Non-fluorinated hydrocarbon polymers are aromatic or aliphatic polymers containing benzene ring structures in the membrane's polymeric backbone or bulky pendant groups attached to the membrane's polymeric backbone. Hydrocarbon membranes are less expensive and readily available in the marketplace. Furthermore, the structure of hydrocarbon membranes allows for the inclusion of polar sites as pendant groups with the ability to uptake water in large amount over a wide temperature range, and the absorbed water is limited to polar groups of polymer chains [38]. Degradation of hydrocarbon polymers can be slowed to some extant with good molecular design, and traditional recycling processes can be used with ease. Despite the fact that hydrocarbon membranes have worse chemical stability than PFSA membranes, sulfonated polyarylenes have become an active research area due to their low cost and high stability. Sulphonation of polymers such as polyarylene ether sulphone (PES), polyarylene ether ether ketone (PEEK), and polyarylene sulfone (PSU) is done directly using concentrated chlorosulfonic acid or sulphuric acid. Smitha *et al* discovered that a sulfonated polyether ether ketone (SPEEK) with a sulphonation level of approximately 50% had an electrochemical performance which is comparable to Nafion at 65 °C, making it an appealing low-cost alternative to Nafion [38].

(c) *Acid–base complexes.* A lot of research has been conducted to develop alternative membranes in terms of excellent MFC performance and low cost. Acid–base polymer complexes are one of the most promising compounds among these possibilities. For proton conductivity, acid–base polymers are created by incorporating an acidic component into an alkaline polymer base [38]. The interactions between ionic cross-linking of acid and base polymers as well as hydrogen bonding bridges play an important role in the control of membrane swelling without a reduction in flexibility. Thus, these membranes have low water uptake, high mechanical strength and flexibility, stability at elevated temperatures, reduced substrate crossover, and high proton conductivity [39]. The phosphoric acid-doped polybenzimidazole (PBI/H_3PO_4), acid–base membranes made of acidic sulfonated polymers (sPPESK, sPBEK, and sPPENK), and basic polyetherimide (PEI) are examples of some commonly used acid–base polymers.

7.3.1.2 Anion exchange membranes

Despite the fact that Nafion and PTFE are widely utilized in the scientific community, they are ineffective in the transmission of hydroxide ions (OH^-).

As a result, when utilized in suitably alkaline FCs, these membranes show reduced efficiency, necessitating the adoption of a type of membrane that meets the needs. The presence of cations (SR_2^+, NR_3^+, NH_4^+, PR_3^+, NR_2H^+, and NHR_2^+) on the polymer matrix of AEMs allows the transfer of anions (OH^-, Cl^-, SO_4^{2-}) in a hydrated state [40–43]. The transfer of cations other than the cation of interest (H^+) in CEMs causes a drop in pH in the anode chamber, which inhibits of microbial activity. As a result of the high pH in the anode chamber, the cathode potential is reduced, and AEMs may readily solve this problem. Fibreglass and nylon, both inexpensive materials with high porosity and permeability, have been evaluated for use as AEMs [27, 44, 45]. When compared to Nafion, Kim *et al* found a 96 mW m^{-2} improvement in power density when utilizing an AEM in MFCs [16]. However, because of the large size of the anions and the decreased hydration capacity of the primary cation group, anionic conductivity is low, resulting in poor electrochemical device performance [46, 47].

7.3.1.3 Bipolar membranes

The combination of a CEM and an AEM creates a BPM. While an AEM or CEM aids in the transport of specific ions, a BPM can transport both protons and hydroxyl ions at the same time due to water splitting at the membrane's interface. Because there is no competition between ion migrations, the biggest difficulty with using this sort of membrane is maintaining a pH gradient. Although the BPM is most commonly used in the electrodialysis process, nothing is known about its applicability in MFCs. Water molecules are split into H^+ and OH^- at the membrane junction when an electric field is applied in the electrodialysis process, after which protons migrate through the CEM and hydroxide ions migrate through the AEM [48]. Kim *et al* compared the performance of the MFC and Cr^{6+} removal in two DC-MFCs with a PEM and a BPM [49]. As a catholyte, they employed hexavalent chromium found in electroplating effluent, and acetate as an electron source for the MFC. The power density of the DC-MFCs operated with PEMs was found to be significantly lower than that of the DC-MFCs operated with a BPM, at 42.7 mW m^{-2} and 150.5 mW m^{-2}, respectively. Furthermore, in the MFC containing a BPM cell, the reduction efficiency of Cr^{6+} to Cr^{3+} was significantly higher. This was due to the fact that the BPM kept the cathodic reaction going in the anodic chamber without causing the pH to drop. In another study, Heijne *et al* [37] employed an MFC with a BPM to achieve ferric iron reduction on a graphite electrode cathode, with a maximum power density of 0.86 W m^{-2} and a η_{coul} of around 90%. In this scenario, the BPM maintained the required pH in the catholyte (less than 2.5) to keep the Fe^{3+} soluble and avoid fouling on the membrane, that may occur if the CEM was employed, since the pH would rise and Fe^{3+} would precipitate on the membrane, reducing the charge transfer process.

7.3.2 Porous membranes

Different types of porous membranes, such as ultrafiltration membranes (UFMs), microfiltration membranes (MFMs), ceramic membranes (CMs), glass wool, and Celgard, have been utilized as an alternative to PEMs in MFCs. The great benefit of using a porous membrane is that it is a low-cost separator in comparison to PEMs [22, 50].

It has been reported that single chamber MFCs using cheap glass wool as separators are found to be more cost effective compared to expensive PEM separators. Another advantage regarding the performance of this type of separator is a low R_{int}, although due to the biofouling, the R_{int} increases throughout the operation time, resulting in a decrease in the MFC performance. The occurrence of this type of phenomenon is higher in porous membranes than in ion-exchange membranes (IEMs) [22]. The major difference between IEMs and porous membranes is that ions (H^+/OH^-) and other molecules of large size can be transferred through the porous membrane, while ion-exchange membranes are highly selective for ions [22]. Because of this character, IEMs are used in the decolourization of azo dyes where the porous structures allow O_2 to diffuse into the anode chambers to degrade the azo bond formed in cleavage during the intermediate reaction.

7.3.2.1 Microfiltration membranes

In wastewater treatment, MFMs have been extensively used as a separator of sludge owing to their stability for long-term operation and filtration efficiency. These excellent properties have made this membrane useful as a separator in MFCs. Like in UFMs, different sorts of ions and molecules with no charge can pass through the pores of the membranes. They are cost effective compared to IEMs, and some commonly used MFMs are nylon mesh, cellulose filters, and polycarbonate filters. Similar to UFMs, this membrane also has problems such as OD, high internal resistance, and fuel crossover [47, 50]. Therefore, before using this membrane, it is necessary to solve these three main problems.

7.3.2.2 Ceramic membranes

Along with applications in different sectors, such as in the pharmaceutical, biotechnological, food, petrochemical, and chemical industries, etc, CMs have been utilized as separators in MFCs, in particular in high temperature FCs [51]. Because of their porosity, permeability, and hydrophobic characteristics, they can play an important role in proton transfer in MFCs. Park and Zeikus reported for the first time the use of CMs as a separator on single-chamber (SC)-MFCs. They utilized different mediators and three electrode settings at pH 7 and sewage sludge as the biocatalyst and found that the maximum current density was 1750 mA m^{-2} and the power density production was 788 mW m^{-2} [28]. Behera *et al* used a 400 ml, off-the-shelf ceramic pot with clay consisting of 58%–68% kaolinite, 15%–26% illite, and 5%–9% smectite and did not utilize any mediators, reporting a power output of 16.8 W m^{-3} [29, 52]. Behera *et al* also conducted another study in the same year using commercial earthen pots and wastewater from rice mills as a substrate to test the efficiency of the pots, both to remove pollutants and to produce electricity at different pH [53, 54]. While the maximum sustainable volumetric power was reported as 2.3 W m^{-3} with 100 Ω external resistance, the chemical oxygen demand removal efficiency (ηCOD) was found to be 96.5% [31, 32]. There are some features that should be considered during the use of CMs. One important feature is thickness. It was found that compared with a thicker membrane, the thinner membrane performed better [52]. Ajayi *et al* studied terracotta pot material with a graphite coating on the outer layer of the membrane for an SC-MFC, where they found a coulombic efficiency (CE) of 21% ± 5%, and the power density obtained was 33.13 mW m^{-2} [55]. Choosing the correct soil/clay for the preparation of the membrane is another important aspect that needs to be considered. The rate of proton transfer through the membrane depends on the total amount of acidic cations (H^+ and Al^{3+}). Following that, the hydraulic conductivity of the clay is determined by the pore throat radii of the material utilized. Hydraulic conductivity is usually low in clay with small pore throat radii. As a result, differing porosities play an important role in proton transfer, in addition to the fuel loss that occurs when MFCs are used. With a commercially accessible CEM, a study was undertaken employing ecofriendly materials such as ceramic and compostable starch-based bags (BioBags) as two ion-exchange materials. With regard to cost and operational stability, starch bags were proved to be an effective material, highlighting porous separators as possibilities for practical MFC implementation [25]. Winfield *et al* looked at the qualities of ceramic and terracotta materials in terms of wall thickness, porosity, and cathode hydration. According to the researchers, cylindrical mud MFCs generated significantly more current and power [56]. The CMs employed in MFCs, according to Winfield *et al*, are cost effective, sustainable, and readily available alternatives that supply the MFC chassis and create the catholyte [29]. They also discovered that a less dense substance produces more current. However, the R_{int} increase with the long-term performance of MFCs and the possible biofouling generated are the main concerns regarding the use of CMs as separators [50].

7.3.2.3 Ultrafiltration membranes

The widespread use of ultrafiltration membranes (UFMs) in wastewater treatment is to separate particulate contaminants on the basis of their molecular weights. As UFMs allow the transfer of both anions and cations, they are also used in DC-MFCs for the separation of anode and cathode electrodes. A study was conducted by Kim *et al* with UFMC membranes (0.5, 1, 3 K daltons) in DC-MFCs where they found that these three membranes had lower oxygen mass transfer coefficients in comparison to NF-117, AEM (AMI-7001), and CEM (CMI-7000), even though two of the UFMs showed relatively higher substrate mass transfer coefficients [16]. Overall, they found lower power densities (8.9%) and lower CEs (44.7%). However, a lot of studies are needed in order to optimize the pore sizes and thickness of UFMs to improve substrate loss and CE. The major drawback of ultrafiltration membranes is the high internal resistance they exhibit in MFCs. Kim *et al* performed a study where they found that reduction of membrane resistance can be done with the addition of polydopamine, which decreased the rate of biofouling and increased the membrane hydrophobicity, resulting in an increase in power density from 137.9 ± 6.0 mW m^{-2} to 159.8 ± 7.7 mW m^{-2} [57].

7.4 The working principle of PEM-based MFCs

The basic working principle of an MFC is a physicochemical oxidation–reduction reaction that is characterized by a flow of electrons where an element or compound releases electrons and becomes oxidized, while another element/compound is reduced by accepting those electrons [58–60]. In a PEM-based MFC, oxidation–reduction reactions are separated by a PEM. The reasons for using this membrane are to prevent the occurrence of oxidation and reduction reactions in the same place, or in other words, to prevent short-circuiting by separating the chemical reactions

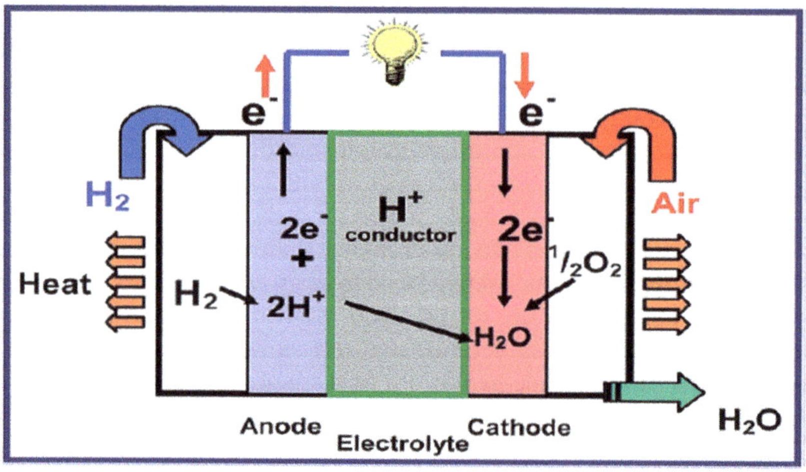

Figure 7.3. The principles of operation of a PEMFC. (Reproduced with permission from [62]. Creative Commons Attribution (CC BY) license.)

and proton conduction generated during oxidation. The operation of a PEM-based MFC requires PEMs, at least two electrodes (anode and cathode) joined via an external circuit, organic compounds as fuel, and the biological activity of micro-organisms as basic elements. In an anode chamber, micro-organisms carry out oxidation of the organic matter used as a substrate and release protons and electrons. The electrons, collected by the anode, travel to the cathode through the external circuit, while the protons are transported to the cathode through the PEM [61]. The transported protons combine with the electrons and an electron acceptor, oxygen, to form water. The working principles of a PEMFC that uses hydrogen as a fuel and oxygen as an oxidant are shown in figure 7.3. H_2 gas is oxidized at the anode, resulting in H^+ ions that are carried to the cathode via a PEM, while electrons flow to the cathode via the external circuit. At the cathode, O_2 gas reacts with H^+ ions and electrons to form water. The anode and cathode reactions are presented in the following equations:

Anode reaction: H_2 (gas) $\rightarrow$ $2H^+ + 2e^-$.

Cathode reaction: $\frac{1}{2}O_2$ (gas) $+ 2H^+ + 2e^- \rightarrow H_2O$ (liquid).

Overall reaction: H_2 (gas) $+ \frac{1}{2}O_2$ (gas) $\rightarrow H_2O$ (liquid).

For the operation and performance of PEM-based MFCs, the proton conduction mechanism plays an important role. Proton conductivity is the first characteristic considered during the selection of the PEM for use in fuel cells to assure high power generation. At the molecular level, in hydrated conditions, the proton conductivity can be described using two principal mechanisms: the Grotthuss mechanism and the vehicular mechanism.

7.4.1 Grotthuss mechanism

This is also known as proton hopping or proton jumping or a diffusion mechanism. In this mechanism, from one hydrolyzed ionic site (SO_3^- H_3O^+), the protons (H^+)

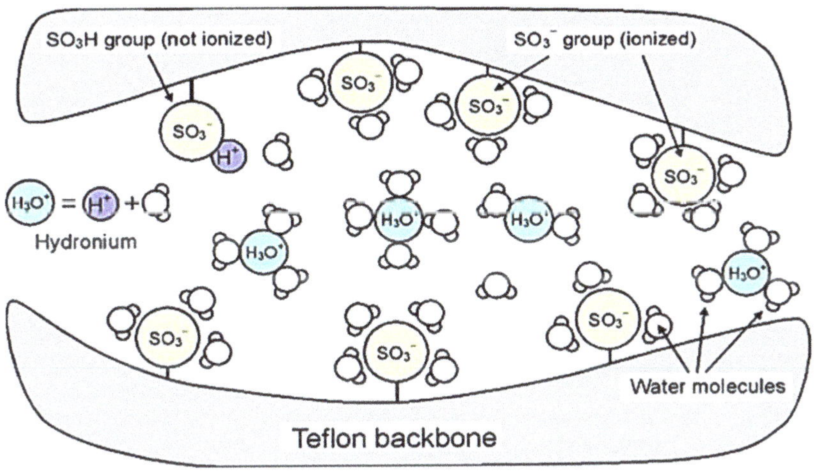

Figure 7.4. A schematic view of the proton hopping mechanism of phosphoric acid (H_3PO_4) in Nafion. Nafion polymer chains self-align into a microchannel structure. The sulfonic acid functional groups along the channel wall enable hydronium conduction. (Reproduced with permission from [66]. Creative Commons Attribution (CC BY) license.)

hop to another hydrolyzed ionic site across the membrane. The proton produced during the oxidation reaction of hydrogen gas at the anode adheres to water molecules to generate provisional hydronium ions. In the next stage, another different proton from the same hydronium ion jumps onto the other water molecule. In the proton hopping mechanism, an ionic region produces specific swelled hydrophilic clusters in the presence of water, allowing the percolation mechanism for protons to transfer [63–65]. The net result of this mechanism is the faster transfer of proton ions than any other dissolved cation. The proton hopping mechanism of phosphoric acid (H_3PO_4) between $H_4PO_4^+$ ions, H_3PO_4 molecules, and $H_2PO_4^-$ ions is shown in figure 7.4.

7.4.2 Vehicular mechanisms

In vehicular mechanisms, water is used as a carrier or vehicle by the proton for diffusion. The protons attach to water and diffusion of the hydrated protons (hydronium ion, H_3O^+) occurs in response to an electrochemical difference in the aqueous medium [63]. The proton transportation rate depends on the presence of free volumes within the polymer backbone and on the vehicular diffusion rate. The design of the vehicular mechanism for proton conductivity in pristine and nano-composite membranes is presented in figures 7.5(a) and (b). In general, these two types of mechanisms are not independent in PEMs and often work together for proton conductivity. In the presence of water, the vehicle-type mechanism is common in all membranes that are attached to sulfonic acid groups, whereas the proton hopping mechanism also plays a role in proton conduction via water molecules through hydrogen bond forming and breaking [62].

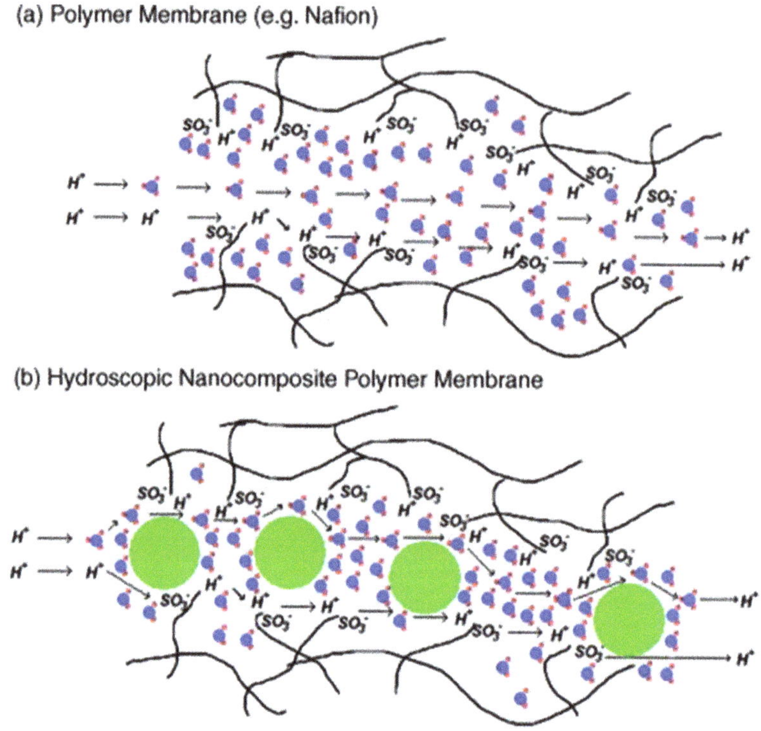

Figure 7.5. The design of the vehicular mechanism for (a) proton conductivity polymer membranes and (b) nanocomposite polymer membranes. (Reproduced with permission from [67]. Copyright 2005 Elsevier.)

7.5 The PEM functions in MFCs

In MFCs, PEMs are the most commonly used separators. PEMs are one of the most significant components in the cell because of their role in proton transfer [1]. For MFC operation, a PEM is needed that serves some basic functions, which are described below.

7.5.1 Conduction of protons

In the anode chamber, protons are produced when the substrate is oxidized by bacteria. The main function of the PEM is to transfer the generated protons from the anode to the cathode, where the electron acceptor oxygen is reduced and combined with the transported protons to form water (figure 7.6). Other ion-exchange membranes transfer anions or both anions and cations, such as the AEM and bipolar membrane, respectively (figure 7.6).

7.5.2 As a separator and an insulator

PEM acts as an electrolyzer because it permits protons to move across the membrane, but this membrane is impermeable to electrons. In the case of Nafion, protons generated from the oxidation of substrates reacts with the sulfonic acid group (SO_3^-) and forms an ion cluster of $SO_3^-H^+$ in the polymer. Ion clusters attract water molecules, and the absorption of water causes the swelling of a plastic membrane

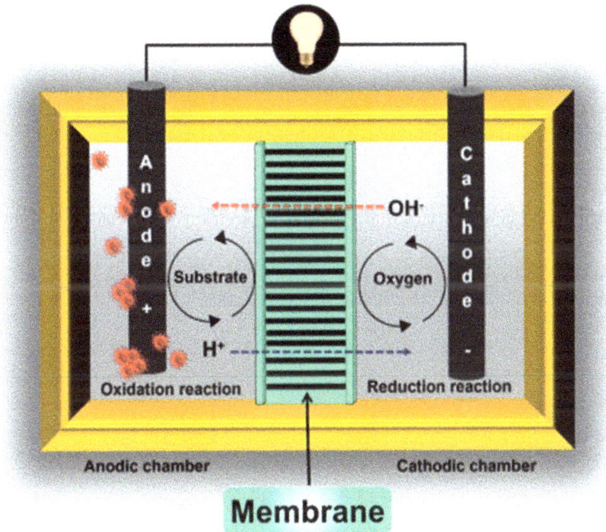

Figure 7.6. The functions of the PEM in MFC operation. (Reproduced with permission from [58]. Creative Commons Attribution (CC BY) license.)

through which protons can move easily by forming bonds with water molecules from one SO_3^- ion to another. Since the plastic membrane has no free electrons, it blocks the transfer of electrons from the anode to the cathode. In this way, PEM acts as an insulator by preventing short-circuiting between the anode and cathode electrodes. The fluids that exists between the anode and cathode chambers are known as anolyte and catholyte, respectively, where two different reactions (oxidation and reduction) occur. The PEM membranes act as separators of anolyte and catholyte and play a vital role in carrying out oxidation and reduction reactions [58, 68].

7.5.3 Prevention of oxygen diffusion and fuel crossover

Another important function of the membrane is to maintain an anaerobic environment in the anodic chamber by preventing the diffusion of oxygen from the cathode to the anode, which is a prerequisite for the anaerobic bacteria to carry out metabolic activity (figure 7.6). The membranes also help to prevent the passage of electron acceptors other than oxygen. For example, ammonia, permanganate, sulphate, ferricyanide, trichloroethene, hydrogen peroxide, perchlorate, nitrate, and certain heavy metals enhance the growth of unwanted bacteria, leading to a reduction in power generation [69–71]. Similar to oxygen diffusion, fuel, including soluble low molecular organic or inorganic compounds, is also prevented by the membrane from crossing from the anode to the cathode electrode.

7.6 Membrane fouling

The proton exchange membrane is an integral part of the MFC system [72]. It acts as a physical spacer and allows proton transfer through the MFC membrane [73]. However, microbial growth on the membrane after a long-term operation is regularly

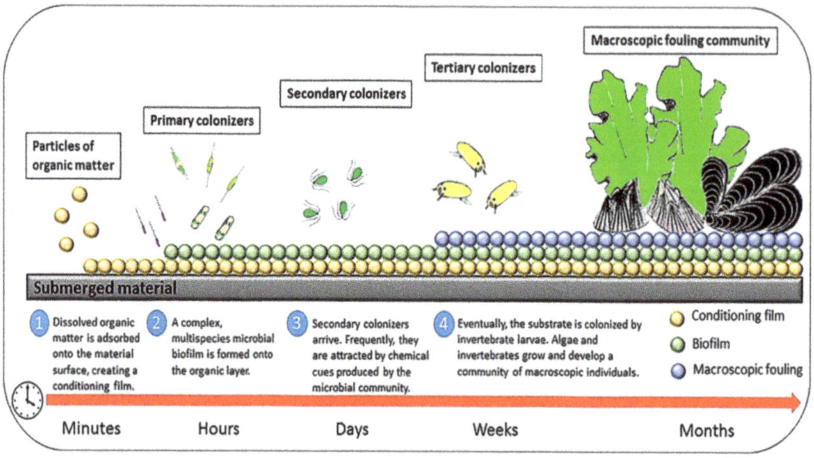

Figure 7.7. Schematic representation of the biofouling process. (Reproduced with permission from [78]. Creative Commons Attribution (CC BY) license.)

encountered, which causes biofouling on the surface of the membrane. Biofouling has been identified as one of the key obstacles to improving the productivity and sustainable operation of MFCs [74]. Organic foulants, such as extracellular polymeric substances (EPSs), micro-organisms, and inorganic salts, facilitate aggregation of residing bacteria and their clustering during the development of biofilms on the surface of the proton exchange membrane, leading to biofouling in MFCs [75]. In addition to EPSs, biopolymers and soluble microbial products (SMP) are also well-recognized foulants, but their production depends on different conditions. For example, recent studies suggest that the absence of an external electron carrier promotes the production of biopolymers and proteins, resulting in membrane biofouling, whereas SMPs are predominantly found in the sludge supernatant [76, 77]. A flow-chart on biofouling development is given in figure 7.7 [78].

7.6.1 Major effects of membrane fouling

7.6.1.1 Biofouling and MFC performance

The biofouling of PEM is regarded as one of the key factors that harms the operation and performance of MFCs [73]. Fouling causes an overall reduction in the cation exchange capacity, electrical conductivity, and cation diffusion coefficients of PEM [79]. Moreover, coulombic efficiencies and maximum power densities have been significantly reduced, leading to reduced power generation [73].

According to imaging and FTIR analyses, micro-organisms encapsulated in extracellular biopolymers and precipitations of inorganic salt were found in the fouling layer linked to PEM [79]. Additionally, a scanning electron microscope (SEM) revealed the presence of bacterial overloads and their metabolic products on the surface of PEM, resulting in a dramatic reduction in the proton transfer from the anode side to the cathode side [73]. Furthermore, the increase in fouling density on the membrane is associated with a reduction in the NF-117 membrane's proton conductivity. This increases membrane resistance due to limits in proton transport across the biofouling membrane [74]. The limited proton transfers in the setting of biofouling occur due to an imbalance in the pH-associated potential of the membrane.

7.6.1.2 Electron transfer and biofouling

Membrane fouling obstructs electron transport and degrades MFC performance [73]. The fouling potential is likely to be considerable in the conditions where no external electron acceptor, nitrate, oxygen, and anode electrode are used. Nevertheless, the type of electron acceptor used in the MFC system influences the composition of biofouling SMPs and EPSs [76].

In an MFC, the pollutants and PEM interact with each other in different manners, including physically, chemically, and biologically. In addition to biofouling, MFC performance, particularly electron transfer, can be affected by chemical changes in the membrane, such as substituting the PEM's proton-binding sites with different cations [80]. Nevertheless, such changes in chemical composition may also be affected by membrane biofouling.

Furthermore, some other metal ions in the medium can quickly replace the proton-binding sites of the sulphonate groups (negatively charged) in PEM to facilitate consistent electron transfers [81]. However, metal ions may promote microbial adherence and biofilm development to the membrane through intermolecular bridging between organic foulants and sulphonate groups [80]. Therefore, careful selection of the type of metal ions aimed for rapid electron transfer in an MFC is essential.

7.6.2 Mitigation of biofouling

Several studies have reported a variety of ways to mitigate the membrane fouling in MFCs. Membrane fouling was recently shown to be reduced when MFC and membrane bioreactor (MBR) were combined, owing to the lower formation of SMPs and EPSs during anodic respiration circumstances [76]. Several solutions have been devised to prevent membrane fouling in MBRs, with the use of an electric field being one of the most popular [82]. By virtue of electrophoresis and electrocoagulation, applying an electric field to an MBR can limit the deposition of foulants on the membrane.

The use of a silver nanoparticle-modified forward osmosis (FO) membrane is currently under development with a view to reducing biofouling formation. The thin-film composite (TFC) polyamide FO membrane modified by silver nanoparticles (AgNPs) demonstrated improved hydrophilicity, negative zeta potential, and antimicrobial properties in the typical tests. Furthermore, *in situ* biocide-mediated modification of the exchange membrane is a powerful approach to prevent bacterial accumulation on the membrane surface [83]. Because of their excellent antibacterial action against a wide range of micro-organisms, AgNPs are extensively utilized as biocides [84].

Membrane biofouling starts to develop after months of ongoing MFC operation. It may arise in a short time if the device remains on standby for a while. Over the

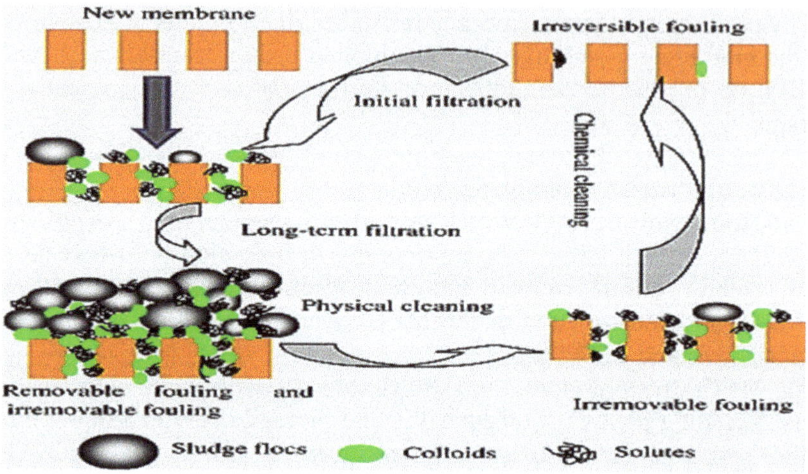

Figure 7.8. Schematic illustration of the fouling process and removal of fouling in a membrane bioreactor. (Reproduced with permission from [85]. Copyright 2009 Elsevier.)

course of six months, the long-term operation of an MFC results in the development of a biofouling layer on the surface of the membrane [77]. The presence of numerous organic and inorganic materials acts as a binding factor for the biofouling layer [74], as shown in figure 7.8. As biofouling of the PEM membrane arises after start-up, it is recommended that the PEM membrane be cleaned every six or seven months to ensure that biofouling no longer occurs [74].

7.7 PEM advances and improvement

Microbial fuel cells have received huge attention worldwide with the hope of finding a sustainable source of clean energy and an ecofriendly method of waste treatment. Several aspects of the MFC system can be improved for better MFC performance. The electron transfer rate is one of the most critical factors that needs attention to improve the MFC performance [86]. Power density, COD removal, and coulombic efficiency are the other factors to evaluate the performance of MFCs [87]. Furthermore, the scope of accelerating MFC performance to generate more energy relies on improving the structure of the reactor, the types of separator materials, the types of microbes in use, and the types of PEMs.

Micro-organisms catalyze the degradation of a substrate present in the MFC and initiate the process of electron transmission, which in turn generates energy. Therefore, selecting suitable bacterial species is of great importance in improving MFC performance. *Shewanella* and *Geobacter* are mainly used for MFC research globally, indicating their suitability for energy generation [88]. In addition, it has been proven that mixed cultures generate six-fold more power than pure cultures by promoting the growth of dominant flora on anode surfaces [89]. A bioengineered *Klebsiella pneumonia–Shewanella oneidensis* microbial consortia was designed using genomic engineering tools for effectively generating power from maize stalk hydrolysate [90]. For microbial adhesion, electrode materials should have strong electron conductivity, a broad surface area, and good biocompatibility [91]. Two aspects need to be considered while choosing electrode material because the types of materials affect the electrical performance of the MFC. These are the roughness and modification of the electrode material. There were numerous examples of modification of the cathode material, such as using a polydimethylsiloxane composite, poly neutral red modified cathode, and modification of the cathode electrode using polyaniline and graphene oxide [92, 93]. Because of their high electrical conductivity and catalytic activity, graphene and carbon nanotube materials are seen as an important path for future study.

The reactor type plays a major role in deciding the performance of an MFC. The types of reactors include single chamber, dual chamber, stacked, and dual tubular. All these types have both advantages and disadvantages. The factors to be considered while designing reactors are the electron transmission rate, mass transfer resistance, assembly, space, and types of connections. Currently, microalgae are considered to have enormous potential in future biofuel production, and their use in MFCs has been established [94]. Salt bridges can also act as alternative separation materials. The MFC was regarded as another viable technique to utilize the sustainable energy of biomass, giving the biomass energy sector a new dimension. Because biomass is an abundant and energy-rich resource, it is projected to be an excellent fuel for MFCs. and MFCs would be an acceptable energy conversion technology for the biomass business. The metabolic process that occurs in MFCs varies depending on the type of biomass and operational conditions, including pH and temperature [95].

According to Tharali *et al* [96] an ideal PEM should be less expensive, improve proton conductivity, and be mechanically strong, heat and chemical resistant, and electronically stable. Nafion, a sulfonated tetrafluoroethylene-based fluoropolymer-copolymer, has drawn wide attention in recent years as a PEM in MFC technology. The salt bridge, another type of PEM, is made up of an ionic salt-like KCl or NaOH, which is melted with agar and put into a cylindrical mould to solidify [96]. However, there is scope to improve PEM performance considering the choice of chemical and biological materials. Incorporating different chemical groups, such as sulphonate, fluoride, and hydroxyl, into polymers as well as doping different chemicals into membranes could be an efficient way to enhance PEM performance [87]. By lowering the crystallinity of the polymer, aromatic sulphonation enhances its ion-exchange capacity (IEC) and hydrophilicity. Sulphonation also increases proton conductivity by increasing the solubility of PEM and facilitating proton transfer [97]. Using naturally abundant polymers for membrane preparation and subsequent modification to improve thermal, chemical, and mechanical properties could provide a better economical approach for energy generation. Chitosan and polyvinyl alcohol are examples [98]. Hickner *et al* (2004) described how the use of graphene oxide and titanium dioxide, along with other chemicals, to fabricate nanocomposites showed better performance than PEMs. One of the most important variables controlling a PEM's effectiveness is its capacity to absorb water as well as its conductivity. Water absorption affects the passage of protons over the membrane, which is essential for fuel cell operation, but it also changes the membrane mechanical characteristics [99].

7.8 Problems associated with PEMs for the commercialization of MFCs

Commercialization is a common goal for MFCs, as many other fields [100]. However, the majority of marketable devices are driven by commercial batteries (e.g. lithium-ion batteries) that have a relatively short and limited lifetime and must be replaced or recharged regularly [101]. In this context, finding self-renewable and efficient energy sources to generate adequate power for distant devices (when battery replacement is not possible and convenient) is a burning question [102]. For the last two decades, PEM-based MFCs (PEMFCs) have emerged as the next generation of environmentally benign and potential renewable energy sources. Self-renewable PEMFCs appear to be a long-lasting alternative power source for distant applications that reduces operational costs and environmental concerns significantly (since they can harvest combustion-free and pollution-free bioelectricity directly from organic biomass) [103]. The PEMs in MFCs are the key requirements for optimum microbiological and electrochemical performance. Because the principal function of this polymer membrane is to avoid redox reactions occurring in the same chamber, i.e. if the oxidation and reduction reactions happen in the same chamber, it will become short-circuited. Conversely, the efficiency of PEMFCs is directly related to the power density (in Wcm^2) linked to the chemistry of the PEM [58]. Nevertheless, depending on their chemical structure and the microbiological and physicochemical properties of the catholyte and anolyte, membranes exhibit several functional limitations. Therefore, those drawbacks must be resolved before it can be considered a sustainable and mature renewable energy technology that is recognized by the scientific community and the stakeholders. First, we have recognized the remaining gaps and challenges regarding PEM as one of the central components of MFCs, and then briefly debated what efforts we require to transform laboratory outcomes into real-world applications aimed at commercialization of PEM-based MFCs. Table 7.1 summarises the current gaps and challenges with PEM in MFCs, as well as their future prospects.

7.9 Conclusions and recommendations

PEMs are most typically employed to separate the anode and cathode compartments in two-compartment MFC systems. In one-compartment MFC techniques, the PEM is directly linked to the cathode, allowing for selective proton transportation in the MFC system. Hydrogen ions formed by oxidizing substrates are delivered selectively from the anode to the cathode via the PEM, where protons combine with oxygen, the ultimate electron acceptor. In comparison to conventional separators such as glass fibres and salt bridges, PEMs have a higher conductivity for protons and a lower interior resistance. The majority of exoelectrogens in the anode compartment grow in anaerobic conditions, resulting in a decrease in coulombic efficacy. As a result, oxygen transport into the anode compartment should be avoided, which PEMs can help with. If protons are not refilled through the PEM due to the lack of a pH stabilizer in an MFC, the pH in the cathode compartment will rise, suggesting that the membrane's role is critical.

On the other hand, the PEM is a major contributor to the high cost of MFC systems. In spite of their selective penetrability to protons, PEMs are regarded as a

Table 7.1. The existing gaps and challenges concerning PEMs in MFCs with future perspectives.

Existing gaps and challenges	Analysis with future outlooks
1. Cost of the PEM	The principal disadvantage of PEM is its high cost, which is not related to MFC performance. The membranes of MFCs could account for nearly 60% of the whole cost [104, 105] of MFCs and are the most significant obstacle that prevents PEMFC devices from being scalable. As the membrane influences the total cost and performance of MFCs, it is suggested that research efforts be focused to develop a new, economically viable membrane that could exhibit promising characteristics and allow for acceptable cell performance during its practical implementation. However, the high cost of the PEM could be minimized as a whole by improving the power density, which could be done by increasing the cathode and anode surface areas. With the integration of novel nanomaterials and nanotechnology following an innovative method or technique for the modification of the electrode surface, the power density and efficiency of PEMFCs could be improved.
2. Higher internal resistance, R_{int}, in the presence of a PEM	The measured voltage from a PEMFC is still less than the theoretical voltage owing to some voltage losses, often termed 'overpotential losses'. The PEM limits its use in MFCs due to the increasing R_{int} resulting from the thickness of the membrane [106] compared to the MFCs without a PEM. The value of R in a PEMFC is mainly contributed by the ohmic resistance (OR). As the membrane thickness increases, the resistance across the membrane increases (leading to a decline in the current, power, and ion-exchange capacity (ICE) of the MFCs), and the proton conductivity decreases (the conductivity of an ion, σ, is decreased with the increase of R, i.e. the lower value of σ is due to the higher value of R). A higher thickness, however, is suitable for restrictive interpenetration of electroactive species. A lower rate of diffusion of protons from the anolyte to the cathode chamber would be reflected in lower MFC performance. These phenomena, such as substrate crossover and biofouling, also play an important role in increasing the R_{int} throughout their prolonged operation. Therefore, in addition to the high cost of the membrane, its thickness effects on the MFC R_{int} is a big challenge that could be overcome by optimizing the thickness of the PEM.
3. Configuration	One of the major problems restricting the application of PEMFCs from commercialization is the perfect engineering design of a scalable architecture. The configuration of the membrane should have the capacity to act as a channel, allowing the selective movement of H^+ produced in the oxidation reaction, and the released electrons at the anode pass through an exterior circuit to the cathode to generate a voltage (ΔV) which is transformed into electrical power [107]. In order to avoid a short-circuit, the design of PEMFCs also needs a separator as a membrane to prevent redox reactions in the same chamber, in particular for a complex multiple-cell system. Therefore, the optimum performance of PEMFCs depends on the perfect configuration and design of the cell system, which must be addressed and optimized by the scientific community and field experts concerned before its commercialization.

4. Oxygen diffusion (OD)	An ideal membrane must be capable of completely preventing OD from the catholyte section to the anolyte chamber. Unfortunately, the PEM partially fails to meet that criterion. That is because oxygen is soluble in water [50] and PEMs need to be hydrated to attain their ion-exchange functionality. Consequently, O_2 diffusion is connected to the requirement for a hydrated membrane. In the case of PEM-based MFCs, O_2 molecules accept electrons at the cathode. The function of a perfect proton selective membrane could maintain a purely anaerobic atmosphere in the anode chamber, but the presence of O_2 in the anolyte inhibits the PEMFC's performance [108] impeding the micro-organism's growth. Because the aerobic reaction occurs at the anode section in the presence of O_2 instead of the anaerobic oxidation process (responsible for releasing electrons), hence the coulombic efficiency, η_{coul} is degraded, inhibiting the MFC's performance [22, 108, 109]. However, even now, it is difficult to synthesize a membrane that resists OD entirely. Therefore, there is a very urgent need to design and synthesize a PEM that could prevent OD entirely to maintain an O_2 free atmosphere in the anode chamber, which is essential for the sustainable survival of micro-organisms that are electroactive [47, 110].
5. Biofouling	Biofouling is a process that is caused by OD, fuel crossover, operating duration, and membrane porosity. When the membrane fouls, it causes a reduction in PEM performance due to a increase in membrane thickness, a decrease in proton conductivity (σ), an increase in R_{int}, and an undesired change in pH. The greater the substrate or O_2 crossover with the operating period, the larger the biofouling thickness will be. This will increase the membrane thickness and, as a direct consequence, Ohmic resistance would be increased. Thus, the performance of the PEMFC would be decreased. It is expressed by the micro-organism's adherence to the surface of the membrane exposed to the inner part of the anode chamber (termed 'anodic biofouling') or the accumulation of small molecular weight organic species as a substrate crossover on the surface of the membrane facing the interior side of the cathode chamber ('cathodic biofouling'); combined with prolonged operation of the MFCs [22, 47]. The crossover of these aerobic and oxidative microenvironments and smaller entities produced on the surfaces of the membrane result in: (i) The higher rate of substrate consumption makes it vulnerable to conversion into electrical energy. (ii) An undesirable O_2 gradient occurs as the demand of aerobic bacteria leads to more O_2 passing through the membrane to the anodic chamber. (iii) An extra barrier such as biofilm sandwiched between the anolyte or catholyte and the membrane increases R_{int} and decreases σ. This decrease of σ favours the increase of acidity in the anode chamber, creating a pH difference between the sections. The summation of all those effects would be reflected in the performance of MFCs regarding power (P_{MFC}) and $\eta_{coul.}$ Therefore, it requires frequent replacement and scheduled cleaning of the PEM to obtain a sustainable performance from PEMFCs.

(Continued)

Table 7.1. (*Continued*)

Existing gaps and challenges	Analysis with future outlooks
6. pH splitting	pH splitting significantly decreases the MFC performance. This phenomenon, associated with the use of membranes, is characterized by the difference in pH between the electrode chambers during the period of MFC operation. It also depends on the selected membrane type and the physicochemical characteristics of the anolyte. The higher cation concentration would compete directly with the diffusion of H^+ ions. Consequently, at lower pH, the rate of oxidation of anaerobic bacteria decreases in the anode chamber as the lower rate of generation of H^+ and electrons increases the pH of the catholyte section. In this context, AEMs could be an alternative choice to eliminate this problem. The rate of transfer of H^+ ions is not restricted, because the AEM associated anions are accountable for the diffusion of protons. Hence, the absence of supplementary cations could not compete with the H^+ ions that stick to the AEM surface. Following this technique, the pH split could be avoided [22, 50] using an AEM.
7. Water loss by evaporation	A dehydration temperature less than 100 °C limits the PEM, which leads to a slower kinetics and lower cost efficiency including CO poisoning at Pt electro-catalysts. The management of water is a difficult issue in PEMFCs, mainly because the hydrated membrane is attracted to the cathode due to the polarization of the electrode. The chemical structure of the membrane makes it partially hydrophilic which favours σ. Moreover, this property eases water evaporation and water transport, making this a design problem. As a result, the management of water is critical to achieve the optimum performance of PEMFCs. If water evaporates very slowly, it will flood the membranes, and accumulating water inside the plates will impede O_2 flow in the PEMFCs. In contrast, if H_2O evaporates too quickly, the membrane becomes dry, increasing R across the membrane. Both cases would cause harm to power output and stability. Therefore, intense research into the design of PEM-based MFCs focusing on the problem is needed immediately, as it is often ignored. In an innovative design, where natural convection aerates the cathode chamber, exposing it to an atmosphere of low humidity at ambient conditions may favour the loss of H_2O molecules through evaporation.
8. Unwanted ion crossing and the performance of the PEM	In nonporous PEM membranes, the permeability to H^+ ions is not 100% efficient regarding the concentration of undesirable ions present in the electrolytic medium. The presence of undesired interfering cations (such as Na^+, NH_4^+, Mg^{2+}, K^+, and Ca^{2+}) competes with protons, and there is an opportunity to diffuse through the membrane, causing a pH gradient. The diffusion of ions, except H^+ ions, creates a problem during the action of the PEM system [50, 111]. Moreover, the size of ionic clusters (volume of ion transport pathways/channel) and

	IEC (ion-exchange capacity) are two vital issues when evaluating the PEM performance. The PEM should function as a proton selective conducting channel, impeding the diffusion of O_2 from the cathode to the anode and of undesired organic compounds between the anode and the cathode chamber. A high σ can be regarded as the second most significant characteristic of the PEM in MFCs, where they would have partial hydrophilic properties to offer ion conduction channels. Furthermore, it must prevent the diffusion of other species such as SO_4^{2-}, MnO_4^-, $[Fe(CN)_6]^{3-}$, H_2O_2, NH_3, ClO_4^-, NO_3^-, some heavy metals, and trichloroethene that can modify the microbial community at the anode, favouring the spread of active micro-organisms which are not electroactive, leading to a reduction in η_{coul} [48, 71].
9. Substrate crossover and degradation	An important function of PEM that happens in PEMFCs is often ignored: avoiding the exchange or crossing of substrates, either as micro-organisms or low molecular weight organic species, among the sections. As a whole, the highly specific transport properties of PEM mainly determine its success in MFCs. The PEM must have the capacity to hinder the transmission of certain entities (such as substrates or O_2) among chambers but let the channel conduct H^+ ions. When the substrate travels from the anolyte to the catholyte chamber, the following effects are detected: (i) The quantity of substrate decreases next to the anode used for the biocatalysts. (ii) The substrate oxidizes on the cathode surface by aerobic bacteria, generating electrons for the oxygen reduction reaction (ORR) at the cathode. (iii) Biofouling limits the transfer of H^+ and changes the pH of the anode chamber, which could translate into inhibiting the potential of bacteria or micro-organisms. (iv) The power (P) is reduced. (v) The η_{coul} decreases as a result of each of those stated activities [50]. In order to escape these effects, the substrate used as an electron source in the anode chamber should be avoided as it crosses the PEM. In this case, the function of the membrane is to prevent soluble low molecular weight organic species from being transferred or diffused through the membranes from the anaerobic anodic section to the aerobic cathodic section [48]. Therefore, the membrane should have good chemical stability to avoid oxidation of the membrane, including microbiological degradation [112, 113]. The degradation could be reduced to some extent by the proper molecular design of the PEM.

drawback to MFCs in this regard. Furthermore, while a PEM's internal resistance is lower than that of other separators, PEMs have their own inner resistance, which reduces power production. Indeed, recent research has demonstrated that the elimination of the PEM in air-cathode MFC systems increases power generation at the cost of decreasing coulombic efficacy due to the PEM's reduced energy density because of this enhanced inside resistance. As a result, there is contradictory evidence about the impact of PEMs on MFC system functioning. As a result, a deeper knowledge of the impact of PEM use on the efficiency of MFC systems is required.

Furthermore, fouling of the PEM is widely thought to be a detriment to MFC performance; it was shown that MFCs with a fouled PEM had higher internal resistance and poorer power production. After fouling, the conductivity, ion interchange capacity, and diffusion coefficients of PEM cations all decreased drastically. Next to membrane fouling, it was discovered that decreasing cation diffusion coefficients and cathodic potential loss were the main causes of MFC performance degradation. Membrane fouling, on the other hand, is unavoidable in MFCs due to the growth of biofilm on the PEM during long-term operation. Other metal ions in the medium can quickly replace the proton-attaching spots of the negatively charged sulphonate moieties in the PEM. Through intermolecular bridging between organic foulants and sulphonate moieties, metal ions may have a role in microbial adherence and biofilm development. To ensure long-term energy production, the fouled PEM must be retrieved or substituted with a fresh one. However, the high price of PEMs is a significant impediment to MFC scalability and potential practical deployment. As a result, PEM fouling not only degrades MFC performance but also increases MFC operating expenses.

During long-term MFC operation, the PEM is easily contaminated by microbes, inorganic salts, and extracellular polymers, lowering the MFC's effectiveness. The major features of the PEM are impacted negatively after fouling. On the one hand, physical cation transfer obstruction causes the current to degrade. On the other hand, the pH-gradient-associated potential deficits were amplified owing to proton transfer limitations. The worsening of the MFC performance is mostly due to cation transfer limitation due to reduced diffusion coefficients and increased cathodic potential loss after PEM fouling. To alleviate PEM fouling difficulties and ensure the long-term stable functioning of MFCs, an anti-fouling PEM with superior cation diffusion coefficients and lower resistance would be developed.

Nafion membranes have been the most extensively used to build up MFCs to date. However, there is a general consensus in the literature that Nafion is costly and, in numerous circumstances, fails to meet the process and user requirements. As a result of these challenges, considerable advancement has been made in the development of alternative materials for membrane manufacturing, including polyester cloth (PC), and ones that rely on the deployment of ionic liquids.

The scale-up of PEMFCs from the laboratory to the market is a challenging step in the development process. There are many problems yet to be solved in the technology to launch the PEMFCs in real-world applications. The underlying facts that restrict the practical implementation of PEMFCs are: design aspects;

electrochemical, technological, physicochemical, and microbiological drawbacks; cathodic electron transfer limitations; insufficient coulombic efficiency (CE); inadequate power output; and high capital costs. Before commercial manufacturing, when the PEM is selected for MFC applications, it must meet some selection criteria that need to be optimized, such as outstanding chemical, mechanical, and thermal stability of the polymer backbone for the reinforcement of the membrane during operation under different operational conditions. Apart from that, the low thickness, conduction of electrons and biofouling, gas impermeability, low levels of swelling, partial hydrophilicity, adequate sulphonation and degree of hydration, high proton conductivity and H^+ ion selectivity, the ability to prevent microbial migration, minimum oxygen and undesirable substrate crossover, long life cycle, low processing cost, and electrical resistance [22, 47, 58] should also be addressed carefully. It is not mandatory to use a membrane in MFCs. However, researchers have determined that a membrane as a separator is highly recommended to confirm a sustainable and efficient MFC application, although it has a direct impact considering its maintenance costs in addition to the primary investment. Excluding PEMs from MFC designs could lower the cost of the technology while simplifying its design, both of which are critical issues for the scale-up and implementation of these devices in real-world scenarios.

Therefore, it is better to focus on continuous searching to synthesize new alternative low-cost membranes as separators with attractive performance in terms of the highest number of optimistic attributes. Any scale-up work from the laboratory-scale, on the other hand, is a difficult task, because it is challenging if we consider the difficulties regarding logistics support, utility supplies, operational complexity, regulatory and safety fulfilment, along with large technical issues. Therefore, with the limited information and knowledge available, the researcher must integrate many different approaches and tailor them to the specific or general situation to accomplish a trade-off solutiom. In this regard, adequate funding should be allocated for collaboration as a team with a variety of expertise in different fields such as chemistry, architecture engineering, electrochemical engineering, physics, materials science, electrical and electronic engineering, fluid and thermodynamics, biology, and microbiology for the transition of PEMFCs from laboratories to the market.

References

[1] Rahimnejad M, Bakeri G, Ghasemi M and Zirepour A 2014 A review on the role of proton exchange membrane on the performance of microbial fuel cell *Polym. Advan. Technol.* **25** 1426–32

[2] Ahn Y and Logan B E 2010 Effectiveness of domestic wastewater treatment using microbial fuel cells at ambient and mesophilic temperatures *Bioresource Technol.* **101** 469–75

[3] Barua E *et al* 2018 Generation of electricity using microbial fuel cell (MFC) from sludge *Bangladesh J. Microbiol.* **35** 23–6

[4] Saha T C *et al* 2019 Microbial fuel cell (MFC) application for generation of electricity from dumping rubbish and identification of potential electrogenic bacteria *Adv. Ind. Biotechnol.* **2** 1–8

[5] Tamakloe R Y 2018 PEM-less microbial fuel cells *Proton Exchange Membrane Fuel Cell* ed T Taner (London: IntechOpen) pp 155–83

[6] Santoro C, Arbizzani C, Erable B and Ieropoulos I 2017 Microbial fuel cells: from fundamentals to applications. a review *J. Power Sources* **356** 225–44

[7] Beyenal H, Chang I S, Mohan S V and Pant D 2021 Microbial fuel cells: current trends and emerging applications *Bioresour. Technol.* **324** 124687

[8] Beylier M R *et al* 2011 Biological nitrogen removal from domestic wastewater *Compr Biotechnol.* **6** 329–40

[9] Ghasemi M, Halakoo E, Sedighi M, Alam J and Sadeqzadeh M 2015 Performance comparison of three common proton exchange membranes for sustainable bioenergy production in microbial fuel cell *Proc. Cirp* **26** 162–6

[10] Stratford J P, Beecroft N J, Slade R C, Grüning A and Avignone-Rossa C 2014 Anodic microbial community diversity as a predictor of the power output of microbial fuel cells *Bioresour. Technol.* **156** 84–91

[11] Esmaeili N, Gray E M and Webb C J 2019 Non-fluorinated polymer composite proton exchange membranes for fuel cell applications—a review *ChemPhysChem* **20** 2016–53

[12] Dicks A and Rand D 2018 Proton-exchange membrane fuel cells *Fuel Cell Systems Explained* 3rd edn (Hoboken, NJ: Wiley) pp 69–133

[13] Mauritz K A and Moore R B 2004 State of understanding of Nafion *Chem. Rev.* **104** 4535–86

[14] Min B, Cheng S and Logan B 2005 Thermodynamic analysis of a single chamber microbial fuel cell *Water Res.* **39** 1675–86

[15] Rabaey K, Clauwaert P, Aelterman P and Verstraete W 2005 Tubular microbial fuel cells for efficient electricity generation *Env. Sci. Technol.* **39** 8077–82

[16] Kim J R, Cheng S, Oh S-E and Logan B E 2007 Power generation using different cation, anion, and ultrafiltration membranes in microbial fuel cells *Env. Sci. Technol.* **41** 1004–9

[17] He Z, Minteer S D and Angenent L T 2005 Electricity generation from artificial wastewater using an upflow microbial fuel cell *Env. Sci. Technol.* **39** 5262–7

[18] Harnisch F, Schröder U and Scholz F 2008 The suitability of monopolar and bipolar ion exchange membranes as separators for biological fuel cells *Env. Sci. Technol.* **42** 1740–6

[19] Vermeiren P, Leysen R, Beckers H, Moreels J and Claes A 2008 The influence of manufacturing parameters on the properties of macroporous Zirfon® separators *J. Porous Mat.* **15** 259–64

[20] Zuo Y, Cheng S and Logan B E 2008 Ion exchange membrane cathodes for scalable microbial fuel cells *Env. Sci. Technol.* **42** 6967–72

[21] Cao X *et al* 2009 A new method for water desalination using microbial desalination cells *Env. Sci. Technol.* **43** 7148–52

[22] Leong J X, Daud W R W, Ghasemi M, Liew K B and Ismail M 2013 Ion exchange membranes as separators in microbial fuel cells for bioenergy conversion: a comprehensive review *Renew. Sust. Energ. Rev.* **28** 575–87

[23] Fan Y, Hu H and Liu H 2007 Enhanced coulombic efficiency and power density of air-cathode microbial fuel cells with an improved cell configuration *J. Power Sources* **171** 348–54

[24] Jana P S, Behera M and Ghangrekar M 2010 Performance comparison of up-flow microbial fuel cells fabricated using proton exchange membrane and earthen cylinder *Int. J. Hydrogen Energy* **35** 5681–6

[25] Winfield J, Chambers L D, Rossiter J and Ieropoulos I 2013 Comparing the short and long term stability of biodegradable, ceramic and cation exchange membranes in microbial fuel cells *Bioresour. Technol.* **148** 480–6

[26] Zhang X, Cheng S, Wang X, Huang X and Logan B E 2009 Separator characteristics for increasing performance of microbial fuel cells *Env. Sci. Technol.* **43** 8456–61

[27] Zhang X, Cheng S, Huang X and Logan B E 2010 The use of nylon and glass fiber filter separators with different pore sizes in air-cathode single-chamber microbial fuel cells *Energy Env. Sci.* **3** 659–64

[28] Park D H and Zeikus J G 2003 Improved fuel cell and electrode designs for producing electricity from microbial degradation *Biotechnol. Bioeng.* **81** 348–55

[29] Winfield J, Gajda I, Greenman J and Ieropoulos I 2016 A review into the use of ceramics in microbial fuel cells *Bioresour. Technol.* **215** 296–303

[30] Niessen J, Harnisch F, Rosenbaum M, Schröder U and Scholz F 2006 Heat treated soil as convenient and versatile source of bacterial communities for microbial electricity generation *Electrochem. Commun.* **8** 869–73

[31] Ucar D, Zhang Y and Angelidaki I 2017 An overview of electron acceptors in microbial fuel cells *Front. Microbiol.* **8** 643

[32] Peighambardoust S J, Rowshanzamir S and Amjadi M 2010 Review of the proton exchange membranes for fuel cell applications *Int. J. Hydrogen Energ.* **35** 9349–84

[33] Lei J *et al* 2021 Under-brine superaerophobic perfluorinated ion exchange membrane with re-entrant superficial microstructures for high energy efficiency of NaCl aqueous solution electrolysis *J. Membrane Sci.* **619** 118801

[34] Prabhu N V and Sangeetha D 2014 Characterization and performance study of sulfonated poly ether ether ketone/Fe_3O_4 nano composite membrane as electrolyte for microbial fuel cell *Chem. Eng. J.* **243** 564–71

[35] Chae K J, Choi M, Ajayi F F, Park W, Chang I S and Kim I S 2008 Mass transport through a proton exchange membrane (Nafion) in microbial fuel cells *Energy Fuel.* **22** 169–76

[36] Rudra R, Kumar V, Nandy A and Kundu P P 2018 Performances of separator and membraneless microbial fuel cell *Microbial Fuel Cell* ed D Das (New York: Springer) pp 125–40

[37] Ter Heijne A, Hamelers H V, De Wilde V, Rozendal R A and Buisman C J 2006 A bipolar membrane combined with ferric iron reduction as an efficient cathode system in microbial fuel cells *Env. Sci. Technol.* **40** 5200–5

[38] Smitha B, Sridhar S and Khan A 2005 Solid polymer electrolyte membranes for fuel cell applications—a review *J. Membrane Sci.* **259** 10–26

[39] Wu D, Xu T, Wu L and Wu Y 2009 Hybrid acid–base polymer membranes prepared for application in fuel cells *J. Power Sources* **186** 286–92

[40] Fornero J J, Rosenbaum M, Cotta M A and Angenent L T 2008 Microbial fuel cell performance with a pressurized cathode chamber *Env. Sci. Technol.* **42** 8578–84

[41] Varcoe J R *et al* 2014 Anion-exchange membranes in electrochemical energy systems *Energy Env. Sci.* **7** 3135–91

[42] Ge X *et al* 2016 Alkaline anion-exchange membranes containing mobile ion shuttles *Adv. Mater.* **28** 3467–72

[43] Elangovan M and Dharmalingam S 2017 Application of polysulphone based anion exchange membrane electrolyte for improved electricity generation in microbial fuel cell *Mater. Chem. Phys.* **199** 528–36

[44] Li W-W, Sheng G-P, Liu X-W and Yu H-Q 2011 Recent advances in the separators for microbial fuel cells *Bioresour. Technol.* **102** 244–52

[45] Guo D, Zhuo Y Z, Lai A N, Zhang Q G, Zhu A M and Liu Q L 2016 Interpenetrating anion exchange membranes using poly (1-vinylimidazole) as bifunctional crosslinker for fuel cells *J. Membrane Sci.* **518** 295–304

[46] Dharmalingam S, Kugarajah V and Sugumar M 2019 Membranes for microbial fuel cells *Microbial Electrochemical Technology: Sustainable Platform for Fuels, Chemicals and Remediation, Biomass, Biofuels, Biochemicals* (Amsterdam: Elsevier) pp 143–94

[47] Scott K 2016 Membranes and separators for microbial fuel cells *Microbial Electrochemical and Fuel Cells: Fundamentals and Applications* (Amsterdam: Elsevier) pp 153–78

[48] Logan B E 2008 *Microbial Fuel Cells* (Hoboken, NJ: Wiley) pp 1–31

[49] Kim C *et al* 2017 Hexavalent chromium as a cathodic electron acceptor in a bipolar membrane microbial fuel cell with the simultaneous treatment of electroplating wastewater *Chem. Eng. J.* **328** 703–7

[50] Dharmalingam S, Kugarajah V and Sugumar M 2019 Membranes for microbial fuel cells *Microbial Electrochemical Technology* (Amsterdam: Elsevier) pp 143–94

[51] Sondhi R, Bhave R and Jung G 2003 Applications and benefits of ceramic membranes *Serono. Sym* **2003** 5–8

[52] Behera M and Ghangrekar M 2011 Electricity generation in low cost microbial fuel cell made up of earthenware of different thickness *Water Sci. Technol.* **64** 2468–73

[53] Behera M, Jana P S and Ghangrekar M 2010 Performance evaluation of low cost microbial fuel cell fabricated using earthen pot with biotic and abiotic cathode *Bioresour. Technol.* **101** 1183–9

[54] Behera M, Jana P S, More T T and Ghangrekar M 2010 Rice mill wastewater treatment in microbial fuel cells fabricated using proton exchange membrane and earthen pot at different pH *Bioelectrochemistry* **79** 228–33

[55] Ajayi F F and Weigele P R 2012 A terracotta bio-battery *Bioresour. Technol.* **116** 86–91

[56] Winfield J, Greenman J, Huson D and Ieropoulos I 2013 Comparing terracotta and earthenware for multiple functionalities in microbial fuel cells *Bioproc. Biosyst. Eng.* **36** 913–21

[57] Kim K-Y, Yang W and Logan B E 2015 Impact of electrode configurations on retention time and domestic wastewater treatment efficiency using microbial fuel cells *Water Res.* **80** 41–6

[58] Ramirez-Nava J *et al* 2021 The implications of membranes used as separators in microbial fuel cells *Membranes* **11** 738

[59] Krauskopf K B and Bird D K 1967 *Introduction to Geochemistry* (New York: McGraw-Hill)

[60] Islam E, Hossain M S, Sarker P K, Towhid S T, Salimullah M and Hashem A 2020 Isolation and characterization of electrogenic bacteria from tannery wastewater *Bangladesh J. Microbiol.* **37** 23–7

[61] Bakonyi P, Koók L, Rózsenberszki T, Tóth G, Bélafi-Bakó K and Nemestóthy N 2020 Development and application of supported ionic liquid membranes in microbial fuel cell technology: a concise overview *Membranes* **10** 16

[62] Zuo Z, Fu Y and Manthiram A 2012 Novel blend membranes based on acid–base interactions for fuel cells *Polymers* **4** 1627–44

[63] Walkowiak-Kulikowska J, Wolska J and Koroniak H 2017 Polymers application in proton exchange membranes for fuel cells (PEMFCs) *Polymer Engineering* ed B Bartosz Tylkowski, K Wieszczycka and R Jastrzab (Berlin: De Gruyter) pp 293–348
[64] Gabbasa M, Sopian K, Fudholi A and Asim N 2014 A review of unitized regenerative fuel cell stack: material, design and research achievements *Int. J. Hydrogen Energy* **39** 17765–78
[65] Seger B, Vinodgopal K and Kamat P V 2007 Proton activity in Nafion films: probing exchangeable protons with methylene blue *Langmuir* **23** 5471–6
[66] Zeis R 2015 Materials and characterization techniques for high-temperature polymer electrolyte membrane fuel cells *Beilstein J. Nanotech.* **6** 68–83
[67] Hogarth W H, Da Costa J D and Lu G M 2005 Solid acid membranes for high temperature (>140 °C) proton exchange membrane fuel cells *J. Power Sources* **142** 223–37
[68] Ogungbemi E *et al* 2019 Fuel cell membranes—pros and cons *Energy* **172** 155–72
[69] Catal T, Bermek H and Liu H 2009 Removal of selenite from wastewater using microbial fuel cells *Biotechnol. Lett.* **31** 1211–6
[70] Wang H and Ren Z J 2014 Bioelectrochemical metal recovery from wastewater: a review *Water Res.* **66** 219–32
[71] Mathuriya A S and Yakhmi J 2014 Microbial fuel cells to recover heavy metals *Environ. Chem. Lett.* **12** 483–94
[72] Du Z, Li H and Gu T 2007 A state of the art review on microbial fuel cells: a promising technology for wastewater treatment and bioenergy *Biotechnol. Adv.* **25** 464–82
[73] Flimban S G, Hassan S H, Rahman M M and Oh S-E 2020 The effect of Nafion membrane fouling on the power generation of a microbial fuel cell *Int. J. Hydrogen Energy* **45** 13643–51
[74] Miskan M, Ismail M, Ghasemi M, Jahim J M, Nordin D and Bakar M H A 2016 Characterization of membrane biofouling and its effect on the performance of microbial fuel cell *Int. J. Hydrogen Energy* **41** 543–52
[75] Kokare C, Chakraborty S, Khopade A and Mahadik K R 2009 Biofilm: importance and applications *Indian J. Biotechnol.* **8** 159–68
[76] Ishizaki S, Papry R I, Miyake H, Narita Y and Okabe S 2019 Membrane fouling potentials of an exoelectrogenic fouling-causing bacterium cultured with different external electron acceptors *Front. Microbiol* **9** 3284
[77] Jiang B *et al* 2020 Impacts of long-term electric field applied on the membrane fouling mitigation and shifts of microbial communities in EMBR for treating phenol wastewater *Sci. Total Env.* **716** 137139
[78] Martín-Rodríguez A J *et al* 2015 From broad-spectrum biocides to quorum sensing disruptors and mussel repellents: antifouling profile of alkyl triphenylphosphonium salts *PLoS One* **10** e0123652
[79] Xu J, Sheng G-P, Luo H-W, Li W-W, Wang L-F and Yu H-Q 2012 Fouling of proton exchange membrane (PEM) deteriorates the performance of microbial fuel cell *Water Res.* **46** 1817–24
[80] Choi M-J *et al* 2011 Effects of biofouling on ion transport through cation exchange membranes and microbial fuel cell performance *Bioresour. Technol.* **102** 298–303
[81] Okada T, Ayato Y, Yuasa M and Sekine I 1999 The effect of impurity cations on the transport characteristics of perfluorosulfonated ionomer membranes *J. Phys. Chem.* B **103** 3315–22
[82] Li H, Xing Y, Cao T, Dong J and Liang S 2021 Evaluation of the fouling potential of sludge in a membrane bioreactor integrated with microbial fuel cell *Chemosphere* **262** 128405

[83] Lu Y, Jia J, Miao H, Ruan W and Wang X 2020 Performance improvement and biofouling mitigation in osmotic microbial fuel cells via *in situ* formation of silver nanoparticles on forward osmosis membrane *Membranes* **10** 122

[84] Liu J and Hurt R H 2010 Ion release kinetics and particle persistence in aqueous nano-silver colloids *Env. Sci. Technol.* **44** 2169–75

[85] Meng F, Chae S-R, Drews A, Kraume M, Shin H-S and Yang F 2009 Recent advances in membrane bioreactors (MBRs): membrane fouling and membrane material *Water Res.* **43** 1489–512

[86] Wu W, Zhang D and Fang P 2020 Advance in improving the electrical performance of microbial fuel cell *IOP Conf. Series: Earth Env. Sci.* **555** 012004

[87] Kim S M, Patel R and Kim J H 2021 Recent advance in microbial fuel cell based on composite membranes *Membrane J.* **31** 120–32

[88] Ha P T *et al* 2017 Syntrophic anaerobic photosynthesis via direct interspecies electron transfer *Nat. Commun.* **8** 1–7

[89] Hassan S H, Kim Y S and Oh S-E 2012 Power generation from cellulose using mixed and pure cultures of cellulose-degrading bacteria in a microbial fuel cell *Enzyme Microb. Technol.* **51** 269–73

[90] Li F *et al* 2019 Engineering microbial consortia for high-performance cellulosic hydrolyzates-fed microbial fuel cells *Front. Microbiol.* **10** 409

[91] Moradian J M, Fang Z and Yong Y-C 2021 Recent advances on biomass-fueled microbial fuel cell *Bioresour. Bioprocess.* **8** 1–13

[92] Zhang F, Saito T, Cheng S, Hickner M A and Logan B E 2010 Microbial fuel cell cathodes with poly (dimethylsiloxane) diffusion layers constructed around stainless steel mesh current collectors *Env. Sci. Technol.* **44** 1490–5

[93] Liu S, Wan G and Ma C 2019 Electrochemical performance of microbial fuel cell with graphene oxide and polyaniline modified cathode *China Env. Sci.* **39** 3866–71

[94] Yang Z, Pei H, Hou Q, Jiang L, Zhang L and Nie C 2018 Algal biofilm-assisted microbial fuel cell to enhance domestic wastewater treatment: nutrient, organics removal and bioenergy production *Chem. Eng. J.* **332** 277–85

[95] Guo Y *et al* 2020 Simultaneous wastewater treatment and energy harvesting in microbial fuel cells: an update on the biocatalysts *RSC Adv.* **10** 25874–87

[96] Tharali A D, Sain N and Osborne W J 2016 Microbial fuel cells in bioelectricity production *Front. Life Sci.* **9** 252–66

[97] Azhar M, Jaafar J, Aziz M, Umar Y, Jafar Mazumder M A and Nazal M K 2021 Mild sulfonated polyether ketone ether ketone ketone incorporated polysulfone membranes for microbial fuel cell application *J. Appl. Polym. Sci.* **138** 50216

[98] González-Pabón M J, Figueredo F, Martínez-Casillas D C and Cortón E 2019 Characterization of a new composite membrane for point of need paper-based micro-scale microbial fuel cell analytical devices *PLoS One* **14** e0222538

[99] Hickner M A, Ghassemi H, Kim Y S, Einsla B R and McGrath J E 2004 Alternative polymer systems for proton exchange membranes (PEMs) *Chem. Rev.* **104** 4587–612

[100] Al Mamun M, Wahab Y A, Hossain M M, Hashem A and Johan M R 2021 Electrochemical biosensors with aptamer recognition layer for the diagnosis of pathogenic bacteria: barriers to commercialization and remediation *TrAC-Trend. Anal. Chem.* **145** 116458

[101] Agubra V and Fergus J 2013 Lithium ion battery anode aging mechanisms *Materials* **6** 1310–25

[102] Donovan C, Dewan A, Peng H, Heo D and Beyenal H 2011 Power management system for a 2.5 W remote sensor powered by a sediment microbial fuel cell *J. Power Sources* **196** 1171–7

[103] Ivars-Barceló F, Zuliani A, Fallah M, Mashkour M, Rahimnejad M and Luque R 2018 Novel applications of microbial fuel cells in sensors and biosensors *Appl. Sci.* **8** 1184

[104] Santoro C *et al* 2020 Urine in bioelectrochemical systems: an overall review *ChemElectroChem* **7** 1312

[105] Shahgaldi S *et al* 2014 Performance enhancement of microbial fuel cell by PVDF/Nafion nanofibre composite proton exchange membrane *Fuel Process. Technol.* **124** 290–5

[106] Sun C Y and Zhang H 2019 Investigation of Nafion series membranes on the performance of iron-chromium redox flow battery *Int. J. Energy Res.* **43** 8739–52

[107] Castellan G W 1983 *Physical Chemistry* 3rd edn (Boston, MA: Addison-Wesley)

[108] Watson V J, Saito T, Hickner M A and Logan B E 2011 Polymer coatings as separator layers for microbial fuel cell cathodes *J. Power Sources* **196** 3009–14

[109] Harnisch F and Schröder U 2009 Selectivity versus mobility: separation of anode and cathode in microbial bioelectrochemical systems *ChemSusChem* **2** 921–6

[110] Hernández-Flores G, Poggi-Varaldo H and Solorza-Feria O 2016 Comparison of alternative membranes to replace high cost Nafion ones in microbial fuel cells *Int. J. Hydrogen Energy* **41** 23354–62

[111] Hernández-Flores G, Poggi-Varaldo H, Romero-Castañón T, Solorza-Feria O and Rinderknecht-Seijas N 2017 Harvesting energy from leachates in microbial fuel cells using an anion exchange membrane *Int. J. Hydrogen Energy* **42** 30374–82

[112] Vélez-Pérez L *et al* 2020 Industrial acid mine drainage and municipal wastewater co-treatment by dual-chamber microbial fuel cells *Int. J. Hydrogen Energy* **45** 13757–66

[113] Rabaey K and Verstraete W 2005 Microbial fuel cells: novel biotechnology for energy generation *Trends Biotechnol.* **23** 291–8

Chapter 8

Biocathode development through a green approach for energy generation in microbial fuel cells

Mustapha Omenesa Idris, Sundas Bahar Yaqoob and Tabassum Parveen

Due to some limitations in conventional microbial fuel cells (MFCs), such as high cost and low efficiency, the development of biocathode MFCs is gaining more interest. The use of biocathodes in MFCs is emerging as a more sustainable option for green energy production. This chapter presents a brief overview of the biocathode progress in MFCs, with an emphasis on suitable biocathode materials. The modification of the biocathode material for improved performance is highlighted. Furthermore, the necessary properties for a biocathode material are discussed. The electron transfer mechanism in biocathode MFCs is also explored. Finally, their application in energy generation is discussed.

8.1 Introduction

The continuous establishment of various industries around the world, combined with an increasing global population, has resulted in a rapid surge in energy demands. This need for energy has resulted in a phenomenal effort to shift the energy structure from a populace that is heavily reliant on conventional fossil fuels to a greener approach to energy with high sustainability. Bioenergy will serve as an efficient and scalable alternative for addressing the increased demand for sustainable energy. Microbial fuel cells (MFCs) have emerged as a promising technology to produce bioenergy with simultaneous bioremediation of toxic substances. MFCs have recently received considerable interest as a modern technology for simultaneous treatment of wastewater and energy production [1]. Various types of MFCs have been evaluated, and they exhibit promising results in bioremediation and waste treatment due to their low operating costs, self-regenerating catalysts, and long-term power supply [2]. A proton exchange membrane (PEM) separates a bioanode and an abiotic cathode in a conventional dual chambered MFC. Biocathode MFCs, on the

doi:10.1088/978-0-7503-4791-4ch8

other hand, are the kinds of MFCs in which the reaction occurring at the cathode chamber is aided by micro-organisms rather than a chemical catalyst. In comparison to abiotic cathode MFCs, microbial biocathode MFCs not only reduce the cost of MFC design and construction, but also expand the operational sustainability of the cathode [3]. Electro-active microbes are used as biocatalysts in biocathode MFCs at both the cathodic and anodic chambers. The system effectively reduces the oxygen sourced into the cathodic chamber by accepting electrons from the cathode via the catalytic activity of the electro-active microbes, associated with a significantly enhanced cathode performance [4]. The biocathode materials and the configuration have a significant impact on biocathode performance because they play a critical role in the oxygen reduction reaction, microbial processes, and the formation of biofilms [5]. Thus one of the major challenges in the development of a biocathode MFC is the identification of the cathode materials that maximise microbial activity for oxygen reduction.

This chapter gives a brief and concise survey of the required properties for biocathode electrodes and the suitable materials for biocathode development. Biocathode material modification strategies for improved performance are also discussed. The electron transfer mechanisms occurring in biocathode MFCs are explained. Finally, highlights on how biocathodes improve the energy performance of MFCs are provided.

8.2 Biocathodes in MFCs

The utilisation of biocathodes in MFCs has gained recognition due to their low build-up and running costs, sustainability, and utilisation of their microbial metabolism to generate valuable industrial products along with the removal of unwanted compounds. According to Freguia *et al* [6] a biocathode is a propitious approach to strengthen the cathode reaction, and it can be used in MFCs to enhance cathode efficiency and effectively replaces the need for synthetic mediators. Clauwaert *et al* [7] are one group among many researchers (table 8.1) who have used biocathodes in MFCs. They used nitrate as an electron acceptor and achieved a power density of about 10 W m^{-3}. When acetate was utilised as the substrate in the anodic chamber of the tubular microbial fuel cell and was integrated with an open air biocathode to generate bioelectricity, the observed maximum power density was about 65 W m^{-3} and 83 W m^{-3} for continuous and batch-fed operations, respectively. As previously stated, the microbial bioactivities in biocathodes can be used to generate valuable products or eliminate undesirable substances. For example, microbial reduction of Fe (III) and Mn (IV), that could serve as terminal electron acceptors in the cathode, is an alternative means of extracting those metals from minerals [8]. Various possible terminal electron acceptors are shown in table 8.2.

Basically, there are two types of biocathodes in MFCs which are categorised based on the element acting as the terminal electron acceptors in the cathode: aerobic biocathodes and anaerobic biocathodes. The terminal electron acceptor (TEA) for the cathode reaction in the aerobic biocathode is oxygen. Oxygen, as a

Table 8.1. Some previous studies on biocathode MFCs.

Anode	Cathode	MFC configuration	Substrate/ source	Power density ($mW\ m^{-2}$)	Reference
Carbon paper	Stainless-steel coated carbon nanotubes	Single chamber	Acetate/sludge	147	[9]
Graphite felt	Graphite felt	Plant MFCs	Aerobic wastewater	240	[10]
Granular activated carbon	Carbon cloth	Constructed wetland MFCs	Sludge	55	[11]
DSA	Stainless-steel plates	Floating MFCs	Seawater	32	[12]
Carbon cloth	Carbon paper	Single chamber	Wastewater/ acetate	500	[13]
Carbon cloth	Graphene oxide	Double chamber	Activated sludge	323	[14]
Carbon cloth	Carbon cloth	Double chamber	Sludge	103	[15]
Carbon felt	Graphite felt	Coupled MFCs system	Synthetic wastewater	144	[16]
Graphite brush	Graphite brush	Double chamber	Activated sludge	72.7 ± 1.2	[17]
Modified graphene	Modified graphene	Double chamber	Wastewater/ glucose	122.4	[18]

Table 8.2. Reactions in the biocathode.

Terminal electron acceptors (TEA)	Biocathode reactions	Reference
Nitrate	$NO_3^- + 2H^+ + 2e^- \rightarrow NO_2^- + H_2O$	[21]
Manganese	$MnO_2 + 4H^+ + 3e^- \rightarrow Mn^{2+} + 2H_2O$	[22]
Iron	$Fe^{2+} \rightarrow Fe^{3+} + e^-$	[23]
Sulphate	$SO_4^{2-} + 4H_2 \rightarrow S^{2-} + H_2OSO_4^{2-} + 4H_2O + 8e^- \rightarrow S^{2-} + 8OH^-$	[24]
Fumarate	Fumarate + $2H^+ + 2e^- \rightarrow$ Succinate	[25]

TAE, is distinguished by its elevated redox potential and large availability in the atmosphere, as well as its low supply cost. Some studies used micro-organisms to help with the oxidation of transition metal compounds such as Mn (II) or Fe (II) for electron transport to oxygen. Furthermore, bacteria in the cathode aid the reaction by supplying oxygen [19]. Anaerobic biocathodes do not require oxygen as a TAE, other elements, such as nitrate, iron, sulphate, manganese, arsenate, selenate,

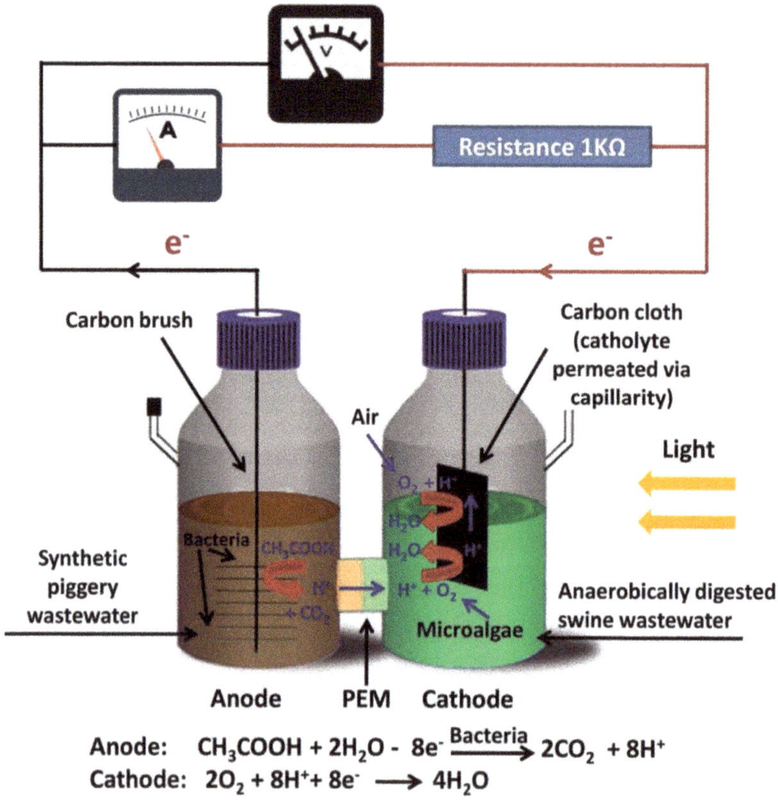

Figure 8.1. Schematic presentation of a biocathode MFC. (Reproduced with permission from [26]. MDPI CC-BY 4.0.)

urinate, carbon dioxide, and fumarate can function as TEAs [20]. Of the frequently commonly occurring compounds, iron, nitrate, and manganese have relative bioactivity close to that of oxygen, whereas sulphate has a comparatively much lower activity. Figure 8.1 shows the schematic representation of a biocathode MFC.

8.3 Properties of biocathode electrodes

The choice of the biocathode electrode material is critical for the MFC system because the material plays a significant role in achieving appropriate results in terms of electron transformation, electrochemical efficiency, and microbial adhesion [27]. In general, low-cost materials with high mechanical strength for use as a biocathode are critical for achieving a sustainable and cost-effective MFC system. Some of the requisite properties which are required in an ideal biocathode electrode are highlighted below.

Conductivity. Because of electron transfer processes, biocathode electrodes must have good conduction. A high conducting biocathode will promote the easy and rapid flow of electrons, increasing the electron transport rate [28]. As a result, conductivity is a critical property of biocathode materials.

Chemical stability. A biocathode material in MFCs must be highly chemically stable. Material swelling, corrosion, and physical strength loss are all risks with biocathode electrodes that lack chemical stability. Furthermore, biocathode materials with a high surface roughness are required to favour bacterial adhesion [29].

Compatibility. The compatibility of the biocathode with bacteria is critical because it aids in the bacterial respiration process. Materials that are prone to corrosion and have a high level of toxicity can inhibit bacterial growth and, as a result, are classified as non-biocompatible for use as biocathode electrodes [30].

Surface area. The surface area provides active sites for bacterial metabolism and improves electro-kinetic efficiency. Microbes are electro-actively immobilised on the surfaces of the biocathode, ensuring that the electron transfer mechanism works properly [31]. The overall performance of the biocathode MFCs is greatly improved by a large surface area [32]. Electrodes in MFCs.

8.4 Biocathode materials

The effectiveness of a biocathode material in an MFC system is generally evaluated by the electrical performance of the MFCs and the bioremediation efficiency of the wastewater [33]. The major hurdles in attempting to build a low-cost biocathode MFC are the material and the design for the biocathode electrode. Various biocathode materials, spanning from metal based to carbon based, have been investigated and examined for suitability as MFC biocathodes [34]. Previous work depicts the various materials used as biocathodes in previous MFC studies [35]. According to previous research, carbon based materials are the most widely used support materials in a variety of forms and modifications. Carbon material has a high mechanical strength and a surface texture, making it ideal for the formation of biofilms on its surfaces [36]. Carbon based materials (table 8.3) are commonly utilised in biocathode electrodes in MFCs and are classified into three types: packed

Table 8.3. Carbonaceous biocathode materials.

Configuration	Carbonaceous materials	Remarks	Study examples
Packed structure	Graphite granules, granular carbon	In recent biocathode MFC studies, graphite granules as biocathodes produce the highest power output.	Freguia *et al* [39]
Plane structure	Carbon cloth, carbon paper, carbon felt, graphite plate, graphite felt	In terms of strength, graphite plate outperforms carbon paper. Even though carbon paper is wafer-thin compared to graphite plate, it is more brittle.	Zhang *et al* [40]
Brush structure	Graphite brush	The graphite brush used as a biocathode in MFCs has a large surface area and high porosity, which improves the adhesion of microbes for electron transport.	Zhang *et al* [41]

structure, plane structure, and brush structure. Carbonaceous bioelectrodes in a packed structure are applied in MFCs to increase the surface area available for bacterial growth. Carbon based biocathodes in the plane structure appear to be thicker with a loose texture, allowing for more room for bacterial growth. Brush structured electrodes are mechanically suitable for biocathode development due to their porous and high surface area, which promote micro-organism attachment for electron transfer [37, 38].

8.5 Modification of biocathode material

Surface modification during electrode fabrication is often done to improve bacterial attachment as well as electron transfer. Surface modification of the biocathode material can significantly improve the material's surface area, porosity, conductivity, and biocompatibility. Physical or chemical surface treatment methods, the addition of highly conducting polymer composites, and the use of metal/graphite composite electrodes are all examples of modification techniques [42]. Chemical oxidation of carbonaceous material surfaces has been identified as an effective treatment method. The acidic treatment of carbon fibre brushes with concentrated sulphuric acid was reported to improve the power output by 8%, and the incorporated acidic and heat surface treatments gave a higher increased power output by 25% [43]. Wang *et al* [44] previously identified some factors associated with the acidic and heat treatment of electrode materials that resulted in improved power generation. The researchers discovered an increase in a specific area of the modified material, which aids in the adhesion of bacteria to surfaces. They also discovered more positive charges on the electrode surface, which promotes bacterial adhesion. Biocathodes now have the significant benefits of being relatively inexpensive, having good stability, and having diverse uses for treating wastewater and biosynthesis. As a result, it has emerged as a rapidly developing topic of interest for researchers in the field of MFCs. The surface properties and active surface area for microbes, which are determined by the character of the electrode materials and their layout, are two major factors that influence the biomass on biocathodes and their efficiency [45]. Recently, researchers were able to modify biocathode material for improved performance of MFCs. Zhuang *et al* [14] developed a graphene-modified bioelectrode that had a higher maximum power density of 0.323 W m^{-2} than the unmodified electrode (0.159 W m^{-2}). The graphene integrated in the biofilm is said to improve electron conductivity between the bacteria and the biocathode. An electrochemical study shows that the graphene-modified biocathode has improved electron transport and lower internal resistance. As a result, the electrocatalytic activity toward oxygen reduction is increased, and electron transfer is aided. Scanning electron microscopy revealed that the modified graphene biocathode has a porous 3D structure that is fitted for substrate mobility and microbial attachment. In a similar study, Chen *et al* [46] created a biocathode from carbon felt and modified it with graphene. Based on the cyclic volammetry, the modified biocathode formed is compared to a control graphite felt biocathode. It was discovered that graphene modification improved power performance. When compared to the conventional biocathode which gave (53.8 mW m^{-2}), the modified biocathode had a greater power density value of (65.4 mW m^{-2}). The improved performance is due to the introduction of graphene, which improved the bioelectrochemical reaction to the electrode as well as the oxygen reduction rate (ORR). The addition of graphene to the graphite felt biocathode increased the ORR, exoelectrogen enrichment, and MFC performance, providing a new insight into the preparation of a microbial self-assembled graphene-modified biocathode for MFC. Aside from graphene-modified biocathodes, Liu *et al* [47] were the first to report the development of a carbon nanotube (CNT) biocathode in an MFC. CNTs are being investigated as a biocathode in MFCs due to their high active surface area, high conductivity, and geared nanostructure potential to adsorb bacteria. In comparison to CNT-modified biocathodes, the use of CNT-modified bioanodes in MFCs has been studied more.

8.6 Electron transfer mechanism in biocathode MFCs

Although there has been much discussion in the literature about the electron transfer mechanism in bioanodes, there have been no adequate studies on the electron transfer mechanism in biocathodes. However, it has been reported that electron acceptors such as oxygen, CO_2, and transition metal compounds are frequently used as electron acceptors in biocathode MFC applications [48]. The two main types of electron transfer observed in biocathode MFCs are direct and indirect (mediated) electron transfer mechanisms as shown in previous work [49]. Whereas the mechanism of microbial electron transfer remains unravelled, many scientists have demonstrated that microbes play an important role in the biocathode electron transfer system [50].

8.6.1 Direct electron transfer mechanism

Direct electron transfer occurs when the surfaces of the biocathode electrode make direct physical contact with the outer layer of the microbial cell. As a result, electrons from the cathode are received directly by the outer layer of the bacterial cell known as cytochromes. When oxygen acts as a direct electron acceptor in a biocathode MFC, bacteria colonise the cathode surface and transform into biofilms that catalyse oxygen reduction. The catalytic activity of the cathode increases, leading to an increase in the on-load cell voltage [51]. Among all the strain types, *Gammaproteobacteria* were found to be the most abundant class of bacteria in terms of division. The presence of microbial biofilms around the cathode of MFCs improves nitrate and oxygen reduction, according to researchers [52]. For the transfer of electrons, the direct electron transfer mechanism does not require the addition of an intermediary or mediator. The key drawback of this mechanism is that bacteria must adhere to the biocathode surface for electrons to be transferred. Most gram-positive bacteria are thought to engage actively in the mechanism via biofilm formation such as through teichoic acid (TA), a compound that enables bacteria to conform to the electrode surface through direct contact [20].

8.6.2 Indirect electron transfer mechanism

Several microbes are unable to transfer electrons directly from their cytochromes to the electrodes because their outer layers are mostly non-conductive. As a result, either exogenous mediators or redox mediators are required to transport electrons between the electrodes and the microbes [53]. Some micro-organisms, for example, are able to release redox-active molecules in order to transfer electrons. Indirect contact between the bacterial cell membrane and the surfaces of the cathode electrode is involved in this mechanism. A redox carrier is used in this method to act as a conduit for electron transfer to the final electron acceptor. These redox mediators, in general, collect electrons from microbe cells, which are then transferred to the anode in a reduced state and oxidised. These oxidised electron steps are free to begin the subsequent electron transfer activity [54]. As previously stated, mediators can be either endogenous or exogenous. Endogenous mediators are redox mediators produced by micro-organisms. Exogenous mediators are redox mediators that can be supplied from outside the body. The main drawbacks of these mediators are their instability and toxicity, which limit their use in MFC systems [55]. There are some characteristics of effective exogenous mediators that have been outlined. Exogenous mediators can accept electrons from electron donors while not interfering with other metabolic processes. A good exogenous mediator should also be cytotoxic and capable of delivering electrons within the cytoplasm for oxidation reactions. The mediator should be able to smoothly hybridise cell membranes and be soluble and stable in both the oxidised and reduced forms. The microbes in the biofilm should not consume it as a nutrient [56, 57].

8.7 Energy performance in applications

Lately, the application of biocathodes in MFC systems has been investigated and improved over time in order to attain optimum performance. The biocathode materials, carbon source for the microbes, reactor configuration, presence of a mediator, and biocathode modification strategies differ between studies. These factors are critical for the overall performance of the MFCs, whether for bioenergy production or bioremediation of wastewater [58]. Sharma *et al* [59] developed a stainless-steel mesh based biocathode of an active carbon fabric electrode to test the performance of a sulphate-reducing microbe. In contrast to the biocathode-based active carbon fabric biocathode that does not have a stainless-steel mesh, it was discovered that stainless-steel mesh combined with activated carbon fabric improved performance. The power density increases by 0.004 79–0.023 W m^{-2} and the current density rises three-fold: 0.075–0.25 A m^{-2}. In a natural marine environment, the biofilm formed on metal surfaces produces a more efficient oxygen reduction cathode. Aelterman *et al* [60] were the first to show the relationship between the electrochemical effectiveness and microbial consortium of a stacked biocathode MFC. They used graphite granules as both the biocathode and the bioanode, and the power density observed by the series and parallel stacks were 228 W m^{-3} and 248 W m^{-3}, respectively. Al-mamun *et al* [61] evaluate the performance of a stake up-flow baffled-reactor MFC based on a biocathode in which autotrophic denitrifying microbes catalysed the cathodic reactions. The system's maximum power density was measured to be 15 W m^{-3} NCV (net cathode volume). The biocathode achieved a maximum current density of 76.5 A m^{-3} NCV and a coulombic efficiency of 97.7%. Xu *et al* [62] investigated a multiple biocathode constructed wetland-microbial fuel cell (CW-MFC) aimed at significantly greater energy production. They discovered that as the number of biocathodes increased the energy losses on both the anode and the cathode decreased significantly. As a result of this improved power performance, the system's maximum power density increased significantly from 12.56 to 26.16 mW m^{-2}. Hence, the authors proposed that the study offers a promising strategy and insight for increased energy production in a biocathode-based CW-MFC. Singh and Kaushik [63] investigate the possibility of using a biocathode to improve energy recovery from wastewater in a double chambered MFC system. Wetland sediments were inoculated in both the anodic and cathodic chambers of the MFCs during their set-up. The power density generated by the biocathode MFCs from wastewater was 722 mW m^{-3}, which was found to be ten times higher than that obtained from the abiotic MFCs. Thus, the new concept of using a biocathode in MFCs shows considerably enhanced recovery of energy without the use of complex and costly catalysts. The characterisation of the anodic and cathodic biofilm revealed how microbial communities of the same inoculum origin varied in their composition and performed separate functions of generating electricity in MFCs.

8.8 Conclusion

Currently, research on MFCs is aimed at addressing the issue of cost effectiveness and large-scale application. The benefits of biocathode MFCs in addressing these challenges are enormous and include, but are not limited to, low cost, reduced secondary pollution, environmental sustainability, uninterrupted operation, and so on. As a result, the development of a biocathode MFC system is aimed at improving the overall MFC performance. Researchers in the field of MFCs have discovered that the material used for biocathode fabrication is critical. These materials are primarily responsible for the biocathode's efficacy. This chapter has discussed some of the properties of biocathode materials. Modification of these materials can result in MFCs with variable energy output. These materials are modified in order to improve the interaction between the biocathode and the microbial biofilms. It is also critical to take into account the material's large surface area and porosity because they improve microbial adhesion. Adhesion is essential for microbe attachment in electron transfer processes. The material conductivity contributes to the improvement of the electron transport system. Biocathode MFCs, like other conventional MFCs systems, use either a direct electron transfer mechanism or a mediated electron transfer mechanism. Conclusively, the biocathode MFCs system is a promising approach for long-term sustainability and, as such, its application in energy generation enhancement merits further investigation.

References

[1] Yaqoob A A, Ibrahim M N M, Yaakop A S, Umar K and Ahmad A 2021 Modified graphene oxide anode: a bioinspired waste material for bioremediation of Pb^{2+} with energy generation through microbial fuel cells *Chem. Eng. J.* **417** 128052

[2] Rahimnejad M, Adhami A, Darvari S, Zirepour A and Oh S-E 2015 Microbial fuel cell as new technology for bioelectricity generation: a review *Alexandria Eng. J.* **54** 745–56

[3] Santoro C, Arbizzani C, Erable B and Ieropoulos I 2017 Microbial fuel cells: from fundamentals to applications. a review *J. Power Sources* **356** 225–44

[4] Zhao J, Li F, Cao Y, Zhang X, Chen T, Song H and Wang Z 2021 Microbial extracellular electron transfer and strategies for engineering electroactive microorganisms *Biotechnol. Adv.* **53** 107682

[5] Wei J, Liang P, Cao X and Huang X 2011 Use of inexpensive semicoke and activated carbon as biocathode in microbial fuel cells *Bioresour. Technol.* **102** 10431–5

[6] Freguia S, Rabaey K, Yuan Z and Keller J 2007 Non-catalyzed cathodic oxygen reduction at graphite granules in microbial fuel cells *Electrochim. Acta* **53** 598–603

[7] Clauwaert P, Van der Ha D, Boon N, Verbeken K, Verhaege M, Rabaey K and Verstraete W 2007 Open air biocathode enables effective electricity generation with microbial fuel cells *Env. Sci. Technol.* **41** 7564–9

[8] Lovley D R 2000 Fe (III) and Mn (IV) reduction *Environmental Microbe–Metal Interactions* (Hoboken, NJ: Wiley) pp 1–30

[9] Zhang Y, Sun J, Hu Y, Li S and Xu Q 2013 Carbon nanotube-coated stainless steel mesh for enhanced oxygen reduction in biocathode microbial fuel cells *J. Power Sources* **239** 169–74

[10] Wetser K, Sudirjo E, Buisman C J and Strik D P 2015 Electricity generation by a plant microbial fuel cell with an integrated oxygen reducing biocathode *Appl. Energy* **137** 151–7

[11] Liu S, Song H, Wei S, Yang F and Li X 2014 Bio-cathode materials evaluation and configuration optimization for power output of vertical subsurface flow constructed wetland—microbial fuel cell systems *Bioresour. Technol.* **166** 575–83

[12] Erable B, Lacroix R, Etcheverry L, Féron D, Délia M-L and Bergel A 2013 Marine floating microbial fuel cell involving aerobic biofilm on stainless steel cathodes *Bioresour. Technol.* **142** 510–6

[13] Cristiani P, Carvalho M, Guerrini E, Daghio M, Santoro C and Li B 2013 Cathodic and anodic biofilms in single chamber microbial fuel cells *Bioelectrochemistry* **92** 6–13

[14] Zhuang L, Yuan Y, Yang G and Zhou S 2012 *In situ* formation of graphene/biofilm composites for enhanced oxygen reduction in biocathode microbial fuel cells *Electrochem. Commun.* **21** 69–72

[15] Wang Z, Zheng Y, Xiao Y, Wu S, Wu Y, Yang Z and Zhao F 2013 Analysis of oxygen reduction and microbial community of air-diffusion biocathode in microbial fuel cells *Bioresour. Technol.* **144** 74–9

[16] Xie S, Liang P, Chen Y, Xia X and Huang X 2011 Simultaneous carbon and nitrogen removal using an oxic/anoxic-biocathode microbial fuel cells coupled system *Bioresour. Technol.* **102** 348–54

[17] Mani P, Fidal V T, Bowman K, Chandra T S, Keshavarz T and Kyazze G 2020 Development of an electroactive aerobic biocathode for microbial fuel cell applications *Env. Microbiol. Rep.* **12** 607–12

[18] Chen J, Hu Y, Tan X, Zhang L, Huang W and Sun J 2017 Enhanced performance of microbial fuel cell with *in situ* preparing dual graphene modified bioelectrode *Bioresour. Technol.* **241** 735–42

[19] He Z and Angenent L T 2006 Application of bacterial biocathodes in microbial fuel cells *Electroanalysis* **18** 2009–15

[20] Prathiba S, Kumar P S and Vo D-V N 2022 Recent advancements in microbial fuel cells: a review on its electron transfer mechanisms, microbial community, types of substrates and design for bio-electrochemical treatment *Chemosphere* **286** 131856

[21] Gregory K B, Bond D R and Lovley D R 2004 Graphite electrodes as electron donors for anaerobic respiration *Env. Microbiol.* **6** 596–604

[22] Rhoads A, Beyenal H and Lewandowski Z 2005 Microbial fuel cell using anaerobic respiration as an anodic reaction and biomineralized manganese as a cathodic reactant *Env. Sci. Technol.* **39** 4666–71

[23] Butler C S, Clauwaert P, Green S J, Verstraete W and Nerenberg R 2010 Bioelectrochemical perchlorate reduction in a microbial fuel cell *Env. Sci. Technol.* **44** 4685–91

[24] Slate A J, Whitehead K A, Brownson D A and Banks C E 2019 Microbial fuel cells: an overview of current technology *Renew. Sustain. Energy Rev.* **101** 60–81

[25] Milner E M, Popescu D, Curtis T, Head I M, Scott K and Eileen H Y 2016 Microbial fuel cells with highly active aerobic biocathodes *J. Power Sources* **324** 8–16

[26] Ling J, Xu Y, Lu C, Lai W, Xie G, Zheng L, Talawar M P, Du Q and Li G 2019 Enhancing stability of microalgae biocathode by a partially submerged carbon cloth electrode for bioenergy production from wastewater *Energies* **12** 3229

[27] Yaqoob A A, Ibrahim M N M and Rodríguez-Couto S 2020 Development and modification of materials to build cost-effective anodes for microbial fuel cells (MFCs): an overview *Biochem. Eng. J.* **164** 107779

[28] Li S, Cheng C and Thomas A 2017 Carbon-based microbial-fuel-cell electrodes: from conductive supports to active catalysts *Adv. Mater.* **29** 1602547

[29] Aryal N, Ammam F, Patil S A and Pant D 2017 An overview of cathode materials for microbial electrosynthesis of chemicals from carbon dioxide *Green Chem.* **19** 5748–60

[30] Katuri K P, Kalathil S, Ragab A a, Bian B, Alqahtani M F, Pant D and Saikaly P E 2018 Dual-function electrocatalytic and macroporous hollow-fiber cathode for converting waste streams to valuable resources using microbial electrochemical systems *Adv. Mater.* **30** 1707072

[31] Yaqoob A A, Mohamad Ibrahim M N, Rafatullah M, Chua Y S, Ahmad A and Umar K 2020 Recent advances in anodes for microbial fuel cells: an overview *Materials* **13** 2078

[32] Angioni S, Millia L, Mustarelli P, Doria E, Temporiti M, Mannucci B, Corana F and Quartarone E 2018 Photosynthetic microbial fuel cell with polybenzimidazole membrane: synergy between bacteria and algae for wastewater removal and biorefinery *Heliyon* **4** e00560

[33] Yaqoob A A, Khatoon A, Mohd Setapar S H, Umar K, Parveen T, Mohamad Ibrahim M N, Ahmad A and Rafatullah M 2020 Outlook on the role of microbial fuel cells in remediation of environmental pollutants with electricity generation *Catalysts* **10** 819

[34] Mao Y, Zhang L, Li D, Shi H, Liu Y and Cai L 2010 Power generation from a biocathode microbial fuel cell biocatalyzed by ferro/manganese-oxidizing bacteria *Electrochim. Acta* **55** 7804–8

[35] Rusli S F N, Bakar M H A, Loh K S and Mastar M S 2019 Review of high-performance biocathode using stainless steel and carbon-based materials in microbial fuel cell for electricity and water treatment *Int. J. Hydrogen Energy* **44** 30772–87

[36] Dumas C, Basseguy R and Bergel A 2008 Microbial electrocatalysis with *Geobacter sulfurreducens* biofilm on stainless steel cathodes *Electrochim. Acta* **53** 2494–500
[37] Wei J, Liang P and Huang X 2011 Recent progress in electrodes for microbial fuel cells *Bioresour. Technol.* **102** 9335–44
[38] Yaqoob A A, Ibrahim M N M and Guerrero-Barajas C 2021 Modern trend of anodes in microbial fuel cells (MFCs): an overview *Env. Technol. Innov.* **23** 101579
[39] Freguia S, Rabaey K, Yuan Z and Keller J 2008 Sequential anode–cathode configuration improves cathodic oxygen reduction and effluent quality of microbial fuel cells *Water Res.* **42** 1387–96
[40] Zhang Y, Sun J, Hu Y, Li S and Xu Q 2012 Bio-cathode materials evaluation in microbial fuel cells: a comparison of graphite felt, carbon paper and stainless steel mesh materials *Int. J. Hydrogen Energy* **37** 16935–42
[41] Zhang G, Zhao Q, Jiao Y, Wang K, Lee D-J and Ren N 2012 Efficient electricity generation from sewage sludge using biocathode microbial fuel cell *Water Res.* **46** 43–52
[42] Yaqoob A A, Ibrahim M N M and Umar K 2021 Electrode material as anode for improving the electrochemical performance of microbial fuel cells *Energy Storage Battery Systems—Fundamentals and Applications* (Rijeka: IntechOpen)
[43] Xie X, Criddle C and Cui Y 2015 Design and fabrication of bioelectrodes for microbial bioelectrochemical systems *Energy Env. Sci.* **8** 3418–41
[44] Wang X, Cheng S, Feng Y, Merrill M D, Saito T and Logan B E 2009 Use of carbon mesh anodes and the effect of different pretreatment methods on power production in microbial fuel cells *Env. Sci. Technol.* **43** 6870–4
[45] Gul M M and Ahmad K S 2019 Bioelectrochemical systems: sustainable bio-energy powerhouses *Biosens. Bioelectron.* **142** 111576
[46] Chen J, Hu Y, Huang W and Zhang L 2017 Enhanced electricity generation for biocathode microbial fuel cell by *in situ* microbial-induced reduction of graphene oxide and polarity reversion *Int. J. Hydrogen Energy* **42** 12574–82
[47] Liu X-W, Sun X-F, Huang Y-X, Sheng G-P, Wang S-G and Yu H-Q 2011 Carbon nanotube/chitosan nanocomposite as a biocompatible biocathode material to enhance the electricity generation of a microbial fuel cell *Energy Env. Sci.* **4** 1422–7
[48] Li J, Li Z, Xiao S, Fu Q, Kobayashi H, Zhang L, Liao Q and Zhu X 2020 Startup cathode potentials determine electron transfer behaviours of biocathodes catalysing CO_2 reduction to CH_4 in microbial electrosynthesis *J. CO_2 Utiliz.* **35** 169–75
[49] Sayed E T and Abdelkareem M A 2017 Yeast as a biocatalyst in microbial fuel cell *Old Yeasts—New Questions* (Rijeka: InTech) pp 41–65
[50] Song H-L, Zhu Y and Li J 2019 Electron transfer mechanisms, characteristics and applications of biological cathode microbial fuel cells—a mini review *Arabian J. Chem.* **12** 2236–43
[51] Hernández-Fernández F, De Los Ríos A P, Salar-García M, Ortiz-Martínez V, Lozano-Blanco L, Godínez C, Tomás-Alonso F and Quesada-Medina J 2015 Recent progress and perspectives in microbial fuel cells for bioenergy generation and wastewater treatment *Fuel Process. Technol.* **138** 284–97
[52] Chen G-W, Choi S-J, Cha J-H, Lee T-H and Kim C-W 2010 Microbial community dynamics and electron transfer of a biocathode in microbial fuel cells *Korean J. Chem. Eng.* **27** 1513–20

[53] Rosenbaum M, Aulenta F, Villano M and Angenent L T 2011 Cathodes as electron donors for microbial metabolism: which extracellular electron transfer mechanisms are involved? *Bioresour. Technol.* **102** 324–33
[54] Ieropoulos I A, Greenman J and Melhuish C 2013 Miniature microbial fuel cells and stacks for urine utilisation *Int. J. Hydrogen Energy* **38** 492–6
[55] Yang Y, Xu M, Guo J and Sun G 2012 Bacterial extracellular electron transfer in bioelectrochemical systems *Process Biochem.* **47** 1707–14
[56] Yang Y, Ding Y, Hu Y, Cao B, Rice S A, Kjelleberg S and Song H 2015 Enhancing bidirectional electron transfer of *Shewanella oneidensis* by a synthetic flavin pathway *ACS Synth. Biol.* **4** 815–23
[57] Verma J, Kumar D, Singh N, Katti S S and Shah Y T 2021 Electricigens and microbial fuel cells for bioremediation and bioenergy production: a review *Environ. Chem. Lett.* **19** 2091–126
[58] Yaqoob A A, Ibrahim M N M, Umar K, Parveen T, Ahmad A, Lokhat D and Setapar S H M 2021 A glimpse into the microbial fuel cells for wastewater treatment with energy generation *Desalin. Water Treat.* **214** 379–89
[59] Sharma M, Jain P, Varanasi J L, Lal B, Rodríguez J, Lema J M and Sarma P M 2013 Enhanced performance of sulfate reducing bacteria based biocathode using stainless steel mesh on activated carbon fabric electrode *Bioresour. Technol.* **150** 172–80
[60] Aelterman P, Rabaey K, Pham H T, Boon N and Verstraete W 2006 Continuous electricity generation at high voltages and currents using stacked microbial fuel cells *Env. Sci. Technol.* **40** 3388–94
[61] Al-Mamun A, Baawain M S, Egger F and Ng H Y 2017 Optimization of a baffled-reactor microbial fuel cell using autotrophic denitrifying bio-cathode for removing nitrogen and recovering electrical energy *Biochem. Eng. J.* **120** 93–102
[62] Xu L, Zhao Y, Wang X and Yu W 2018 Applying multiple bio-cathodes in constructed wetland-microbial fuel cell for promoting energy production and bioelectrical derived nitrification-denitrification process *Chem. Eng. J.* **344** 105–13
[63] Singh A and Kaushik A 2021 Removal of Cd and Ni with enhanced energy generation using biocathode microbial fuel cell: insights from molecular characterization of biofilm communities *J. Clean. Prod.* **315** 127940

Chapter 9

The effect of organic substrates on the generation of electrons in microbial fuel cells

Collins N Nwaokocha, Solomon O Giwa, Abayomi T Layeni, Oluwakemi S Fasinu, Modaser Momin, Sunday Oyedepo, Olusegun D Samuel and Mohsen Sharifpur

The emerging climate issues associated with fossil fuel derived energy use serve as a motivation to adopt sustainable energy sources. Energy remains a major socio-economic parameter for the developmental strides of nations, hence the search for eco-friendly solutions. Microbial fuel cells (MFCs) are a novel approach for generating clean energy from waste material. This pilot experiment examined the use of starch effluent from a Nigerian species of corn as substrate for a double-chamber fuel cell set-up. Iron and lead electrodes served as a means of electron transfer. The results (current, voltage, current density, and voltage density) present improved results compared to the open literature. This study proves viability for a sustainable off-grid energy source, in particular post-Covid-19.

9.1 Introduction

The global need for sustainable energy sources continues to gain prominence as a result of the associated challenges of pollution, climate change, and global warming resulting from the continuous use of petroleum based energy products [1–3]. Humans and their environment continue to be affected negatively by the resultant emission [4–6]. Population growth and technological advancements also increase energy demands, making the world even more energy dependent. Energy is still the major parameter for socio-economic growth [3]. Thus there is a need to develop green energy sources such as bio-hydrogen [7–9], biomass [3, 10], geothermal [11, 12], refuse derived fuel [3, 10, 13], solar dryers [14, 15], tidal energy [16, 17], wind energy [18], etc. To achieve a sustainable energy source that will rival and replace the capacity and use of conventional fossil fuel based energy resource, alternative energy hybrids will do a good job, such as biomass–solar hybrids, wind turbine–photovoltaic hybrids, solar–hydrogen fuel cells, photovoltaic–wind–tidal hybrids, wind turbine–photovoltaic–fuel cell hybrids,

doi:10.1088/978-0-7503-4791-4ch9

solar–wind hybrids, photovoltaic–wind–hydro–battery hybrids, etc [19–23]. Upgrading of grid systems is currently not feasible because of the high cost, hence off-grid hybrid power systems which incorporate non-conventional energy sources that are renewable, sustainable, and eco-friendly are a viable solution [1–3, 16, 19, 24, 25].

Wastewater is an effluent emanating from human activities, which could be domestic, communal, or industrial, with contaminant properties. Effluents lead to pollution and environmental issues. The disposal of effluent is cost-intensive, either on a domestic or industrial scale, such as sludge and wastewater treatments [1, 26, 27]. Hence, the need arises to determine the possibility of a sustainable and economically viable option, with microbial fuel cells (MFCs) being a novel clean energy harvesting method, comparable to traditional fuel cells. MFCs use the biocatalytic potential of viable micro-organisms by transforming the energy stored in the chemical bonds of biodegradable substrates to produce bio-energy [2, 28, 29]. MFCs make it possible to convert biodegradable materials directly into off-grid energy and concurrently treating effluent using the fuel cells [27, 28, 30], biofuel cell set-ups [27, 31, 32], bioenergy [29, 32], and cultures [1, 31]. Potter first pioneered the idea of an MFC in 1911 [30, 33, 34] and this was later confirmed by Cohen in 1931 [28, 30, 33–35]. The approaches used were electrode designs [36], electron transfer in biofuel cells [33, 34], bacterial cultures [34, 35], etc.

The US National Aeronautics and Space Administration (NASA) confirmed the viability of transforming domestic waste into bioenergy for use in space flight [37–39]. Over the years, energy has been generated from resources such as oil, gas, coal, etc, which release greenhouse gases (GHGs) and particulate matter (PM) which are harmful to human health, and these also have a firm link with global warming [3, 10, 40]. Energy sufficiency remains the key economic parameter for the development and growth of nations [3, 25]. Thus there arises the need for a novel method using MFCs to transform effluents into useful bioenergy. The use of the effluent of a Nigerian species of corn as substrate for MFCs in generating bio-energy is scarcely reported in the open literature. In addition to gain knowledge and provide a contribution to zero-carbon energy, an investigation was conducted to examine the viability of underused corn wastewater as substrate for generating bio-energy, adopting MFCs, as reported in this chapter. This experimental investigation was conducted in the Clean Energy Research Laboratory of the Mechanical Engineering Department, Olabisi Onabanjo University, Ago-Iwoye, Nigeria.

9.2 Materials and methods

9.2.1 MFC set-up

Effluent samples were sourced from a corn processing plant for this pilot experimental study. The plant is located within the Ibogun community where the College of Engineering and Environmental Studies Campus of Olabisi Onabanjo University is located in the Ifo Local Government Area of Ogun State, Nigeria. A two-chambered fuel cell set-up was used for a working volume of 1000 ml for both the anode and cathode, as depicted in figure 9.1. The two-chambered container was developed using two plastic containers with a diameter of 100.5 mm and a height of 120 mm. A salt bridge of PVC pipe was the connecting medium for the two

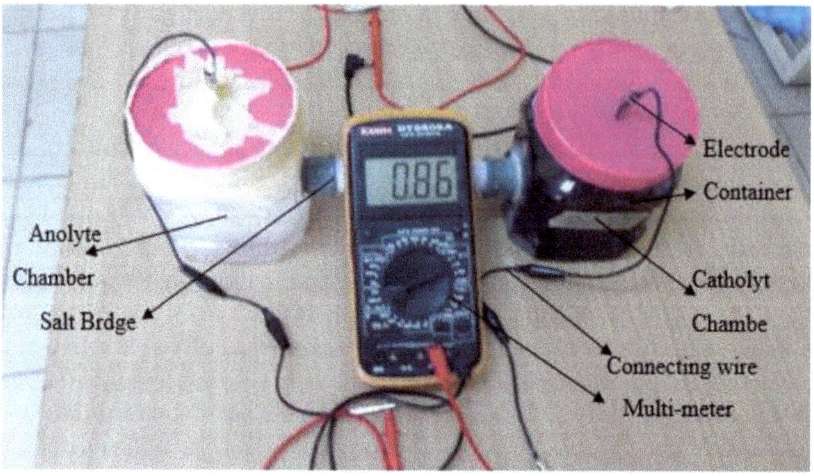

Figure 9.1. A pilot experimental set-up for a double chamber MFC.

chambers. A hole with a diameter of 25 mm and a of height 50.4 mm from the top of each container was drilled to position the salt bridge. In the anodic set-up, 1000 ml of the anodic inoculation was applied and the container was made air-proof. In the cathodic chamber, 1000 ml of potassium dichromate solution was applied. Pipe fittings were used to improve the proper sealing of the proton exchange membrane (PEM) casing to the plastic containers, and a digital multimeter (DT-9205A) was connected for measuring the output current voltages (OCVs).

9.2.2 Preparation of electrodes

Two experimental set-ups were designed using an iron electrode and a soldering lead electrode as anode and cathode, respectively, for each set-up. The electrodes were connected with copper wire for electron transfer between the cathode and anode. The electrodes are locally available and comparatively cheap. The electrodes (iron and soldering lead) were pre-treated in pure ethanol for 30 min and cleaned in 1 M HCl. Then the electrodes were recleaned in a 1 M sodium hydroxide solution for an hour to deactivate and remove inorganic contaminants. Thereafter, they were stored in distilled water preceding use. The choice of electrode material has significant effect on the performance of MFC set-ups.

9.2.3 Preparation of the effluent sample

The anodic chamber which contains 1000 ml of the prepared anodic inoculation (substrate and biocatalyst micro-organisms) was kept under anaerobic conditions. The cathodic compartment contained 1000 ml of potassium dichromate solution, which was also controlled under aerobic conditions with a salt bridge connecting the cathode chamber to the anodic chambers, thus allowing free ion movement. A digital multimeter (DT-9205A) was used for measuring the OCVs.

9.2.4 Preparation of the substrate

First, 85 g of wet starch was dissolved in 75 cm^3 of distilled water and heated to a temperature range of 80 °C–100 °C for 30 min under controlled stirring. A dried sample of sodium chloride of 5 g was then added to the mixture, with continued stirring for another 5 min to make for an even mixture. Sodium chloride was added to enhance the electrical conductivity of the wet starch, while the prepared hot gel was poured instantly into a PVC pipe 25 mm in diameter and 15 cm long, and later allowed to air-cool and set. Thus the PEM was ready for usage in the experimental MFC set-ups.

9.2.5 Preparation of the anolyte

First, 177.42 g of the substance was mixed with 1000 ml of buffer solution and kept airtight. The solution underwent controlled stirring to produce an even mixture, controlled at a fixed pH of 7.2 and a concentration of 177.40 g dm^{-3} buffer solution. The buffer solution was a mix of 4.6 g dm^{-3} dipotassium hydrogen photo-phosphate (KH_2PO_4) solution (base) and 2.70 g dm^{-3} potassium di-hydrogen photo-phosphate (KH_2PO_4) solution (acid), with 1000 cm^3 of each and 1000 ml of distilled water. The anode chamber was covered and made airtight to prevent atmospheric oxygen from coming into the chamber, allowing the bacteria to function in absolute anaerobic conditions.

9.2.6 Preparation of the catholyte

The catholyte solution contained 32.93 g dm^{-3} of potassium ferry cyanide ($K_3Fe(CN)_6$), a conducting material, and was poured into a completely aerated cathodic enclosure, as the cathode will require air with oxygen as the oxidant.

9.2.7 Experimental set-up and procedure

To prevent impurities in the constituents of the substrate, the wastewater used for the experiment was collected on the first day of the experiment and daily twelve-hourly readings were recorded for a designated period of 14 days.

For the set-up using an iron electrode, the readings were recorded for nine (9) days due to a corrosion effect leading to electrode breakage in the cathode chamber on the ninth day, so readings were paused. The set-up using the lead electrode was monitored for 14 days. The current and voltage readings were taken using a digital multimeter (DT-9205A). The values for voltage densities, current densities, and columbic efficiencies were estimated from the experimental datasets using the relevant equations. Tests for pH, chemical oxygen demand (COD), and biological oxygen demand (BOD) were carried out on the substrate.

9.2.8 Calculations

The readings of the measured current and voltage were adopted to estimate power density P (mW m^{-2}) using the equation [33]

$$\text{Power density, } P = \frac{\text{Current} \times \text{Voltage}}{\text{Surface area of the projected anode}}. \tag{9.1}$$

Current density C (mA m^{-2}) was estimated using [41]

$$\text{Current density, } C = \frac{\text{Current produced}}{\text{Surface area of projected anode}}. \tag{9.2}$$

Columbic efficiency E_c was determined using the following equations [38, 41]:

$$E_c = \frac{C_p}{C_n} \times 100, \tag{9.3}$$

where C_p is the total coulombic efficiency estimated by integrating the current over time, and C_n is the theoretical coulombic efficiency produced from the cell,

$$C_n = \frac{F \times b_{si} \times S_i}{M_i}, \tag{9.4}$$

where F, b_{si}, S_i, and M_i are the Faraday constant (96 485 coulombs/mol-electron), moles of electrons/moles of substrate, substrate concentration, and the molecular weight of the substrate (basically carbohydrate), respectively.

Table 9.1 shows the design parameters, while table 9.2 presents the characterization of the substrates.

Table 9.1. Design parameters for a double chamber MFC set-up.

Fuel cell configuration	Parameter
Capacity of cathode and anode chamber	1000 ml
Quantity of corn starch wastewater	1000 ml
Nature of electrode (cathode and anode)	Lead and iron
Mediator-cathode	Air
Mediator-anode	Nil
Projected electrode length	10.0 cm
Projected cathode surface area	0.001 138 m^2
Membrane	Wastewater
Effluent membrane	15 cm PVC pipe (25 mm diameter)

Table 9.2. Characterization of substrate.

	Iron electrode	Lead electrode
Relative humidity	90% ± 5%	90% ± 5%
Operating temperature	27 °C ± 2 °C	27 °C ± 2 °C
Operating pH	8.05	7.09
Biological oxygen demand (BOD)	13.33 mg l^{-1}	228.60 mg l^{-1}
Chemical oxygen demand (COD)	Nil	475.30 mg l^{-1}

9.3 Results and discussion

9.3.1 Characterization of substrate

The characterization of the substrate (corn effluent) was performed using standard methods. Laboratory readings for the characterization of the substrate are shown in table 9.2. The set-up with the iron electrode had a pH value of 8.05 at the start of the experiment. This value is higher than the pH 7.4 reported for domestic effluent [42] and a little less than the pH 8.3 measured for paper industry effluent [43]. The BOD was 13.330 mg l^{-1} and COD could not be measured due to inadequate reagents at the point of characterization.

The set-up with the lead electrode had a pH value of 7.09 prior to the experiment, which was less than that of the iron electrode set-up. This pH reading depicts the relatively neutral (basic) nature of this wastewater. The estimated BOD and COD readings for corn wastewater were slightly lower than the values of 290 mg l^{-1} (BOD) and 654 mg l^{-1} (COD) published for domestic wastewater [42]. The pH was also slightly lower compared to the pH of 8.00 obtained for tapioca wastewater produced using pure starch [44] and slightly lower than the pH of 8.3 reported for paper industry effluent [43]. The difference in the pH values may be due to the chemical composition of both wastewaters.

9.3.2 Current generation

For the MFC set-ups, the generated current at a stable ohmic resistance (100 Ω) was read with a digital multi-meter (DT-9205A). The MFC set-up was controlled at fixed internal resistance with the surrounding temperature not exceeding 27 °C ± 2 °C. For the MFC set-up with iron electrodes, an average daily current of 1.13 ± 0.03 mA was achieved, with a standard deviation of 0.20 mA. Minimum and maximum currents of 0.79 mA and 1.43 mA, respectively, were achieved in this investigation. Figure 9.2

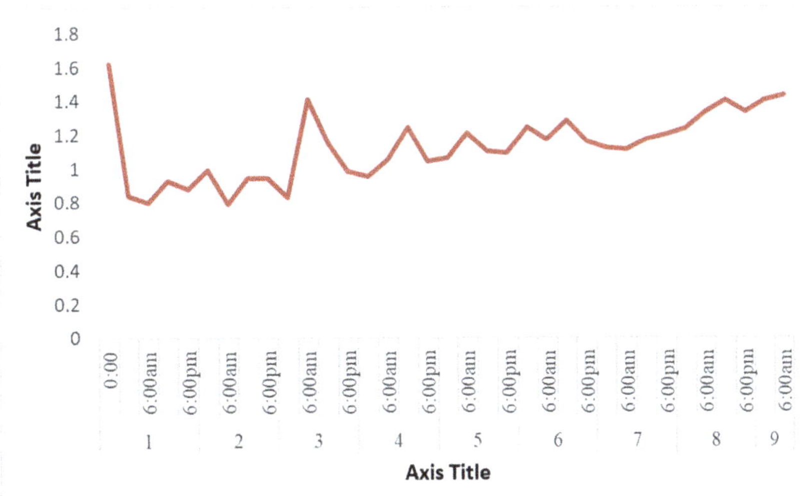

Figure 9.2. Plot of current against time (iron electrode).

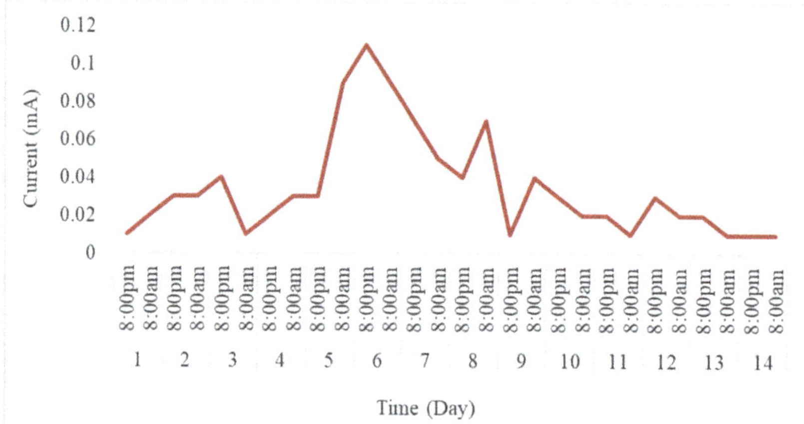

Figure 9.3. Plot of current against time (lead electrode).

depicts the graph of the average output currents for 9 days at 12 h intervals. The initial measured current was constant at 1.62 mA on the first day, but decreased later. The MFC average output current improved steadily from day 3 to the middle of day 9 before the iron electrode was damaged. The highest current obtained in this investigation is quite a bit higher than the highest current of 0.15 mA reported for poultry droppings [45].

For the MFC set-up with lead electrodes, figure 9.3 presents the daily average current output. The readings increased from day 1 to a maximum value on day 6, and thereafter began to decrease gradually until the designed day 14. The daily average current was $0.03 \pm 5.18 \times 10^{-3}$ mA with am SD of 2.74×10^{-2} mA. The highest current attained was 0.11 mA, which is a little lower than the highest current of 0.15 mA reported for animal-based waste [45].

9.3.3 Voltage generation

For the MFC set-up with iron electrodes, an average daily voltage of 0.84 ± 0.01 V was recorded, with a maximum reading of 0.97 V on day 2. Figure 9.4 shows a seemingly stable voltage for the duration of nine days of readings before the iron electrode was damaged. The voltage readings are remarkably high compared to the 0.203 V achieved for artificial wastewater [29], and a bit more than the highest voltage of 0.85 V obtained for potato extract effluent [46]. The voltage readings obtained in this investigation are greater when compared to previous investigations, probably due to the type of electrode used, the surface area of the electrode, etc.

For the MFC set-up with the lead electrodes, an average voltage of 0.67 ± 0.02 V was obtained for the 14 day experiment, shown in figure 9.5. The highest voltage of 0.78 V and lowest voltage of 0.42 V were achieved. The highest voltage recorded was slightly lower than the maximum voltage of 1.1 V recorded for a substrate of poultry droppings [45]. The graph of voltage with time in figure 9.5 shows an initial value of 0.74 V for day 1 and a steady pattern until day 7, thereafter there is an undulating

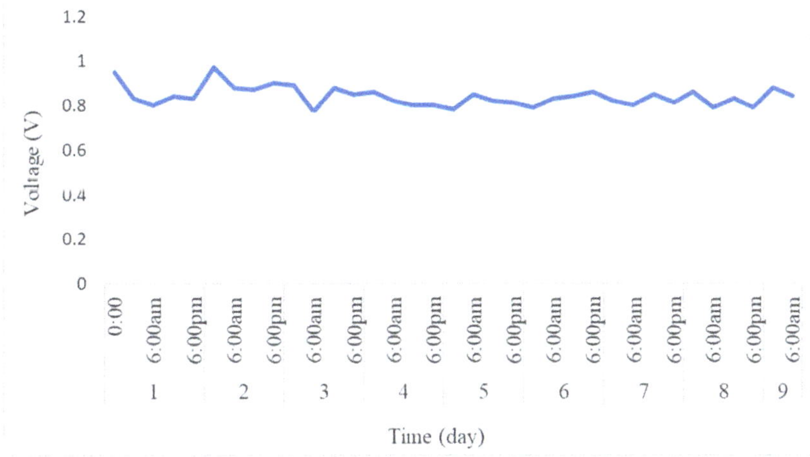

Figure 9.4. Plot of voltage against time (iron electrode).

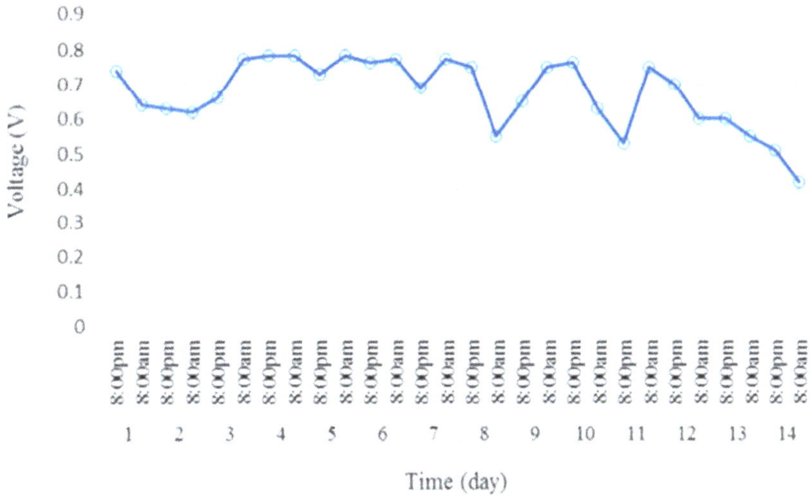

Figure 9.5. Plot of voltage against time (lead electrode).

trend until day 14. This could be attributed to the cell's internal stability and steady microbial activities.

9.3.4 Current and power densities

Current and power densities are estimated by dividing the values of current and power by the area of the projected anode. For the MFC set-up with iron electrodes, the evaluated daily mean of current and power density was 4.69 ± 0.15 mA cm^{-2} and 5.59 ± 0.17 mW cm^{-2}, respectively. The highest values of estimated current and power densities were 8.10 mA cm^{-2} and 7.70 mW cm^{-2}, respectively. The estimated current density is higher than the value of 0.0893 mA cm^{-2} obtained using cassava starch effluent as substrate in an MFC for an internal resistance of 120 Ω [47]. The obtained power density is considerably higher than the 0.0536 mW cm^{-2} for artificial wastewater [29]. Figure 9.6 shows a plot of current and power densities against time. Both parameters displayed a similar pattern throughout the duration of the experiment. This presents a sustainable energy solution for societal use, as advanced in the literature [3, 7, 10, 27, 32, 48, 49].

For the MFC set-up with lead electrodes, the evaluated current and power densities are shown in figure 9.7. Maximum current and power densities of 14.00 mA cm^{-2} and 10.46 mW cm^{-2}, respectively, were estimated. Both the evaluated current and power densities were observed to be higher than the current and power density values of 0.015 mA cm^{-2} and 0.005 mW cm^{-2}, respectively, as reported when oil refinery wastewater [50] was used as the substrate. Also, the estimated power density is considerably higher than that of fish market wastewater as a substrate in a range of 3.2–5.5 mW m^{-2} [51]. Figure 9.7 shows a similar sinusoidal pattern for the current

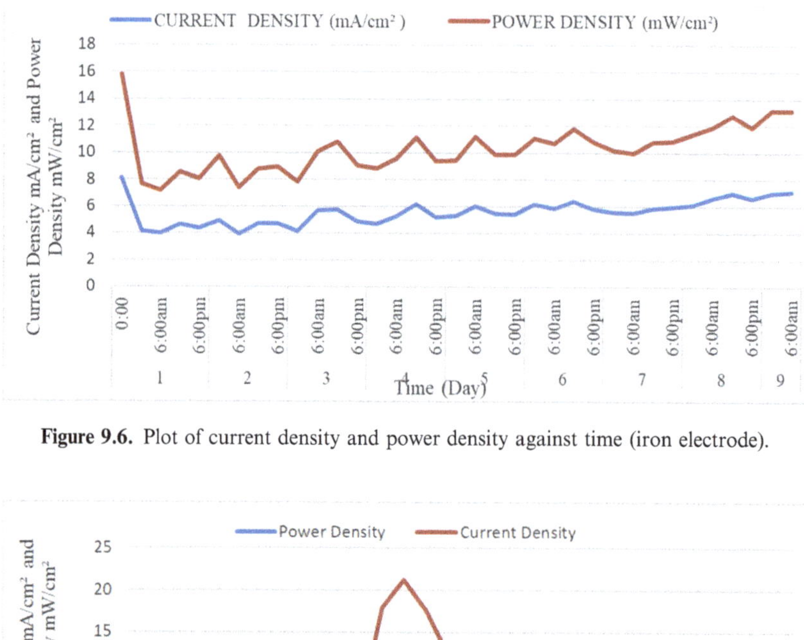

Figure 9.6. Plot of current density and power density against time (iron electrode).

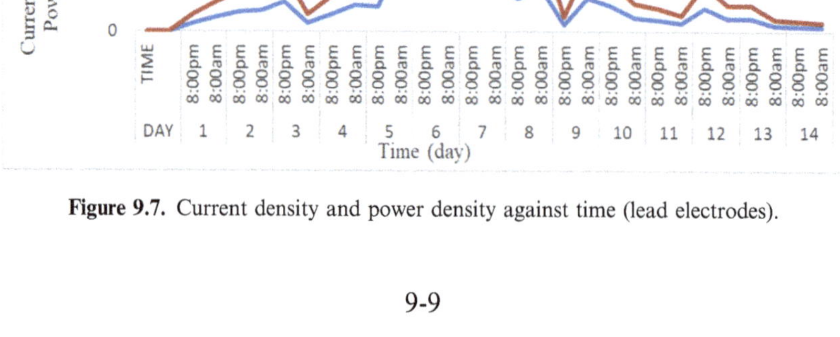

Figure 9.7. Current density and power density against time (lead electrodes).

Table 9.3. ANOVA—iron electrode MFC set-up.

Source of variation	SS	df	MS	$F_{observed}$	P-value	$F_{critical}$
Between groups	394.522	4	98.6306	23.8075	6.498E-15	2.4387
Within groups	559.282	135	4.1428			
Total	953.81	139				

Table 9.4. ANOVA—lead electrode MFC set-up.

Source of variation	SS	df	MS	$F_{observed}$	P-value	$F_{critical}$
Between groups	443.74	4	110.934	28.6507	3.078E-17	2.4387
Within groups	522.71	135	3.8719			
Total	966.45	139				

density and power density against time, with intermittent increases and decreases in the readings for both parameters.

9.3.5 Statistical analysis of data

The result for the analysis of variance (ANOVA) test of the iron electrode MFC set-up is depicted in table 9.3. The value of $F_{critical}$ (2.4387) is moderately less than that of $F_{observed}$ (23.8075), denoting that the data are independent of one another, i.e. statistically unequal. A P-value of $\ll 0.05$ (95% confidence limit) depicts a significant ANOVA result. This demonstrates that the datasets obtained are statistically dissimilar and quite significant for the experiment. The correlation existing between the voltage and current yielded −0.3846, which depicts a negative and weak relationship existing between the current and voltage.

The ANOVA test result for the lead electrode MFC set-up was carried out on the experimental dataset and is shown in table 9.4. The $F_{critical}$ of 2.4387 was observed to be lower than the $F_{observed}$ of 28.6507, which denotes a statistical independence between the data for current and voltage. This fact was confirmed by a P-value $\ll 0.05$ at a 95% confidence level, indicating the significance of the result. The correlation between voltage and current was 0.5717, which implies that a positive and moderately strong correlation exists between the current and voltage.

9.4 Conclusion

In this experimental investigation, the feasibility of corn wastewater as substrate for MFC technology for the generation of off-grid energy using iron and lead electrodes was examined. With technological advancements, it is imperative that our dependence on energy advances as well. The iron electrode MFC set-up produced a high current potential up to 1.62 mA (on the first day at a voltage of 0.95 V) with a daily average of 0.84 ± 0.01 V and 1.13 ± 0.03 mA. Comparatively high values for current, voltage, power density, and current density were obtained relative to the values in the open literature. The lead electrode MFC set-up showed considerable potential with a high voltage of about 0.78 V with a very low current potential ranging from 0.01 to 0.11 mA. It was observed that the principle aspect in an MFC for the removal of toxicity is the availability of biodegradable compounds present in the effluent sample. This investigation revealed the possibility of effluent treatment and simultaneous generation of off-grid energy using Nigerian corn effluent with the aid of MFC solution, which confirms a sustainable mode for obtaining eco-friendly, clean, and cheap energy, with focus on rural dwellers in developing nations at this time and post-Covid-19.

References

[1] Nwaokocha C N *et al* 2021 Microbial fuel cell: bio-energy production from Nigerian corn starch wastewater using iron electrodes *Mater. Today Proc.* **46** 5565–9

[2] Nwaokocha C *et al* 2021 Energy generation from corn starch effluent using microbial fuel cell using lead electrodes *Advances in Electrical and Electronics Engineering* ed S A Kale (Pune: Grinrey Publications) pp 87–96

[3] Nwaokocha C N *et al* 2019 The viability of refuse-derived fuel as a green household cooking fuel *Int. J. Recent Technol. Eng.* **8** 1031–4

[4] Giwa S, Sulaiman M and Nwaokocha C 2017 Inventory of greenhouse gases emissions from gasoline and diesel consumption in Nigeria *Nigerian J. Technol. Dev.* **14** 1–12

[5] Giwa Solomon O 2019 Noise and emission characterization of off-grid diesel-powered generators in Nigeria *Manag. Env. Qual.* **30** 783–802

[6] Giwa S O *et al* 2019 Gas flaring attendant impacts of criteria and particulate pollutants: a case of Niger Delta region of Nigeria *J.King Saud Univ. Eng. Sci.* **31** 209–17

[7] Giwa S *et al* 2017 Hydrogen production from alternative aqueous sources: a feasibility study *Int. J. Technol.* **8** 867–77

[8] Chen Y, Yin Y and Wang J 2022 Biohydrogen production using macroalgal biomass of *Laminaria japonica* pretreated by gamma irradiation as substrate *Fuel* **309** 122179

[9] Fu Q *et al* 2021 Towards hydrogen production from waste activated sludge: principles, challenges and perspectives *Renew. Sustain. Energy Rev.* **135** 110283

[10] Nwaokocha C N and Giwa S O 2016 Investigation of bio-waste as alternative fuel for cooking *Proc. of the 3rd Int. Conf. on African Development Issues (CU-ICADI 2016) (Covenant University, Ota, Nigeria, May 9–11, 2016)* 548–51

[11] Hu Y, Cheng H and Tao S 2021 Opportunity and challenges in large-scale geothermal energy exploitation in China *Crit. Rev. Env. Sci. Technol.* **52** 3813–34

[12] Lund J W and Boyd T L 2016 Direct utilization of geothermal energy 2015 worldwide review *Geothermics* **60** 66–93

[13] Yang Y *et al* 2021 Gasification of refuse-derived fuel from municipal solid waste for energy production: a review *Env. Chem. Lett.* **19** 2127–40

[14] Nwaokocha C N *et al* 2017 Performance evaluation of a domestic passive solar food dryer *Renew. Energy Syst.* 67

[15] Nwaokocha C N *et al* 2018 Solar energy and food preservation in Nigeria *Advances in Renewable Energy and Sustainable Systems* (New York: Nova Science) pp 97–105

[16] Nwaokocha C N, Waheed M A and Layeni A T 2015 Estimation of tidal power potential of Nigeria coastal area *Renewable Energy and Sustainable Development* (New York: Nova Science) pp 163–71
[17] Chowdhury M *et al* 2021 Current trends and prospects of tidal energy technology *Env. Dev. Sust.* **23** 8179–94
[18] Sadorsky P 2021 Wind energy for sustainable development: driving factors and future outlook *J. Clean. Prod.* **289** 125779
[19] Debnath D and Ray S 2022 Hybrid energy system for an academic institution: a case study *Renewable Energy Optimization, Planning and Control* (Berlin: Springer) pp 31–9
[20] Hosseinalizadeh R *et al* 2016 Economic sizing of a hybrid (PV–WT–FC) renewable energy system (HRES) for stand-alone usages by an optimization-simulation model: case study of Iran *Renew. Sustain. Energy Rev.* **54** 139–50
[21] Alagar K and Thirumal S 2021 Standalone PV-wind-DG-battery hybrid energy system for zero energy buildings in smart city Coimbatore, India *Advanced Controllers for Smart Cities: An Industry 4.0 Perspective* (Cham: Springer) pp 55–63
[22] Sharma C, Singh M P and Chaudhary S 2021 Cost analysis of solar-wind based hybrid renewable energy system *Advances in Automation, Signal Processing, Instrumentation, and Control* (Berlin: Springer) pp 2733–46
[23] Karthick A *et al* 2021 Optimization of PV–wind hybrid renewable energy system for health care buildings in smart city *Hybrid Renewable Energy Systems* ed U Sahoo (Hoboken, NJ: Wiley) pp 213–28
[24] Olatomiwa L *et al* 2016 Energy management strategies in hybrid renewable energy systems: a review *Renew. Sustain. Energy Rev.* **62** 821–35
[25] Nwaokocha C and Oyedepo S O 2018 The potentials of minimal hydropower for sustainable energy development in Nigeria *Advances in Energy Research* ed M J Acosta (New York: Nova Science) pp 55–91
[26] Ali A *et al* 2019 Comparison between conventional and modified microbial fuel cell for wastewater treatment and electricity generation *Int. J. Env. Sci. Technol.* **16** 8141–50
[27] Suresh R *et al* 2022 Current advances in microbial fuel cell technology toward removal of organic contaminants—a review *Chemosphere* **287** 132186
[28] Slate A J *et al* 2019 Microbial fuel cells: an overview of current technology *Renew. Sustain. Energy Rev.* **101** 60–81
[29] Du Z, Li H and Gu T 2007 A state of the art review on microbial fuel cells: a promising technology for wastewater treatment and bioenergy *Biotechnol. Adv.* **25** 464–82
[30] Liu H, Ramnarayanan R and Logan B E 2004 Production of electricity during wastewater treatment using a single chamber microbial fuel cell *Env. Sci. Technol.* **38** 2281–5
[31] Rahimnejad M *et al* 2009 Low voltage power generation in a biofuel cell using anaerobic cultures *World Appl. Sci. J.* **6** 1585–8
[32] Hoang A T *et al* 2022 Microbial fuel cells for bioelectricity production from waste as sustainable prospect of future energy sector *Chemosphere* **287** 132285
[33] Rabaey K *et al* 2005 Microbial phenazine production enhances electron transfer in biofuel cells *Env. Sci. Technol.* **39** 3401–8
[34] Logan B E 2008 *Microbial Fuel Cells* (New York: Wiley)
[35] Cohen B 1931 The bacterial culture as an electrical half-cell *J. Bacteriol.* **21** 18–9
[36] Park D H and Zeikus J G 2003 Improved fuel cell and electrode designs for producing electricity from microbial degradation *Biotechnol. Bioeng.* **81** 348–55

[37] Canfield J, Goldner B and Lutwack R 1963 Research on applied biochemistry *NASA Technical Report* Magna Corporation, Anaheim, CA

[38] Schröder U 2011 Discover the possibilities: microbial bioelectrochemical systems and the revival of a 100-year-old discovery *J. Solid State Electrochem.* **15** 1481–6

[39] Pal M, Giri R and Sharma R K 2022 Production of biofuels in a microbial electrochemical reactor *Biofuels and Bioenergy* (Amsterdam: Elsevier) pp 303–19

[40] Nwaokocha C N *et al* 2016 An experimental investigation of LPG as an alternative fuel for a retrofitted spark ignition engine *Int. J. Energy, Env. Econ.* **24** 121

[41] Liu H, Cheng S and Logan B E 2005 Production of electricity from acetate or butyrate using a single-chamber microbial fuel cell *Env. Sci. Technol.* **39** 658–62

[42] Mahendra B and Mahavarkar S 2013 Treatment of wastewater and electricity generation using microbial fuel cell technology *Int. J. Res. Eng. Technol.* **1** 277–82

[43] Mathuriya A S and Sharma V 2010 Bioelectricity production from various wastewaters through microbial fuel cell technology *J. Biochem. Technol.* **2** 133–7

[44] Rachma R M *et al* 2010 Electricity generation from tapioca wastewater using a microbial fuel cell (MFC) *Southeast Asian Water Env.* **4** 115

[45] Oladejo D *et al* 2015 Evaluation of electricity generation from animal based wastes in a microbial fuel cell *Int. J. Sci. Technol. Res.* **4** 85–90

[46] Herrero-Hernández E, Smith T and Akid R 2013 Electricity generation from wastewaters with starch as carbon source using a mediatorless microbial fuel cell *Biosens. Bioelectron.* **39** 194–8

[47] Lu N *et al* 2009 Electricity generation from starch processing wastewater using microbial fuel cell technology *Biochem. Eng. J.* **43** 246–51

[48] Dwivedi G and Sharma M 2014 Impact of cold flow properties of biodiesel on engine performance *Renew. Sustain. Energy Rev.* **31** 650–6

[49] Verma P, Sharma M and Dwivedi G 2016 Impact of alcohol on biodiesel production and properties *Renew. Sustain. Energy Rev.* **56** 319–33

[50] Majumder D *et al* 2014 Electricity generation and wastewater treatment of oil refinery in microbial fuel cells using *Pseudomonas putida Int. J. Mol. Sci.* **15** 16772–86

[51] Lee C-Y, Chen J-H and Cai Y-Y 2010 Bioelectricity generation and organic removal in microbial fuel cells used for treatment of wastewater from fish-market *J. Env. Eng. Manag.* **20** 173–80

IOP Publishing

Microbial Fuel Cells
Emerging trends in electrochemical applications
Mohamad Nasir Mohamad Ibrahim, Asim Ali Yaqoob and Akil Ahmad

Chapter 10

The effects of operating factors on the electrochemical performance of microbial fuel cells

Ansa Kanwal, Affia Siddique and Showkat Ahmad Bhawani

Microbial fuel cells (MFCs) are revolutionary devices that generate power through the oxidation of organic molecules in a substrate. It is critical to comprehend the factors that affect MFCs in order to obtain the desired results. The substrate, micro-organisms and their metabolism, electron transfer mechanism in the anodic chamber, electrode material and shape, type of membrane, operating conditions such as temperature, pH, and salinity, electron acceptor in the cathodic chamber, and geometric design of MFCs are considered to be the most important factors among the numerous factors affecting MFC performance. Among the many substrates, wastewater is a sustainable, rich medium that MFCs can address. Exoelectrogenic bacteria, which can oxidize organic materials and transfer electrons to the anode without the use of mediators, are found in wastewater. Optimal pH and temperature, as for other microbial systems, promote bacterial growth, which can increase MFC efficiency. Despite the fact that salt inhibits microbial growth, increased salinity and ionic strength can increase substrate conductivity and hence improve MFC effectiveness. The scaling up of MFCs is a contentious topic that necessitates a thorough grasp of these issues. A more cost-effective design for scalable wastewater treatment and high power production may be realized by employing novel affordable materials for the electrodes and membranes in MFC manufacture. MFCs are also a viable option for the bioremediation of polluted groundwater. In the current chapter, these parameters and their influence on MFC performance are studied.

10.1 Introduction

The availability of clean water is now a global challenge, and it is increasing day by day with the increase in the world population, climate change, and a higher demand

doi:10.1088/978-0-7503-4791-4ch10

for clean water [1]. At the moment, water treatment through the activated sludge process is considered to be the most successful, common, and widely applied biological treatment technology. The process is mostly concerned with the treatment of organic waste using aerobic micro-organisms to improve the quality of water [2]. Membrane technology is a relatively new technique that can be utilized in water treatment. The use of non-renewable energy resources such as fossil fuels makes this technique expensive and harmful to the environment due to the release of greenhouse gases during burning. Thus it is essential to develop an energy and cost-efficient process for water treatment to meet this global challenge effectively [2]. An innovative type of bioreactor with the potential to meet the requirements in this case is the microbial fuel cell (MFCs). Unfortunately, the future of the said technology is uncertain in terms of its efficiency and development at the industrial level [3]. The major benefit of MFCs is the decrease in the cost of energy for wastewater treatment. Logan claimed that if all of the energy is retrieved as electricity, a population of 100 000 people that produces around 16.4×10^6 (l/year) could potentially run a 2.3 MW power plant from wastewater treatment [4]. However, MFCs have shown a low power density, so the more feasible power from a wastewater treatment plant is 0.5 mW [4].

A standard MFC is made up of two chambers separated by a proton exchange membrane with an anaerobic anodic chamber and aerobic cathodic chamber [5]. The treatment of wastewater takes place in the anodic chamber, and electricity generation during the process of aerobic treatment of wastewater using biocathodes has been achieved in some cases [6, 7]. The removal of oxygen from the cathode chamber and the application of an additional small voltage to the set-up results in the production of hydrogen gas at cathode. This type of biofuel cell is known as a bioelectrochemically assisted microbial reactor [8]. A microbial desalination cell (MDC) is a new type of bioelectrochemical system that has recently been devised by inserting two membranes between the cathode and anode in MFCs [9, 10]. The voltage generated by the bacteria limits the desalination performance of MDCs. In order to desalinate saline water and concurrently produce hydrogen gas, in some instances microbial electrodialysis cells have been set up [11]. The application of MCDs results in the simultaneous process of saline water desalination, the generation of electricity, and wastewater treatment [12]. This chapter reviews MFCs and summarizes the effects of the operating parameters on their performance for wastewater treatment and efficient electricity generation.

10.2 The history and evolution of MFCs

Galvani was the first to reveal the correlation between biology and electricity [13]. Potter was able to use platinum electrodes to generate electricity from *Escherichia coli* in 1910 [14]. An initial result of electricity generation was documented by Cohen who reported the generation of 35 V fusing serial MFCs [15]. Even though this potential application of micro-organisms was known, the idea of using micro-organisms as catalysts in fuel cells was not assessed until the 1970s [16]. In 1960 the

USA space program developed an interest in MFCs for the establishment of a power generation and waste management system in space flight [17]. In 1991 MFCs for domestic wastewater treatment were introduced. It was demonstrated that waste-water treatment can be carried out using MFCs along with power generation [18]. Some of the latest developments in MFCs have been reviewed in various reports [5, 16]. Few reports classify biofuel cells as microbially and enzymatically catalyzed cells [17, 27]. In 2005 Vestraete and Rebaey published a review of bacterial metabolism in MFCs [16]. Rebaey and Keller reviewed some of the latest improvements in MFC cathode research [19]. In 2005 Logan *et al* reported various techniques and materials used for the manufacture of MFCs and methods to assess their efficiency [4].

10.3 Performance affecting factors for MFCs

The chemical energy stored in an organic matter substrate is converted to electrical energy using MFCs [20]. This conversion occurs when bacteria in the anode compartment oxidize the organic substrate (oxygen, sulphur, or nitrate species acting as electron donors) and yield protons and electrons. The anode absorbs the electrons and transfers the electrons to the cathodes via a power user or resistor in order to reduce the electron acceptors at the cathode [5, 16]. In order to balance the charge density cations, ideally protons, move from the anode to the cathode compartment via an ion selective membrane such as a salt bridge or a proton exchange membrane (PEM). On the other hand, water is formed by the combination of oxygen and protons in the cathode chamber [19]. Carbon dioxide is produced in the anaerobic anode compartment as the oxidation product. A diagrammatic representation of a two-chamber MFC is presented in figure 10.1.

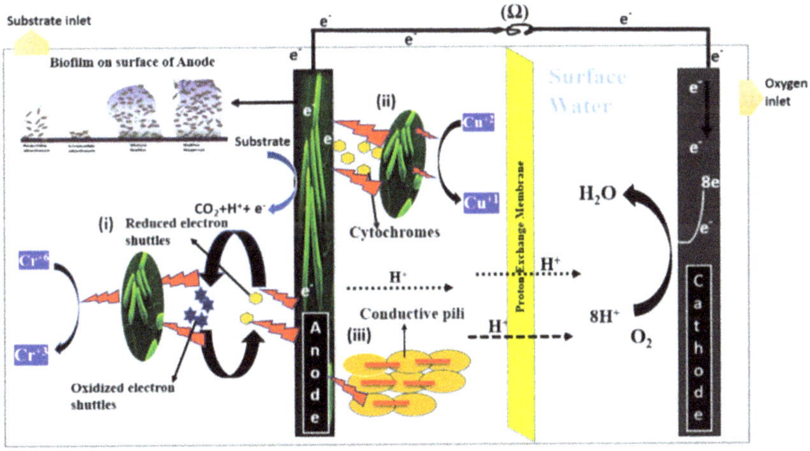

Figure 10.1. Diagrammatic representation of two-chamber MFCs. (Reproduced with permission from [20]. Copyright 2021 Elsevier.)

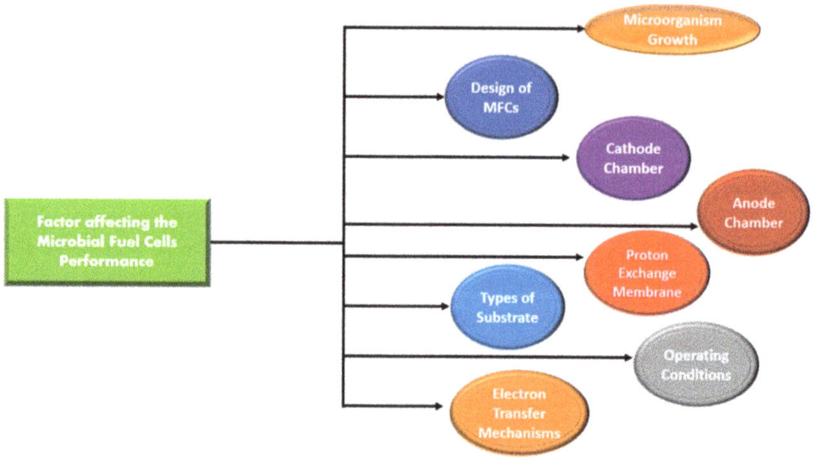

Figure 10.2. Parameters affecting the efficiency of MFCs.

There are numerous factors that affect the efficiency of MFCs and the resultant power generation during the process of wastewater treatment. The recognition and consideration of these conditions is very important in order to develop an efficient MFC. In the anodic chamber, the micro-organisms are significant because of their metabolic activities and the mediators they use to transfer electrons to the anode. Different substrates can be oxidized by micro-organisms and act as a source of electron donors in MFCs. Electricity generation can be influenced greatly by the working parameters, such as the temperature, pH, ionic strength of the medium, and the material and technique used to design the anode, cathode, and membranes used. Figure 10.2 presents a summary of these parameters.

10.4 Mechanism of electron transfer

Electron mediators or shuttles should be used to transport electrons produced during the activity of micro-organisms in the anode compartment. In some MFCs the micro-organisms used (such as *E. coli* and *Saccharomyces*) have an outer non-conductive membrane of lipids, lipopolysaccharides, and peptidoglycans that delay the transfer of electrons to the anode [5]. Dyes such as neutral red, methylene blue, methyl viologen, thionine, and humic acid are generally applied as mediators for electron transfer to the anode [5, 21]. These dyes must possess some properties in order to be used as mediators, such as the potential to penetrate the bacterial membrane and access the reductive species in bacteria and be reduced by the metabolic activity of micro-organisms. The redox potential of the reductive metabolite and dyes should match [22]. The mediator must not interact with the other substances present in the bacteria. The mediator in its reduced form must diffuse out of the cell membrane easily and pass to the anode where it will transfer the electrons to the anode and become oxidized. The electrochemical kinetics of this oxidation process at the anode should be very fast [22]. In the electrolyte solution, the mediator must be chemically stable in either its reduced or oxidized state. It must also stay in the medium and should not be adsorbed on the anode or bacterial surface [23]. In some cases, a higher efficiency is reported with the application of a mixture of two mediators in the medium [24]. In one case, anthraquinone-2-sulphonate and riboflavin were used as mediators in an MFC. The set-up was investigated for the biodegradation of petroleum mixture treated with anaerobic sludge. The assessment of this set-up indicated a 30-fold increase in the power generation capacity with riboflavin as the mediator as compared to the mediatorless MFC set-up [25]. Most mediators are quite expensive and some of the water-soluble phenolic compounds are also toxic so, despite the benefits of mediators in MFCs, they may lead to the development of an expensive and not-so-environmentally-friendly set-up. As a result of this, the industrialization of MFCs appears to be challenging [26]. Some of the bacteria can potentially use their specific metabolites, known as endogenous mediators or pillis, and nanowires to transport the electrons to the anode. These particular kinds of bacteria are known as electrogenesis bacteria. Fe(III)-reducing micro-organisms have been discovered to be electrochemically active due to the presence of cytochromes in their outer membranes. These micro-organisms can be applied in mediator free MFCs [27, 28]. *Shewanella* and *Geobacter* species have been reported to produce nanowires which show a highly conductive nature and can develop biofilms on the surface of the anode [29]. Also, the thermophilic fermentative bacteria and oxygenic photosynthetic cyanobacteria can develop electrically conductive nanowires [29]. A comparison of the electron transfer mechanism and similar working conditions in three types of MFCs was carried out by Ieropoulos *et al.* They reported a higher efficiency for MFCs developed with anodophillic *Geobacter sulfurreducens* without a soluble mediator and the sulphate-reducing species *Desulfovibrio desulfuricans* using the intrinsic mediating character of sulphate/sulphide, compared to the efficiency of MFC based on *E. coli* with synthetic redox mediators [30]. This showed that the application of electrogenesis bacteria was more favourable. Figure 10.1 presents a schematic view of these mechanisms.

10.5 Cell potential and microbial metabolism

The primary factor in defining the cell potential is the metabolic pathway of the micro-organisms and the resultant anode potential. In MFCs, the catabolic activities of bacteria act as a rate limiting process [31]. The oxidation of organic matter provides energy to the heterotrophic organisms. Because of the presence of exogenous oxidants or the external terminal electron acceptors, fermentation and the respiratory chain are two significant metabolic channels that occur in the anode chamber [5, 16, 32]. In respiratory chain metabolic process, the electrons are released by the oxidation of the substrate and transported to the external terminal electron acceptor through a redox system. With a particular substrate acting as the electron donor, the higher redox potential of the external terminal electron acceptor produces a higher energy gain for the organism [32],

$$G^{\varphi} = nF[E^{\varphi}(\text{substrate}) - E^{\varphi}(\text{anode})], \tag{10.1}$$

where E^{φ} is the biological standard potential and G^{φ} indicates the free energy change under standard biological conditions [32].

The electrical potential of MFCs is determined by the potential difference of the two electrodes. The electrode potential at the anode is determined by a complex respiratory chain mechanism that transfers the electrons to the anode through the activity of micro-organisms on the organic substrate. As different micro-organisms show different respiratory chain mechanisms for the transfer of electrons and a change in the growth conditions for a particular strain also changes the resultant voltage, it is not possible to determine the ideal cell voltage [5]. Anaerobic respiration produces most of the energy but in MFCs reactions in the anode chamber are carried out by particular anaerobic micro-organisms that use other extracellular, organic, or inorganic substances as the terminal acceptors of electrons in order to carry out anaerobic respiration. Generally, these terminal electron acceptors are sulphates, nitrates, metal ions (Fe^{3+}), or fumarate [32].

Iron and manganese based respiratory process in *Geobacter* and *Shewanella* species are found to be the processes for the transfer of electrons to the cell surface [33]. The most noteworthy oxidation mechanism of an organic substrate to generate carbon dioxide is the Fe(III) reduction [34]. Activated sludge in an MFC contains some 3% of the dissimilatory iron-reducing strains as one of the many exoelectrogenic bacteria [35, 36]. The dispersion of iron and manganese in groundwater, aquatic sediments, and submerged soil are mainly affected by the dissimilatory Fe (III) and Mn(IV) reduction process. As a result, these MFCs can be utilized in areas with increased organic flux, such as beneath deep rivers and other water bodies where there are suitable anaerobic conditions and organic substrate for the oxidation in the MFCs. Furthermore, iron-reducing bacteria can potentially remove odours from swine wastewater [35].

Fermentation is the primary process taking place in the absence of sulphates or nitrates as electron acceptors. Only 7%–8% of the glucose energy content is used by the micro-organisms during the process of fermentation, resulting in the production of acetate or butyrate [37]. This partial oxidation of the substrate results in the loss of major energy content [16, 32]. In 2011 Rismani-Yazdi *et al* [38] reported the dominance of fermentation over anaerobic respiration with cellulose being used as the organic substrate. As a conclusion, it was suggested that a fermentable substrate is suitable for anaerobic MFC while nonfermentable substrates are effectively oxidized in anaerobic MFCs. Chen *et al* [39] demonstrated that the anaerobic MFCs had a higher voltage than the aerobic MFCs. This was attributed to the presence of more dissolved oxygen in the anode compartment that results in the production of water and decreases the quantity of electrons being transferred to the cathode [39, 40].

10.6 Micro-organisms

In biofuel cells, both enzymes and micro-organisms can be applied but using micro-organisms offers multiple enzymes, and thus it can be used to oxidize multiple substrates or mixtures of various substrates. High purity enzymes can be applied for particular and characterized reactions; these types of biofuel cells are being devised concurrently with biosensor technology [17]. As discussed previously, two types of bacterial strains have been applied in MFCs: bacteria which require a mediator and bacteria that are categorized as mediatorless bacteria. *E. coli* [30, 41], *Proteus mirabilis* [42], *Saccharomyces cerevisiae* [33], *Actinobacillus succinogenes* [32], *Pseudomonas fluorescens*, and *Proteus vulgaris* need a mediator to work, while *Rhodoferax ferrireducens*, a family of *Geobacteraceae* and *Geobacter* [30, 43], *D. desulfuricans* [30], and *Shewanella putrefaciens* [27] do not need any mediators, and they have received more interest recently [26, 29, 43]. Kim *et al* [27] reported electrochemical activities for an anaerobically cultured strain of *S. putrefaciens*, whereas no activity was recorded for aerobically grown *S. putrefaciens* cells. Either pure or mixed cultures of bacteria can be inoculated in MFCs. Mixed cultures, because of their stress resistance and adaptability to various nutrients, outperform pure cultures [45]. The use of a vibrant and diversified bacterial source such as activated sludge, wastewater, sediments, or soil in mediatorless MFCs are quite beneficial in wastewater treatment. This is because of the increased power density and variety of bacteria present in the medium including electrogenic bacteria [46]. The exoelectrogenic bacteria found in wastewater have the ability to oxidize the organic substrate and electron transfer to the anode [47]. MFCs require at least one electrogenic bacterium in the medium to oxidize the organic matter and convert it to electricity. The use of these bacteria in MFCs is more beneficial due to the increased columbic efficiency and longer sustainability which saves the energy required for growth and maintenance in the electron transfer to the anode. The preliminary oxidation of the organic substrate is carried out by methanogenic or fermentative micro-organisms when mixed cultures are in use. Later, the presence of at least one electricigen will be necessary to retrieve the electrons from the product of the activity of nonelectricigens and attain higher electron recoveries to generate electricity [46]. Mixed cultures that have been applied in many studies as inoculum include domestic wastewater, activated sludge, and anaerobic sludge [48, 49]. Yates *et al* [50] assessed the use of swamp sediments and wastewater as inoculums in MFCs. They reported that the application of swamp sediments generated higher energy compared to wastewater inoculated MFCs in the first 20 cycles but after that the power density and voltage were almost similar. The presence of methanogens, fermentative bacteria, and sulphate-reducing bacteria in anaerobic sewage sludge makes it a better choice as inoculum for MFCs. Repeated observation of methane generation with anaerobic sludge as the inoculum indicated that the micro-organisms used the anaerobic mechanism of oxidative metabolism [37, 65]. Certain enrichment approaches were recommended in such cases to decrease the growth of methanogens in the anode chamber while increasing the growth of electrochemically active bacteria [37]. Mathuriya reported that waste removal and power generation are affected greatly by the type of inoculum used in MFCs [46]. Mathuriya compared the activated sludge, self-microbial flora from plain wastewater, and mixed and individual electrogenic micro-organisms taken from a self-tannery environment. The report concluded a higher energy production for the mixed electrogenic strain sample along with an increased efficiency of waste removal [46]. The most commonly used micro-organisms in the anode chamber of MFCs are *Proteobacteria* [40, 51]. *Clostridium* and *Propionibacterium* were reported as the most common bacteria for two-chamber MFCs when soil was used as the inoculum [52]. Moreover, they also reported the detection of some common bacterial strains correlated to *Pseudomonas* in the first week of activity, which implied that these micro-organisms acted as oxygen scavengers [52]. Rismani-Yazdi *et al* used ruminal micro-organisms as inoculums for the purpose of power generation from cellulose [51]. They applied *Sedimentibacter* and *Clostridium* as major strains in biofilms connected to an anode with a comparable sequence to that of *Ruminococcus* and *Desulfotomaculum*, whereas *Comamonas* were abundant in suspended consortia [51]. Chen *et al* [40] deduced that *Delftia acidovorans, Citrobacter freundii*, and *Ochrobactrum intermedium* are potentially novel bioelectricity generating species from proteobacteria species. According to an analysis of the microbial community for MFCs, *Thiobacillus* was prominent in biofilms attached to the anode while sulphate-reducing strains such as *Desulfobacteraceae* were abundant in suspended micro-organisms in the treatment of wastewater containing a high sulphate content [53]. Nor *et al* [54] reported a higher maximum power generation and current density with inoculation of a pure culture of *Pseudomonas aeruginosa* in a double chambered MFC membrane in comparison to the application of anaerobic palm oil mill effluent sludge as the MFC inoculum.

10.7 Substrate

A wide range of substrates have been used in MFCs ranging from simple, pure compounds to mixtures of complex organic matter [26]. Pure compounds such as glucose [34, 48, 55], butyrate [56], cellulose, proteins [39, 51, 57], acetate [49, 56, 63], lactate [52], glycine, glycerol [40], and cysteine [58] have been applied. Acetate is a nonfermentable substance and effective electron donor used for iron-reducing bacteria that produces up to 66% more power compared to butyrate [56]. Among the various media that can be treated with MFCs, wastewater is a sustainable and rich medium. Various reports have documented the effective power generation from complex organic wastewater such as tannery wastewater [46], municipal wastewater [59], dairy wastewater [60], swine wastewater [61], rice mill wastewater [62], chemical wastewater [63], azo dye [64, 65], brewery wastewater [66], molasses wastewater [67], sulphide-rich wastewater [68], winery wastewater [69], slaughterhouse wastewater [70], cassava mill wastewater [71], refinery wastewater [67], food waste leachate [72], landfill leachate [73], and solid waste such as rice straw [74]. Simultaneous removal of nitrate and sulphide from sulphate rich wastewater and synthetic wastewater can be achieved through MFCs [75, 76]. Marine sediments and soil are organic matter that are made up of extremely complex and heterogenous mixtures that can be effectively applied for electricity generation [67, 77]. MFCs, acting as an inexhaustible source of terminal electron acceptors, can be effectively applied in the biotreatment of groundwater pollutants and aquatic sediment [78]. MFCs have been utilized successfully for the treatment of diesel range organic hydrocarbons [78], hydrocarbon-contaminated sediments [79], nitroaromatic compounds [80], ammonium- and benzene-contaminated groundwater [81], alkanes, chloroethane, pentachlorophenol, pyridine, phenol, and indoles [82]. The power generation capacity of the MFCs depends on the nature of the substrate and the construction of MFCs along with certain other conditions that cannot be compared. Another important factor that must be considered is substrate concentration. Jiang and Li reported a boost in power generation form 0.2 up to 1.2 W m^{-3} by increasing the concentration of the substrate from 100 to 850 mg l^{-1}. However, no change in the output was recorded at the higher concentration of 1000–1500 mg l^{-1} [83].

10.8 Anode

The most significant properties of the anode material include the stability, conductivity, surface area, biocompatibility, and non-corrosive nature. Furthermore, the techniques for the design and development of the anode and MFC architecture also have a considerable effect on the efficiency of the MFC. Copper, despite having very good conductivity, cannot be used as an anode because of its toxicity for micro-organisms. On the other hand, carbon is a suitable material for bacterial adhesion although it has low conductivity. Carbon is available as fibres, carbon felt, foam, cloth, and paper [27, 84]. Very little power (2 mW m^{-2}) is generated with a copper anode compared to carbon cloth (CC; 880 mW m^{-2}) or stainless-steel (12 mW m^{-2}) anodes. This can be attributed to the increased specific surface area of the CC when compared to the metal mesh [85]. Graphite plates or rods are more efficient for use as anodes because of their ease-of-handling, cost effectiveness, and specific surface area [27]. When we compare other materials for their use as anodes in MFCs, CCs show a higher porosity and mechanical strength, but they are economically unviable because of their higher cost [60]. Graphene based electrodes have gained a lot of attention recently as a promising material due to their active electrolyte–electrode–microbe interaction, higher specific area, and efficient conductive nature [86]. The use of ammonia gas treatment with a CC anode increased the charge density on the electrode surface and resulted in an improved MFC performance. The power generation was improved by 20% compared to the untreated anode [49]. The core brush of a graphite fibre brush is made up of a non-corrosive metal such as titanium or stainless steel. Because of their configuration, they offer a higher surface area and efficient electron terminal, making them suitable for a wide range uses in MFCs [84, 87–89]. Granular activated carbon (GAC), which is a low-cost packing material, is widely utilized for water and wastewater treatment because of its higher surface area, and is also a potentially suitable material to make anodes for MFCs [83, 90]. It can be applied in packed or fluidized bed conditions coupled with anode current terminal. In the case of a fluidized bed, the power generation density was 951 mW m^{-2}, while it was 813 mW m^{-2} in the case of a packed bed and about 525 mW m^{-2} without GAC particles and no stirring [90]. A 34% increase in power generation (form 1020 to 1370 mW m^{-2}) was recorded when carbon fibre material was treated with a simple acid and heat was used [88]. Saito *et al* [91] reported that the application of modified CC anodes with 4(*N*,*N*-dimethylamino)benzene diazonium tetrafluoroborate improves the MFC efficiency as it provides increased nitrogenous functional groups at the surface of the anode. Some of the other techniques that can be applied to increase the nitrogen content of the electrode surface include heating, or ammonia or diazonium treatment that basically increase the adhesive capacity of the electrode surface towards bacteria without any adverse effect on the electron transfer to the surface or microbial activity [88, 89, 91]. When an anode made up of carbon felt coated with manganese dioxide was applied, it produced 24.7% higher power density and better anode conductivity compared to pure carbon felt as the anode [92]. A stainless-steel mesh coated with graphite was used as an anode by Mardanpour *et al* as it is cost effective and offers a better porosity for the formation of biofilms [60]. According to Jiang and Li, GAC is a better alternative to CC [83]. An innovative category of nanomaterials is carbon nanotubes (CNTs) that offer exceptional structural and electrical characteristics for their application in fuel cells [93]. However, they demonstrate a microbial toxicity and can hinder growth which leads to cell death. Hence a customized CNT–polyaniline (PANI) composite was a potential alternative material for the anode that improved the power generation capacity of MFCs [93]. Ci *et al* demonstrated a simple technique to modify the anode material made up of nitrogen-doped CNTs using bamboo-like nanostructures (bamboo-NCNTs). The efficiency of these NCNTs was significantly higher than those of non-doped CNTs and CC [94]. Zou *et al* [95] used a CNT composite coated with polypyrrole (PPy) as an anode material and reported an improved efficiency compared to plain carbon paper. A considerable increase of 38% in the power generation capacity of an MFC was recorded with PPy–CNTs composites in comparison to plain carbon fibre. In contrast to an unmodified CC anode, a polypyrrole-coated polyacrylonitrile–CNT nanofibre anode material demonstrated the highest power generation capacity of an MFC by 40% [96]. A comparison of four different anode materials was carried out by Heijne *et al* [97]. They reported a decrease in the activity of the anode the an order graphite > Pt-coated titanium > flat graphite > uncoated titanium [97]. Ganesh and Jambeck investigated the activity of a biochar anode and activated carbon (AC) anode in batch cycles with landfill sludge as the substrate and no inoculum added. A considerably higher carbon oxygen demand (COD) removal (74.7% ± 5.5%) was recorded for the AC anode [97]. Similarly, another comparison for electrode materials was performed by Chen *et al* [42] where they compared the activity of CC, aluminium-alloy mesh composite carbon cloth (AAMCC), and aluminium-alloy mesh (AAM). The appearance of an electrical double layer in the AAMCC offered the highest adhesion properties for the surface and the smallest resistance. Also, the power generation capacity of an MFC with AAMCC was found to be higher than any of the others consisting of AAM or CC electrodes, i.e. 1497 and 132.4 times, respectively. Wu *et al* [98] demonstrated that the CNT–Au–TiO_2 electrode is a best suited anode for the MFCs as it has the benefit of properties of nanocomposites, such as better adhesions of microbes to the surface, higher specified surface area, and higher conductivity. However, different types of electrode materials are presented in figure 10.3.

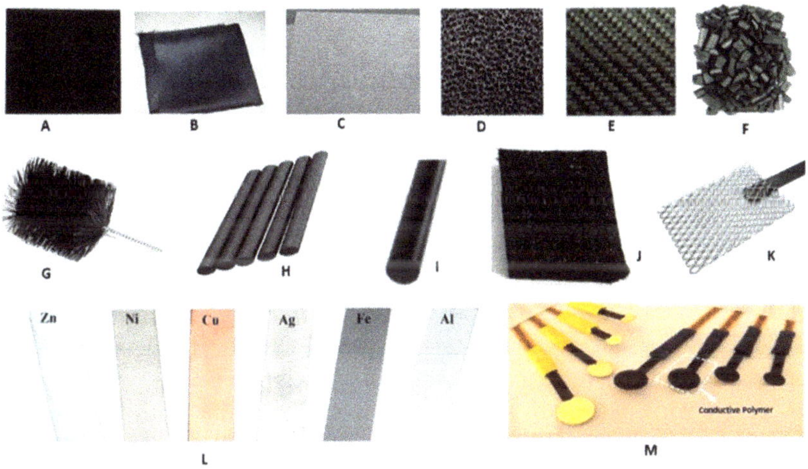

Figure 10.3. Commonly used electrodes: (a) carbon paper, (b) carbon cloth, (c) carbon fibre, (d) reticulated vitrified carbon, (e) carbon mesh, (f) graphitic granular, (g) carbon brushes, (h) graphite rod, (i) polycrystalline graphite, (j) carbon felt, (k) platinum mesh, (l) different metal electrode strips, and (m) conductive polymer-based strips. (Reproduced with permission from [24]. MDPI CC-BY 4.0.)

10.9 Catholyte and cathode

10.9.1 Catholyte

In the cathode compartment there is a cathode, electron acceptor, and a buffer solution (sometimes) present. Species that can act as electron acceptors in MFCs include ferric iron, hexacyanoferrate [99], oxygen, ferricyanide [99, 101], nitrobenzene, manganese, and H_2O_2. Among all these electron acceptors, oxygen is most sustainable, easily accessible, and eco-friendly electron acceptor for practical applications. The electrolyte in the cathode compartment of an MFC must be oxygenated in order to deliver oxygen to cathode [102–104]. However, in terms of both substrate removal and power generation, a ferricyanide catholyte has a higher efficiency than an oxygenated catholyte. This could be due to the benefit of the efficient mass transfer process taking place in the case of a ferricyanide catholyte [105]. In the case of single-chamber MFCs, one side is immersed in the anolyte while the other is in air. Hence, single chambered cathodes are quite practical for use in MFCs owing to their higher power generation capacities, simple configuration, sustainability, and cost effectiveness due to the exclusion of the aeration process [103–106].

10.9.2 Cathode

The material and strategies used to construct the cathode have a significant effect on the efficiency of MFCs. Most commonly, platinum electrodes are applied as a catalyst for the oxidation reduction. Although copper electrodes outperform carbon and graphite electrodes, they are very costly and unsustainable due to their poisoning of the microbial solution when used as aerated biocathodes [6, 107]. Using platinum as the electrode material elevates the cost of the MFCs so the costs of MFC construction can be reduced by reducing the need for Pt. In certain cases, commercially developed cathodes with a Pt loading of 0.5 mg cm^{-2} were used [48, 103]. It was reported that 0.1 mg cm^{-2} Pt loading in the electrolyte of an air cathode does not have any effect on the power density of the MFC. Santoro *et al* reported that at the start of the process, a higher power generation capacity was recorded with Pt-based cathodes (330 mW m^{-2}) compared to the Pt-free cathodes (253mW m^{-2}), however, this contrast disappeared after several weeks of working [108]. Some of the other, less expensive materials that can be used as alternatives to platinum are nitrogen functionalized CNTs, carbon materials [109], AAMCC, AAM [42], mesoporous nitrogen-rich carbon [110], and co-tetramethyl phenyl porphyrin [48, 111]. In order to avoid leakage of solution from an open air cathode, it should be hydrophobic and have a high capacity to diffuse oxygen at the open air end. However, on the water side, it must be hydrophilic [112]. The Pt-free cathode, made up of hydrophobic CC (a gas diffusive layer) in open air with a microporous layer in the solution was also tested by Santoro *et al* [108]. Zhang *et al* [113] suggested that a low-cost AC air cathode can be applied instead of a Pt-catalyzed electrode for the purpose of oxygen reduction in MFCs. Cold pressing of AC and a polytetrafluoro-ethylene (PTFE) binder and subsequent pressing on a stainless-steel or nickel mesh was performed to develop an AC air cathode. PTFE is used as a binder alternative to Nafion (perfluoro-sulphonate) with Pt to reduce the cost. These cathodes demonstrate comparable or mildly increased maximum power generation capacity in comparison to CC and pt/C catalyst cathodes [113, 114]. Unfortunately, the formation of a thick biofilm on the catalytic surface of air cathodes is reported to decrease the efficiency and stability of air cathodes in MFCs. Yao *et al* [115] integrated a broad-spectrum fluoroquinolone antibiotic known as enrofloxacin in the catalytic layer of an AC air cathode to prevent the growth of biofilms on the cathode. Another important parameter that greatly affects the power generation capacity is cathode surface area [87, 116]. Lou *et al* [104] demonstrated an increased power generation by doubling the size of the anode and cathode in domestic wastewater treatment. The power-up was about 62% for the cathode and 12% for the anode.

10.9.3 Biocathodes

Lately, there has been some research on biological materials to be used as cathodes that are cost-effective and durable alternatives to the synthetic cathodes that reduce the oxygen reduction potential with micro-organisms [107, 116]. Various micro-organisms and enzymes can work as biological catalysts for oxygen reduction. Some of the micro-organisms can act as catalysts with certain mediators such as chromium [117, 118] and manganese [119]. The biocathodes can be utilized to remove water pollutants such as NO_3, 2-chlorophenol, chloroethenes, ClO_4, Cr (VI), and U(VI) in some cases [118]. Clauwaert *et al* [153] conducted a study where an MFC with both a biocathode and bioanode performed concurrent degradation of an organic substrate. This MFC was also applied to completely denitrify the substrate without depending on an external power source or hydrogen formation. Oxygen mass transfer and charge transfer limits the activity of biocathodes. In MFCs the transfer of oxygen in a dissolved form as electron acceptor is a major limitation. It diffuses from the batch solution to the diffusion layer and afterwards diffuses to a stagnated water layer connected with the biofilm and finally out of the biofilm itself. Oxygen transfer across the layer connected to the batch solution can be reduced by reducing the layer thickness and it can be achieved by increasing the linear flow rate [116]. Another factor that limits the efficiency of biocathodes is the transfer of electrons from the surface of the cathode to the micro-organisms [118].

10.10 Membrane

In two-chamber MFCs, a salt bridge or PEM is present between the anode and cathode compartments to split the two chambers and avoid oxygen diffusion to the anode chamber [84]. A PEM is most commonly applied because it is selectively permeable for protons and demonstrates higher electrical conductivity [5, 120]. Some cheaper and more efficient composite membranes have been used recently as alternatives to Nafion. Recently, a few innovative separators have been evaluated for their use in MFCs, including disulphonated poly(arylene ether sulphone) [121], Nafion–activated carbon nanofibre composite, Nafion–carbon nanofibre composite [122], earthen ceramic [62], and sulphonated poly(ether ether ketone) (SPEEK) in poly(ether sulphone) (PES) membranes with different compositions of SPEEK [123]. Single-chamber MFCs can also work with air cathodes containing ultrafiltration (UF) and microfiltration membranes of varying molecular weight cut-offs such as 1K (UF-1K), 5K (UF-5K), and 10K (UF-10K) [99]. Another substitute for PEM is nanocomposite membranes as they are relatively cheaper, show higher resistance towards fouling, and are capable of proton transfer [124]. Applying the same working conditions, Rahimnejad *et al* recorded a maximum power density of 20 mW m^{-2} and a current of 148 mA m^{-2} with a novel synthetic Fe_3O_4–PES nanocomposite membrane compared to Nafion that generated 15.4 mW m^{-2} and 112 mA m^{-2}, respectively [124]. An even more improved performance was recorded for a titanium dioxide (TiO_2)-quaternized poly(vinyl alcohol) (QAPVA) hybrid anion exchange membrane compared with Nafion in a study carried out by Tao *et al* [125]. In some studies, glass fibre and glass wool were used as membranes and displayed a very good performance with reduced electrode spacing and also lowered the cost of the membranes [84, 126]. Glass fibre, being a non-biodegradable material, exhibits lower ohmic resistance, higher proton transfer, and a lower oxygen transfer coefficient [89, 127]. The efficiency of glass fibre was also compared with a textile separator prepared using 54% polyester and 46% cellulose in a single-chamber air cathode. The textile separator displayed a maximum voltage of 0.65 V whereas it was just 0.18 V for glass fibre due to the short circuiting between the cathode and anode [128]. Sevda *et al* [129] compared two other membranes, called Fumasep and Zirfon, for their efficiency as separators in an air cathode MFC. The highest power density recorded for Fumasep was

38.03 mW m^{-2} while for Zirfon it was 424.50 mW m^{-2}. Philamore *et al* [130] worked with two new materials, Tangoplus 3D-printed acrylic-based photopolymer and natural rubber latex as ion exchange membranes in an MFC. They provided a rapid, cost-effective fabrication and substantially increased the ion exchange surface area compared to the currently used membranes made up of Nafion [130]. Some MFCs, such as single-chamber MFCs (SCMFCs) and sediment MFCs, do not require any membrane. Without the PEM, oxygen transfers to the anode chamber at a higher rate, resulting in increased aerobic bacterial growth in the anode compartment [27, 103].

10.11 Working conditions for MFCs

The efficiency of MFCs is mainly affected by the factors that control the bacterial growth, such as pH, temperature, salinity, and ionic strength. Generally, MFCs work well in moderate reaction conditions such as atmospheric pressure and temperature and neutral pH [21].

10.11.1 pH and salinity

The optimal pH for bacterial growth is around neutral pH at 7, but in some cases the pH of the wastewater is acidic and greatly influences the MFC performance. Rice mill wastewater at a pH around 3–4 was used as an organic substrate by Behera *et al.* They reported that a weakly basic condition of around 7.5 pH was favoured for increased power generation and decomposition of organic substrate [62]. The lower concentration of protons at neutral pH increases the internal resistance of the cell compared to the chemical fuel cells that work in basic or acidic conditions [130]. Increasing the ionic strength of the medium can increase the conductivity of the solution and reduced the internal resistance without affecting the pH of the solution [131]. MFCs can be effectively used to treat the around 5% of the world's wastewater that is identified as highly saline wastewater. Higher salinity results in higher power generation as conductivity is improved. Improved conductivity promotes the proton transfer and hence reduces the internal resistance of the MFCs [132]. Lefebvre *et al* [132] also reported that the addition of 20 g l^{-1} of NaCl improved the efficiency of the MFC as it decreased the internal resistance by 33% and increased the highest power generation capacity by 30%. However, a higher concentration of NaCl had an adverse effect on the efficiency of MFCs. The efficiency of MFCs for power generation and substrate treatment improved with an increase in the salinity of the medium up to 1.5% w/v, but it decreased four-fold when there was a 2.5% w/v increase in salinity [26].

10.11.2 Temperature

The temperature of the medium affects the rate of Pt-catalyzed oxygen reduction at the cathode and proton transfer across the solution. It also influences the bacterial growth in the medium. As a result, temperature has a considerable impact on the performance of MFCs [131]. Generally, MFCs are tested for their operation at room temperature or a slightly higher temperature of around 20 °C–35 °C.

It is reported that in the lower temperature range of 4 °C–30 °C, a higher start-up time is required for repeated power cycles and better efficiency. At a low temperature of 15 °C or below, no considerable power generation was recorded even after working for a month. However, starting MFC operation at a 30 °C temperature, it is possible to achieve repeated cycles of power generation even at the two lowest temperature points of 4 °C and 10 °C [63]. A two-fold increase in the highest power generation capacity of MFCs was recorded at 40 °C compared to 30 °C. However, an increase in temperature to 50 °C decreased the power generation capacity four-fold [26].

10.11.3 MFC architecture

The system architecture is one of the major elements that determine the power generation capacity of MFCs [9]. Du *et al* [5] and Logan *et al* [26] have outlined some of the MFC designs published up to 2007. Generally, MFCs consist of two chambers in the typical 'H' shape, but they generate less power density [26]. The output of the MFC can be changed by altering the material type and construction techniques of the electrodes, the PEM applied in two-chambered MFCs, and the surface area, etc. The complex model of two-chamber MFCs offers challenges in the upgrade of MFCs but these are still a more common topic of investigation than any other type. It is important to seal the anode chamber to avoid the diffusion of oxygen that may result in the decrease of MFC performance [20]. Logan probed the impact of the size of electrodes in two-chambered MFCs that contain PEMs with three varying surface areas. They demonstrated that the surface area of the PEMs limits the power generation capacity of the MFCs by elevating the internal resistance in cases where the PEMs had a smaller surface area compared to that of the anode and cathode [133]. There are a few models for such cases where different shaped compartments can be used for MFCs. Most MFCs operate in batch, fed-batch, or continuous modes with the purpose of wastewater treatment rather than power generation. In the case of continuous mode configuration, a novel fuel cell called an up-flow microbial fuel cell (UMFC) was constructed by incorporating the benefits of an up-flow anaerobic sludge bioreactor (UASB) system and dual chamber MFCs. This fuel cell produced a maximum power density of 0.17 W m^{-2} [99]. Another UMFC system containing an anode chamber with a U-shaped cathode inside the anode was investigated. It demonstrated a maximum volumetric power density of 29.2 W m^{-3} when using a volumetric loading rate of 3.40 kg COD/(m^3 day) [134]. Construction of a dual anode chambered MFC is achieved by placing two anode chambers with a common air cathode in an electrically parallel and geometrically stacked pattern [135]. This kind of MFC demonstrated maximum power generation of approximately 24 W m^{-3} that is similar to that of single electrode (anode and cathode) chambered MFCs. However, the difference is in the maximum current generation, where dual anode chambered MFCs displayed a 3.66 mA (at 50 Ω) current which is twice the 1.69 mA (at 100 Ω) current for a single anode chambered MFC working at almost identical volumetric current density [135]. The cathode chamber could be eliminated by revealing the cathode in a single-chamber MFC to air. It consists of only an anode chamber and also aeration is no longer required [103, 136]. Rabaey *et al* [101] designed a single chambered, tubular, continuous MFC that contained granular graphite matter as the anode and ferricyanide solution as the cathode. Some other types of SCMFCs are low-cost MFCs as they are membraneless. These MFCs, however, can potentially lose the substrate because of the aerobic oxidation in the anode chamber by the activity of bacteria [103]. Liu *et al* [137] designed a cylindrical SCMFC made up of eight graphite anodes and one air cathode. They operated this fuel cell in continuous mode for wastewater treatment. The result was the generation of a maximum power of 26 mW m^{-2} and removal of up to 80% of wastewater COD. Another significant parameter that influences the power generation capacity of MFCs is the spacing of the electrodes [83, 131, 138]. In a separator electrode assembly (SEA), there should be less distance between the two electrodes and there must be separators placed between the two electrodes to avoid direct contact of the electrodes which can cause short circuiting [67, 89, 127, 139]. In this way, the internal resistance is reduced, and the capacity of power generation is enhanced. However, in the case of SEAMFC, where closely spaced electrodes are used, oxygen can crossover to the anode from the cathode where it can affect and reduce the power production capabilities of exoelectrogenic bacteria [139, 140]. The negative effect of oxygen crossover on the power generation capacity can be reduced by using thick anodes as it will retain the anaerobic conditions in the anode compartment [141]. Ahn and Logan designed another SEAMFC with several graphite fibre brush anodes and one air cathode working in continuous and fed-batch mode [128]. Liao *et al* [142] recorded an elevated mass transfer and subsequently a higher power density of up to 1.4 times higher when a tubular rotating carbon brush anode was applied in a cell compared to those with a non-rotating anode. Jiang and Li [83] established a granular activated carbon single-chamber MFC (GAC-SCMFC) system that displayed higher and more stable power generation in comparison with the trivial dual chamber MFCs. The effect of spacing among the electrodes and the concentration of granular activated carbon was also assessed. It showed that a decrease in the distance between the electrodes decreases the internal resistance. With an increase in the concentration of the GAC, the capacity of power generation was elevated because of the higher quantity of appended biomass. Several studies associated with biocathodes have already been stated here, but the majority of the MFCs applied in these reports are traditional two-chambered MFCs. An innovative integrated MFC–membrane bioreactor system was developed by Wang *et al* where the aeration tank of the bioreactor served as the cathode chamber [143]. A stacked MFC that can be used for continuous mode operation is a parallel series of MFCs [144, 154]. The currents and voltages of the system can be increased by using this type of MFC. A different kind of MFC is the sediment MFC that generates electrical power through the bioremediation of sediment and polluted groundwater. In this type of MFC the cathode is kept in aerobic seawater above and the anode is implanted in sediment and the two electrodes are electrically connected [145].

10.12 The latest MFC designs for wastewater treatment

Although there have been various studies on MFCs, only a few works are applied for significant wastewater treatment at a large scale. The design of MFCs is one of the major impediments to their integration in industrial scale applications. As discussed earlier, other factors of significant importance in MFC construction are the material and configuration of electrodes, the separation membrane between the two chambers, and the spacing between the electrodes. Power generation capacity and wastewater treatment are also influenced by operating parameters such as temperature, pH, mediators, salinity, and micro-organisms. Jiang *et al* [145] developed a multi-anode/cathode microbial fuel cell (MAC-MFC) consisting of 12 anodes/cathodes for power generation through wastewater treatment. The MAC-MFC was made up of a square cathode box at the top and a cylindrical anode chamber at the bottom and had a liquid volume of 20 l. In this MFC, an increase in the number of electrodes led to improved power generation. It also demonstrated a higher biological to chemical oxygen demand (COD/BOD) removal efficiency (more than 70%) at hydraulic retention times (HRT) of 10 h and 20 h. Traditional aerobic and aerobic techniques of wastewater treatment can also be applied in combination with MFCs. Zhang *et al* [67] established an inclusive set-up of an up-flow anaerobic sludge blanket reactor–MFC–biological aerated filter (UASB–MFC–BAF) for electrical power generation and simultaneous treatment of molasses wastewater. The COD removal and sulphate reduction reaction occurred in the UASB reactor unit whereas the energy production occurred in the MFC unit via oxidation of the generated sulphide. The degradation of phenol derivatives and removal of coloured substances was the responsibility of the BAF unit. The treatment of high strength molasses wastewater of 127 500 mg l^{-1} COD resulted in the maximum power density of 1410.2 mW m^{-2} and 53.2% COD removal efficiency. A 90 l prototype MFC with five conveniently stacked modules was designed for brewery wastewater treatment. This design had minimum spacing between the electrodes and a double layered textile separator which was inserted between the cathode and anode to avoid short circuiting. The anode was made up of six carbon brushes fabricated on titanium wire. The overall efficiencies for COD removal and suspended solid were 87.6% and 86.3%, respectively [145]. A comparison of the treatment for chemical wastewater using a bioelectrochemical treatment (BET) system and anoxic treatment (AxT) was devised. The results showed COD removal efficiency of 47% and 90% for AxT and BET, respectively [91].

10.13 Future aspects

MFCs are an excellent choice for the treatment of harmful wastewater, such as wastewater from industries containing toxic substances and heavy metal ions. Saline wastewater can also be treated using MFCs depending on their adjustments with the salinity of the anode compartment. Hence MFCs could be a prospective method for the treatment of all types of wastewaters. MFCs do not produce a significant amount of energy and the main outcome of the process is wastewater treatment. It is critical to consider all the discussed parameters in the correct dimensions to attain the desired results for power generation and wastewater treatment [3]. If the main purpose of the process is power generation, then parameters can be adjusted to enhance the cell potential such as the application of engineered bacteria to elevate the rate of bacterial metabolic activity. Combining electricigen bacteria with simple substrates such as acetate, as well as taking the design parameters into account, can result in a higher potential. However, these engineered bacteria cannot be applied for the process where wastewater treatment is the main purpose as activated sludge is mainly applied as inoculum in MFCs in such cases [23]. The expansion of MFCs to a larger scale is a contentious issue because of the lack of understanding regarding MFCs. Innovative techniques, materials, and a better understanding and handling of the parameters affecting the efficiency of the MFCs will pave the way for the industrial scale application of MFCs for wastewater treatment. All the basic details in the design of the MFCs, such as the material, surface area, and spacing of the electrodes and membrane, and the adaptability of the system need consideration in order to upgrade them for the wider scale application [23]. To summarize, it can be stated that the development of a cost efficient MFC for optimized wastewater treatment and higher power generation capacity is a preference.

10.14 Conclusion

MFCs are a smart choice when considering the treatment of wastewater or the removal of pollutants from groundwater resources along with power generation. MFCs can offer increased removal of COD. As various factors influence the efficiency of MFCs, it is possible to design and develop a desired MFC. Generally, the MFCs used for the treatment of wastewater contain aerobic or anaerobic activated sludge and there is no need to apply a mediator. MFC performance can be enhanced by maintaining the optimum temperature and pressure required for the bacterial growth, just as for any other microbial system. Also, the salinity and ionic strength of the medium are critical factors to consider. A medium with high salinity and ionic strength can improve the substrate conductivity and increase the efficiency of the MFCs even though the high salinity has an adverse effect on microbial growth. Material and fabrication techniques for electrodes and membranes is also critical and should be given special consideration when designing an MFC relevant to their activity, efficiency, and cost.

References

[1] Yaqoob A A, Parveen T, Umar K and Ibrahim M N M 2020 Role of nanomaterials in the treatment of wastewater: a review *Water* **12** 495

[2] Oh S T, Kim J R, Premier G C, Lee T H, Kim C and Sloan W T 2010 Sustainable wastewater treatment: how might microbial fuel cells contribute *Biotechnol. Adv.* **28** 871–81

[3] Yaqoob A A, Ibrahim M N and Rodríguez-Couto S 2020 Development and modification of materials to build cost-effective anodes for microbial fuel cells (MFCs): an overview *Biochem. Eng. J.* **164** 107779

[4] Logan B E 2005 Simultaneous wastewater treatment and biological electricity generation *Water Sci. Technol.* **52** 31–7

[5] Du Z, Li H and Gu T 2007 A state of the art review on microbial fuel cells: a promising technology for wastewater treatment and bioenergy *Biotechnol. Adv.* **25** 464–82

[6] Mohan S V, Saravanan R, Veer Raghavulu S, Mohanakrishna G and Sarma P N 2008 Bioelectricity production from wastewater treatment in dual chambered microbial fuel cell (MFC) using selectively enriched mixed microflora: effect of catholyte *Bioresour. Technol.* **99** 596–603

[7] Yaqoob A A, Ibrahim M N, Yaakop A S, Umar K and Ahmad A 2021 Modified graphene oxide anode: a bioinspired waste material for bioremediation of Pb^{2+} with energy generation through microbial fuel cells *Chem. Eng. J.* **417** 128052

[8] Aghababaie M, Farhadian M, Jeihanipour A and Biria D 2015 Effective factors on the performance of microbial fuel cells in wastewater treatment: a review *Env. Technol. Rev.* **4** 71–89

[9] Cao X *et al* 2009 A new method for water desalination using microbial desalination cells *Env. Sci. Technol.* **43** 7148–52

[10] Luo H, Xu P, Roane T M, Jenkins P E and Ren Z 2012 Microbial desalination cells for improved performance in wastewater treatment, electricity production, and desalination *Bioresour. Technol.* **105** 60–6

[11] Lou H, Jenkins P E and Ren Z 2011 Concurrent desalination and hydrogen generation using microbial electrolysis and desalination cells *Env. Sci. Technol.* **45** 340–4

[12] Jacobson K S, Drew D M and He Z 2011 Use of a liter-scale microbial desalination cell as a platform to study bioelectrochemical desalination with salt solution or artificial seawater *Env. Sci. Technol.* **45** 4652–7

[13] Bresadola M 1998 Medicine and science in the life of Luigi Galvani (1737–1798) *Brain Res. Bull.* **46** 367–80

[14] Potter M C 1911 Electrical effects accompanying the decomposition of organic compounds *R. Soc.* **84** 260–76

[15] Cohen B 1931 The bacterial culture as an electrical half cell *J. Bacteriol.* **21** 18–9

[16] Rabaey K and Verstraete W 2005 Microbial fuel cells: novel biotechnology for energy generation *Trends Biotechnol.* **23** 291–8

[17] Bullen R A, Arnot T C, Lakemanc J B and Walsh F C 2006 Biofuel cells and their development *Biosens. Bioelectron.* **21** 2015–45

[18] Habermann W and Pommer E H 1991 Biological fuel cells with sulphide storage capacity *Appl. Microbiol. Biotechnol.* **35** 128–33

[19] Yaqoob A A, Ibrahim M N, Rodríguez-Couto S and Ahmad A 2021 Preparation, characterization, and application of modified carbonized lignin as an anode for sustainable microbial fuel cell *Pro. Safe Env. Protect.* **155** 49–60

[20] Yaqoob A A, Ibrahim M N M and Guerrero-Barajas C 2021 Modern trend of anodes in microbial fuel cells (MFCs): an overview *Env. Technol. Innov.* **23** 101579

[21] Yaqoob A A, Khatoon A, Mohd Setapar S H, Umar K, Parveen T, Mohamad Ibrahim M N, Ahmad A and Rafatullah M 2020 Outlook on the role of microbial fuel cells in remediation of environmental pollutants with electricity generation *Catalyst* **10** 819

[22] Shukla A K, Suresh P, Berchmans S and Rajendran A 2004 Biological fuel cells and their applications *Curr. Sci.* **87** 455–68

[23] Yaqoob A A, Mohamad Ibrahim M N, Rafatullah M, Chua Y S, Ahmad A and Umar K 2020 Recent advances in anodes for microbial fuel cells: an overview *Material* **13** 2078

[24] Tanaka K, Vega C A and Tamamushi R 1983 Thionine and ferric chelate compounds as coupled mediators in microbial fuel cells *Bioelectrochem. Bioenerg.* **11** 289–97

[25] Adelaja O, Keshavarz T and Kyazze G 2015 The effect of salinity, redox mediators and temperature on anaerobic biodegradation of petroleum hydrocarbons in microbial fuel cells *J. Hazard. Mater.* **283** 211–7

[26] Fatemi S, Ghoreyshi A A, Najafpour G and Rahimnejad M 2012 Bioelectricity generation in mediatorless microbial fuel cell: application of pure and mixed cultures *Iran J. Energy Env.* **3** 104–8

[27] Kim H J, Park H S, Hyun M S, Chang I S, Kim M and Kim B H 2002 A mediatorless microbial fuel cell using a metal reducing bacterium, *Shewanella putrefaciens Enzyme Microb. Technol.* **30** 145–52

[28] Asim A Y, Mohamad N, Khalid U, Tabassum P, Akil A, Lokhat D and Siti H 2021 A glimpse into the microbial fuel cells for wastewater treatment with energy generation *Desal. Water Treat.* **214** 379–89

[29] Gorby Y A *et al* 2006 Electrically conductive bacterial nanowires produced by *Shewanella oneidensis* strain MR-1 and other microorganisms *Proc. Natl Acad. Sci. USA* **103** 11358–63

[30] Ieropoulos I A, Greenmana J, Melhuish C and Hart J 2005 Comparative study of three types of microbial fuel cell *Enzyme Microb. Technol.* **37** 238–45

[31] Sammes N 2006 *Fuel Cell Technology Reaching Towards Commercialization* (London: Springer)

[32] Yaqoob A A, Mohamad Ibrahim M N, Umar K, Bhawani S A, Khan A, Asiri A M, Khan M R, Azam M and AlAmmari A M 2021 Cellulose derived graphene/polyaniline nanocomposite anode for energy generation and bioremediation of toxic metals via benthic microbial fuel cells *Polymer* **13** 135

[33] Richter K, Schicklberger M and Gescher J 2012 Dissimilatory reduction of extracellular electron acceptors in anaerobic respiration *Appl. Env. Microbiol.* **78** 913–21

[34] Lovley D R 1991 Dissimilatory Fe(III) and Mn(IV) reduction *Microbiol. Rev.* **78** 259–87

[35] Kim J R, Dec J, Bruns M A and Logan B E 2008 Removal of odors from swine wastewater by using microbial fuel cells *Appl. Env. Microbiol.* **74** 2540–3

[36] Kim J R, Min B and Logan B E 2005 Evaluation of procedures to acclimate a microbial fuel cell for electricity production *Appl. Microbiol. Biotechnol.* **68** 23–30

[37] Thauer R K, Jungermann K and Decker K 1977 Energy conservation in chemotropic anaerobic bacteria *Bacteriol. Rev.* **41** 100–80

[38] Rismani-Yazdi H, Christy A D, Carver S M, Dehority Z Y A and Tuovinen O H 2011 Effect of external resistance on bacterial diversity and metabolism in cellulose-fed microbial fuel cells *Bioresour. Technol.* **102** 278–83

[39] Chen C-Y, Chen T-Y and Chung Y-C 2014 A comparison of bioelectricity in microbial fuel cells with aerobic and anaerobic anodes *Env. Technol.* **35** 286–93

[40] Ringeisen B R, Ray R and Little B 2007 A miniature microbial fuel cell operating with an aerobic anode chamber *J. Power Sources* **165** 591–7

[41] Chen Y-M, Wang C-T, Yang Y-C and Chen W-J 2013 Application of aluminum-alloy mesh composite carbon cloth for the design of anode/cathode electrodes in *Escherichia coli* microbial fuel cell *Int. J. Hydrogen Energy* **38** 11131–7

[42] Choi Y, Jung E, Kim S and Jung S 2003 Membrane fluidity sensoring microbial fuel cell *Bioelectrochemistry* **59** 121–7

[43] Bond D R and Lovley D R 2003 Electricity production by *Geobacter sulfurreducens* attached to electrodes *Appl. Env. Microbiol.* **69** 1548–55
[44] Mathuriya A S 2013 Inoculum selection to enhance performance of a microbial fuel cell for electricity generation during wastewater treatment *Env. Technol.* **34** 1957–64
[45] Yaqoob A A, Ibrahim M N and Umar K 2021 Biomass-derived composite anode electrode: synthesis, characterizations, and application in microbial fuel cells (MFCs) *J. Env. Chem. Eng.* **9** 106111
[46] Cheng S, Liu H and Logan B E 2006 Power densities using different cathode catalysts (Pt and CoTMPP) and polymer binders (Nafion and PTFE) in single chamber microbial fuel cells *Env. Sci. Technol.* **40** 364–9
[47] Cheng S and Logan B E 2007 Ammonia treatment of carbon cloth anodes to enhance power generation of microbial fuel cells *Electrochem. Commun.* **9** 492–6
[48] Yates M D *et al* 2012 Convergent development of anodic bacterial communities in microbial fuel cells *ISME J.* **6** 2002–13
[49] Rismani-Yazdi H, Christy A D, Dehority B A, Morrison M, Yu Z and Tuovinen O H 2007 Electricity generation from cellulose by rumen microorganisms in microbial fuel cells *Biotechnol. Bioeng.* **97** 1398–407
[50] Futamata H, Bretschger O, Cheung A, Kan J, Owen R and Nealson K H 2013 Adaptation of soil microbes during establishment of microbial fuel cell consortium fed with lactate *J. Biosci. Bioeng.* **115** 58–63
[51] Sangcharoen A, Niyom W and Suwannasilp B B 2015 A microbial fuel cell treating organic wastewater containing high sulfate under continuous operation: performance and microbial community *Process. Biochem.* **50** 1648–55
[52] Yaqoob A A, Ibrahim M N, Yaakop A S and Rafatullah M 2022 Utilization of biomass-derived electrodes: a journey toward the high performance of microbial fuel cells *Appl. Water Sci.* **12** 1–20
[53] Kim K Y *et al* 2011 Enhanced coulombic efficiency in glucose-fed microbial fuel cells by reducing metabolite electron losses using dual-anode electrodes *Bioresour. Technol.* **102** 4144–9
[54] Liu H, Cheng S and Logan B E 2005 Production of electricity from acetate or butyrate using a single-chamber microbial fuel cell *Env. Sci. Technol.* **39** 658–62
[55] Hassan S H A, Kim Y S and Oh S-E 2012 Power generation from cellulose using mixed and pure cultures of cellulose-degrading bacteria in a microbial fuel cell *Enzyme Microb. Technol.* **51** 269–73
[56] Logana B E, Murano C, Scott K, Gray N D and Head I M 2005 Electricity generation from cysteine in a microbial fuel cell *Water Res.* **39** 942–52
[57] Liu G, Yates M D, Cheng S, Call D F, Sun D and Logan B E 2011 Examination of microbial fuel cell startup times with domestic wastewater and additional amendments *Bioresour. Technol.* **102** 7301–6
[58] Yaqoob A A, Serrà A, Ibrahim M N M and Yaakop A S 2021 Self-assembled oil palm biomass-derived modified graphene oxide anode: an efficient medium for energy transportation and bioremediating Cd (II) via microbial fuel cells *Arabian J. Chem.* **14** 103121
[59] Min B, Kim J R, Oh S E, Regan J M and Logan B E 2005 Electricity generation from swine wastewater using microbial fuel cells *Water Res.* **39** 4961–8

[60] Behera M, Jana P S, More T T and Ghangrekar M M 2010 Rice mill wastewater treatment in microbial fuel cells fabricated using proton exchange membrane and earthen pot at different pH *Bioelectrochem.* **79** 228–33
[61] Mohan S V, Mohanakrishna G, Reddy B P, Saravanan R and Sarma P N 2008 Bioelectricity generation from chemical wastewater treatment in mediatorless (anode) microbial fuel cell (MFC) using selectively enriched hydrogen producing mixed culture under acidophilic microenvironment *Biochem. Eng. J.* **39** 121–30
[62] Hou B, Sun J and Hu Y 2011 Simultaneous Congo red decolorization and electricity generation in air-cathode single-chamber microbial fuel cell with different microfiltration, ultrafiltration and proton exchange membranes *Bioresour. Technol.* **102** 4433–8
[63] Kumru M, Eren H, Catal T, Bermek H and Akarsubas A T 2012 Study of azo dye decolorization and determination of cathode microorganism profile in air-cathode microbial fuel cells *Env. Technol.* **33** 2167–75
[64] Mshoperi E, Fogel R and Limson J 2011 Application of carbon black and iron phthalocyanine composites inbioelectricity production at a brewery wastewater fed microbial fuel cell *Electrochim. Acta* **196** 1103–6
[65] Zhang B, Zhao H, Zhou S, Shi C, Wang C and Ni J 2009 A novel UASB–MFC–BAF integrated system for high strength molasses wastewater treatment and bioelectricity generation *Bioresour. Technol.* **100** 5687–93
[66] Zhang L *et al* 2013 Effect of the chemical oxidation demand to sulfide ratio on sulfide oxidation in microbial fuel cells treating sulfide-rich wastewater *Env. Technol.* **34** 269–74
[67] Sciarria T P, Merlino G, Scaglia B, D'Epifanio A, Mecheri B and Borin S 2015 Electricity generation using white and red wine lees in air cathode microbial fuel cells *J. Power Sources* **274** 393–9
[68] Katuri K P, Enright A-M, O'Flaherty V and Leech D 2012 Microbial analysis of anodic biofilm in a microbial fuel cell using slaughterhouse wastewater *Bioelectrochemistry* **87** 164–71
[69] Kaewkannetra P, Chiwes W and Chiu T Y 2011 Treatment of cassava mill wastewater and production of electricity through microbial fuel cell technology *Fuel* **90** 2746–50
[70] Rikame S S, Mungray A A and Mungray A K 2012 Electricity generation from acidogenic food waste leachate using dual chamber mediator less microbial fuel cell *Int. Biodeterior. Biodegrad.* **75** 131–7
[71] Gálvez A, Greenman J and Ieropoulos I 2009 Landfill leachate treatment with microbial fuel cells; scale-up through plurality *Bioresour. Technol.* **100** 5085–91
[72] Hassan S H A *et al* 2014 Electricity generation from rice straw using a microbial fuel cell *Int. J. Hydrogen Energy* **39** 9490–6
[73] Cai J, Zheng P, Zhang J, Xie Z, Li W and Sun P 2013 Simultaneous anaerobic sulfide and nitrate removal coupled with electricity generation in microbial fuel cell *Bioresour. Technol.* **129** 224–8
[74] Yaqoob A A, Ibrahim M N, Yaakop A S and Ahmad A 2021 Application of microbial fuel cells energized by oil palm trunk sap (OPTS) to remove the toxic metal from synthetic wastewater with generation of electricity *Appl. Nanosci.* **11** 1949–61
[75] Hong S W, Kim H S and Chung T H 2010 Alteration of sediment organic matter in sediment microbial fuel cells *Env. Pollut.* **158** 185–91
[76] Fadzli F S, Rashid M, Yaqoob A A and Ibrahim M N 2021 Electricity generation and heavy metal remediation by utilizing yam (*Dioscorea alata*) waste in benthic microbial fuel cells (BMFCs) *Biochem. Eng. J.* **172** 108067

[77] Morris J M and Jin S 2012 Enhanced biodegradation of hydrocarbon contaminated sediments using microbial fuel cells *J. Hazard. Mater.* **213–214** 474–7
[78] Li J, Liu G, Zhang R, Luo Y, Zhang C and Li M 2010 Electricity generation by two types of microbial fuel cells using nitrobenzene as the anodic or cathodic reactants *Bioresour. Technol.* **101** 4013–20
[79] Wei M, Harnisch F, Vogt C, Ahlheim J, Neu T R and Richnow H H 2015 Harvesting electricity from benzene and ammonium-contaminated groundwater using a microbial fuel cell with an aerated cathode *RSC Adv.* **5** 5321–30
[80] Huang L, Gan L, Zhao Q, Logan B E, Lu H and Chen G 2011 Degradation of pentachlorophenol with the presence of fermentable and non-fermentable co-substrates in a microbial fuel cell *Bioresour. Technol.* **102** 8762–8
[81] Jiang D and Li B 2009 Granular activated carbon single-chamber microbial fuel cells (GAC-SCMFCs): a design suitable for large-scale wastewater treatment processes *Biochem. Eng. J.* **47** 31–7
[82] Yaqoob A A, Ibrahim M N and Yaakop A S 2021 Application of oil palm lignocellulosic derived material as an efficient anode to boost the toxic metal remediation trend and energy generation through microbial fuel cells *J. Cleaner Prod.* **314** 128062
[83] Zhu X and Logan B E 2014 Copper anode corrosion affects power generation in microbial fuel cells *J. Chem. Technol. Biotechnol.* **89** 471–4
[84] Yuan H and He Z 2015 Graphene-modified electrodes for enhancing the performance of microbial fuel cells *Nanoscale* **7** 7022–9
[85] Logan B E, Cheng S, Watson V and Estadt G 2007 Graphite fiber brush anodes for increased power production in air cathode microbial fuel cells *Env. Sci. Technol.* **41** 3341–6
[86] Feng Y, Yang Q, Wang X and Logan B E 2010 Treatment of carbon fiber brush anodes for improving power generation in air–cathode microbial fuel cells *J. Power Sources* **195** 1841–4
[87] Zhang X, Cheng S, Liang P, Huang X and Logan B E 2011 Scalable air cathode microbial fuel cells using glass fiber separators, plastic mesh supporters, and graphite fiber brush anodes *Bioresour. Technol.* **102** 372–5
[88] Liu J, Zhang F, He W, Zhang X, Feng Y and Logan B E 2014 Intermittent contact of fluidized anode particles containing exoelectrogenic biofilms for continuous power generation in microbial fuel cells *J. Power Sources* **261** 278–84
[89] Saito T *et al* 2011 Effect of nitrogen addition on the performance of microbial fuel cell anodes *Bioresour. Technol.* **102** 395–8
[90] Zhang C, Liang P, Jiang Y and Huang X 2015 Enhanced power generation of microbial fuel cell using manganese dioxide-coated anode in flow-through mode *J. Power Sources* **273** 580–3
[91] Qiao Y, Li C M, Bao S J and Bao Q L 2007 Carbon nanotube/polyaniline composite as anode material for microbial fuel cells *J. Power Sources* **170** 79–84
[92] Ci S, Wen Z, Chen J and He Z 2012 Decorating anode with bamboo-like nitrogen-doped carbon nanotubes for microbial fuel cells *Electrochem. Commun.* **14** 71–4
[93] Roh S-H and Woo H-G 2015 Carbon nanotube composite electrode coated with polypyrrole for microbial fuel cell application *J. Nanosci. Nanotechnol.* **15** 484–7
[94] Roh S-H 2015 Electricity generation from microbial fuel cell with polypyrrole-coated carbon nanofiber composite *J. Nanosci. Nanotechnol.* **15** 1700–3

[95] Heijne A, Hamelers H V M, Saakes M and Buisman C J N 2008 Performance of nonporous graphite and titanium-based anodes in microbial fuel cells *Electrochim. Acta* **53** 5697–703
[96] Wu Y, Zhang X, Li S, Lv X, Cheng Y and Wang X 2013 Microbial biofuel cell operating effectively through carbon nanotube blended with gold–titania nanocomposites modified electrode *Electrochim. Acta* **109** 328–32
[97] He Z, Minteer S D and Angenent L T 2005 Electricity generation from artificial wastewater using an upflow microbial fuel cell *Env. Sci. Technol.* **39** 5262–7
[98] Oh S E, Min B and Logan B E 2004 Cathode performance as a factor in electricity generation in microbial fuel cells *Env. Sci. Technol.* **38** 4900–4
[99] Rabaey K, Clauwaert P, Aelterman P and Verstraete W 2005 Tubular microbial fuel cells for efficient electricity generation *Env. Sci. Technol.* **39** 8077–82
[100] Zhang X, Sun H, Liang P, Huang X, Chen X and Logan B E 2011 Air-cathode structure optimization in separator-coupled microbial fuel cells *Biosens. Bioelectron.* **30** 267–71
[101] Liu H and Logan B 2004 Electricity generation using an air-cathode single chamber microbial fuel cell in the presence and absence of a proton exchange membrane *Env. Sci. Technol.* **38** 4040–6
[102] Luo Y, Zhang F, Wei B, Liu G, Zhang R and Logan B E 2011 Power generation using carbon mesh cathodes with different diffusion layers in microbial fuel cells *J. Power Sources* **196** 9317–21
[103] Venkata Mohan S, Saravanan R, Veer Raghavulu S, Mohanakrishna G and Sarma P N 2008 Bioelectricity production from wastewater treatment in dual chambered microbial fuel cell (MFC) using selectively enriched mixed microflora: effect of catholyte *Bioresour. Technol.* **99** 596–603
[104] Liu H, Cheng S, Huang L and Logan B E 2008 Scale-up of membrane-free single-chamber microbial fuel cells *J. Power Sources* **179** 274–9
[105] Clauwaert P *et al* 2007 Open air biocathode enables effective electricity generation with microbial fuel cells *Env. Sci. Technol.* **41** 7564–69
[106] Santoro C, Lei Y, Li B and Cristiani P 2012 Power generation from wastewater using single chamber microbial fuel cells (MFCs) with platinum-free cathodes and pre-colonized anodes *Biochem. Eng. J.* **62** 8–16
[107] Tharamani C N, Kundu S B, Muhler M and Schuhmann W 2010 Nitrogen-doped carbon nanotubes as a cathode catalyst for the oxygen reduction reaction in alkaline medium *Electrochem. Commun.* **12** 338–41
[108] Ahn Y, Ivanov I, Nagaiah T C, Bordoloi A and Logan B E 2014 Mesoporous nitrogen-rich carbon materials as cathode catalysts in microbial fuel cells *J. Power Sources* **269** 212–5
[109] Zuo Y, Cheng S, Call D and Logan B E 2007 Tubular membrane cathodes for scalable power generation in microbial fuel cells *Env. Sci. Technol.* **41** 3347–53
[110] Guerrini E, Grattieri M, Faggianelli A, Cristiani P and Trasatti S 2015 PTFE effect on the electrocatalysis of the oxygen reduction reaction in membraneless microbial fuel cells *Bioelectrochemistry*
[111] Zhang F, Cheng S, Pant D, Bogaert G V and Logan B E 2009 Power generation using an activated carbon and metal mesh cathode in a microbial fuel cell *Electrochem. Commun.* **11** 2177–9
[112] Zhang F, Pant D and Logan B E 2011 Long-term performance of activated carbon air cathodes with different diffusion layer porosities in microbial fuel cells *Biosens. Bioelectron.* **30** 49–55

[113] Liu W, Cheng S, Sun D, Huang H, Chen J and Cen K 2015 Inhibition of microbial growth on air cathodes of single chamber microbial fuel cells by incorporating enrofloxacin into the catalyst layer *Biosens. Bioelectron.* **72** 44–50

[114] Heijne A, Strike D P B T B, Hamelers H V M and Buisman C J N 2010 Cathode potential and mass transfer determine performance of oxygen reducing biocathodes in microbial fuel cells *Env. Sci. Technol.* **44** 7151–6

[115] Tandukar M, Huber S J, Onodera T and Pavlostathis S 2009 Biological chromium(VI) reduction in the cathode of a microbial fuel cell *Env. Sci. Technol.* **43** 8159–65

[116] Huang L, Chai X, Chen G and Logan B E 2011 Effect of set potential on hexavalent chromium reduction and electricity generation from biocathode microbial fuel cells *Env. Sci. Technol.* **45** 5025–31

[117] Rhoads A, Beyenal H, Lewandowski A, Huber S J, Onodera T and Pavlostathis S G 2009 Microbial fuel cell using anaerobic respiration as an anodic reaction and biomineralized manganese as a cathodic reactant *Env. Sci. Technol.* **43** 8159–65

[118] Rahimnejad M, Jafary T, Haghparast F, Najafpour G D and Ghoreyshi A A 2010 Nafion as a nanoproton conductor in microbial fuel cells *Turk. J. Eng. Env. Sci.* **34** 289–92

[119] Choi T H *et al* 2012 Electrochemical performance of microbial fuel cells based on disulfonated poly(arylene ether sulfone) membranes *J. Power Sources* **220** 269–79

[120] Ghasemi M, Shahgaldi S, Ismail M, Yaakob Z and Daud W R W 2012 New generation of carbon nanocomposite proton exchange membranes in microbial fuel cell systems *Chem. Eng. J.* **184** 82–9

[121] Lim S S, Daud W R W, Jahim J M, Ghasemi M, Chong P S and Ismail M 2012 Sulfonated poly(ether ether ketone)/poly(ether sulfone) composite membranes as an alternative proton exchange membrane in microbial fuel cells *Int. J. Hydrogen Energy* **37** 11409–24

[122] Rahimnejad M *et al* 2011 Synthesis, characterization and application studies of self-made Fe_3O_4/PES nanocomposite membranes in microbial fuel cell *Electrochim. Acta* **85** 700–6

[123] Tao H-C, Sun X-N and Xiong Y 2015 A novel hybrid anion exchange membrane for high performance microbial fuel cells *RSC Adv.* **5** 4659–63

[124] Ghangrekar M M and Shinde V B 2007 Performance of membraneless microbial fuel cell treating wastewater and effect of electrode distance and area on electricity production *Bioresour. Technol.* **98** 2879–85

[125] Zhang X, Liang P, Shi J, Wei J and Huang X 2013 Using a glass fiber separator in a single-chamber air-cathode microbial fuel cell shortens start-up time and improves anode performance at ambient and mesophilic temperatures *Bioresour. Technol.* **130** 529–35

[126] Ahn Y and Logan B E 2012 A multi-electrode continuous flow microbial fuel cell with separator electrode assembly design *Appl. Microbiol. Biotechnol.* **93** 2241–8

[127] Sevda S, Dominguez-Benetton X, Vanbroekhoven K, Sreekrishnan T R and Pant D 2013 Characterization and comparison of the performance of two different separator types in air–cathode microbial fuel cell treating synthetic wastewater *Chem. Eng. J.* **228** 1–11

[128] Philamore H, Rossiter J, Walters P, Winfield J and Ieropoulos I 2015 Cast and 3D printed ion exchange membranes for monolithic microbial fuel cell fabrication *J. Power Sources* **289** 91–9

[129] Sevda S, Dominguez-Benetton X, Vanbroekhoven K, Sreekrishnan T R and Pant D 2013 Characterization and comparison of the performance of two different separator types in air–cathode microbial fuel cell treating synthetic wastewater *Chem. Eng. J.* **228** 1–11

[130] Lefebvre O, Tan Z, Kharkwal S and Ng H Y 2012 Effect of increasing anodic NaCl concentration on microbial fuel cell performance *Bioresour. Technol.* **112** 336–40

[131] Oh S E and Logan B E 2006 Proton exchange membrane and electrode surface areas as factors that affect power generation in microbial fuel cells *Appl. Microbiol. Biotechnol.* **70** 162–9

[132] He Z, Wagner N, Minteer S D and Angenent L T 2006 An upflow microbial fuel cell with an interior cathode: assessment of the internal resistance by impedance spectroscopy *Env. Sci. Technol.* **40** 5212–7

[133] Kim M H, Iwuchukwu I J, Wang Y, Shin D, Sanseverino J and Frymier P 2011 An analysis of the performance of an anaerobic dual anode-chambered microbial fuel cell *J. Power Sources* **196** 1909–14

[134] Du Z, Li Q, Tong M, Li S and Li H 2008 Electricity generation using membrane-less microbial fuel cell during wastewater treatment *Chin. J. Chem. Eng.* **16** 772–7

[135] Liu H, Ramnarayanan R and Logan B E 2004 Production of electricity during wastewater treatment using a single chamber microbial fuel cell *Env. Sci. Technol.* **38** 2281–5

[136] Ahn Y, Hatzell M C, Zhang F and Logan B E 2014 Different electrode configurations to optimize performance of multielectrode microbial fuel cells for generating power or treating domestic wastewater *J. Power Sources* **249** 440–5

[137] Wei B, Tokash J C, Zhang F, Kim Y and Logan B E 2013 Electrochemical analysis of separators used in single-chamber, air-cathode microbial fuel cells *Electrochim. Acta* **89** 45–51

[138] Zhang F, Xia X, Luo Y, Sun D, Call D F and Logan B E 2013 Improving startup performance with carbon mesh anodes in separator electrode assembly microbial fuel cells *Bioresour. Technol.* **133** 74–81

[139] Ahn Y and Logan B E 2013 Altering anode thickness to improve power production in microbial fuel cells with different electrode distances *Energy Fuels* **27** 271–6

[140] Liao Q, Zhang J, Li J, Ye D, Zhu X and Zhang B 2015 Increased performance of a tubular microbial fuel cell with a rotating carbon-brush anode *Biosens. Bioelect.* **63** 558–61

[141] Wang Y P *et al* 2012 A microbial fuel cell–membrane bioreactor integrated system for cost-effective wastewater treatment *Appl. Energy* **98** 230–5

[142] Rahimnejad M, Ghoreyshi A A, Najafpour G D, Younesi H and Shakeri M 2012 A novel microbial fuel cell stack for continuous production of clean energy *Int. J. Hydrogen Energy* **37** 5992–6000

[143] Tender L M *et al* 2008 The first demonstration of a microbial fuel cell as a viable power supply: powering a meteorological buoy *J. Power Sources* **179** 571–5

[144] Jiang D *et al* 2011 A pilot-scale study on utilizing multi-anode/cathode microbial fuel cells (MAC MFCs) to enhance the power production in wastewater treatment *Int. J. Hydrogen Energy* **36** 876–84

[145] Dong Y *et al* 2015 A 90-liter stackable baffled microbial fuel cell for brewery wastewater treatment based on energy self-sufficient mode *Bioresour. Technol.* **195** 66–72

IOP Publishing

Microbial Fuel Cells
Emerging trends in electrochemical applications
Mohamad Nasir Mohamad Ibrahim, Asim Ali Yaqoob and Akil Ahmad

Chapter 11

Influence of micro-organisms on the electrochemical process of microbial fuel cells

Rizwana Kausar and Faiz Ur Rahman

Micro-organisms act as a biological catalyst in microbial fuel cells (MFCs) to generate electrical energy by oxidizing an organic substrate. Biofilm formation is the key characteristic of micro-organisms, which has a profound influence on the MFCs' performance. Extracellular polysaccharides affect greatly the formation, development, and regulation of biofilms and quorum sensing in micro-organisms. The use of biofilm-producing bacteria and consortia versus pure cultures shows the production of high-density current in MFCs. The microbial metabolism is another important factor in MFC electrogenesis. Micro-organisms with various ways of transferring extracellular electrons (the basis of electricity generation in MFCs) without utilizing artificial means are of great importance in improving the MFC performance. The presence of specific proteins and genes in micro-organisms is related to the various mechanisms for the transfer of electrons. Similarly, the choice of the substrate, such as a coastal lagoon, anaerobic sludge of wastewater, rice paddy soil, etc, is also shown to have an impact on the amount of energy production by MFCs. This chapter discusses the formation, development, and regulation of biofilm and the factors influencing it. It also discusses the impact of microbial metabolism in electrogenesis, and various processes of extracellular electron transfer in MFCs which can influence the bioelectrochemical process in MFCs.

11.1 Introduction

Microbial fuel cells (MFCs) use micro-organisms as biological catalysts to generate electrical energy by oxidizing an organic substrate. Protons and electrons are released from the organic substrate as a result of oxidation by the micro-organisms. An electrical current is generated when micro-organisms take these free electrons and transfer them to an anode which is then transferred to a cathode, hence

doi:10.1088/978-0-7503-4791-4ch11

generating an electrical current [1]. This electron transfer is carried out by micro-organisms by producing electron shuttles [2], electron mediators [3], or numerous filaments, also called microbial nanowires [4]. Simultaneously, at the cathode protons combine with an electron acceptor such as oxygen and electrons and produce water [1].

MFC technology is a sustainable way of producing bioelectricity as it offers the availability and flexibility of using diverse groups of micro-organisms which can utilize a vast range of organic waste. MFC technology may be helpful not just for producing bioenergy to satisfy rising global energy demands in an era of limited non-renewable energy resources, but also for organic waste biodegradation [5]. As a consequence, it is a promising energy harvesting technology with a wide range of applications, including bioremediation, sensors, and powering electronic monitoring equipment [6]. The basics of MFC design and operation are well understood at this time, but technical issues, particularly microbiological ones, are still under investigation (figure 11.1). MFCs have not yet achieved extremely high power densities, limiting their real-world application. The design and microbiological composition of these devices may be improved further. The most significant microbiological factors that influence an MFC's efficiency are biofilm formation, bacterial metabolism, and bacterial electron transfer.

It has been found that a range of micro-organisms (mostly bacteria) from different taxonomic groups may generate electricity without the need for a mediator in MFCs. Electrical current has been seen in five phyla of Proteobacteria, Firmicutes, and Acidobacteria, as well as using microalgae, yeasts, and fungi as the substrate or to assist the cathode or anode in MFCs. There is not a specific nomenclature at present, and the micro-organisms that assist exogenously in transporting of electrons from the anode with no need of an artificial mediator are known by different names [7]. Some of the terms used by different authors for

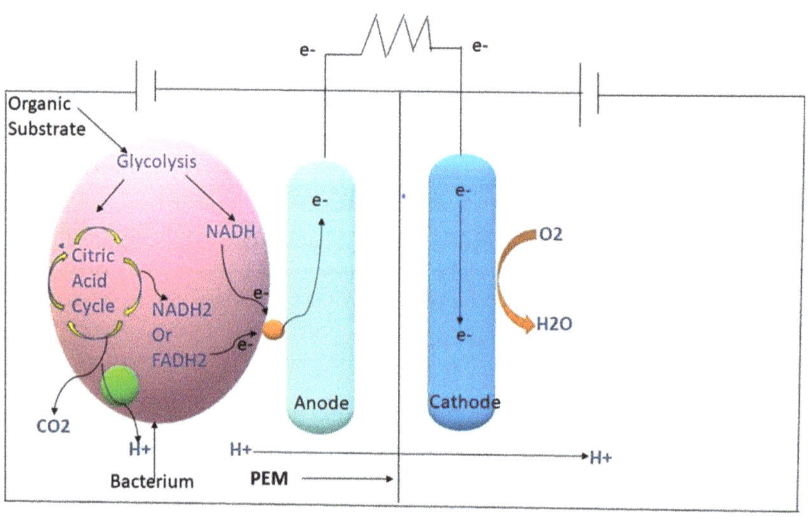

Figure 11.1. Overall principles of microbial fuel cells.

these micro-organisms are anode-respiring bacteria, exoelectrogens, electricigens, electrogenic micro-organisms, anodophiles, and electrochemically active bacteria. For example, based on their functions in MFCs, iron-reducing and sulphate-reducing bacteria may be called iron reducers and sulphate reducers [7]. Electrode reducers provide electrons to the electrode (anode) in MFCs, while electrode oxidizers remove electrons away from the electrodes. The most common bacterial species that generate electricity in MFCs are *Geobacter* spp., *Shewanella* spp., *Clostridium butyricum*, *Rhodoferax ferrireducens*, *Aeromonas hydrophila*, and *Pseudomonas aeruginosa* [8–14]. Microalgae have been used as biocathodes or substrates in MFCs [15]. Excellent current is produced in an MFC by *Hansenula anomala* and *Saccharomyces cerevisiae* [16]. However, it does not seem that the utilization of yeasts for MFC based energy generation has been investigated extensively. *Synechocystis* sp., a phototrophic oxygenic cyanobacterium, can produce electricity in MFCs and nanowires [17]. The micro-organisms that can transfer electrons to the anode because of complete oxidation of organic compounds at a faster rate may make a significant contribution to power generation. The increase in current density has been observed when micro-organisms in a biofilm on an anode directly transfer electrons between themselves and the anode. The micro-organisms in a mixed culture biofilm can produce current densities higher than a biofilm of pure cultures [18]. For example, a pure culture of *Brevibacillus* sp. in an MFC produced less electricity compared to when it was grown with *Pseudomonas* sp. [19]. Micro-organisms with a capacity for dissimilatory metal reduction may produce power efficiently in a mediatorless MFC. Such bacteria can interact directly with cytochromes in their outer membrane or excrete electron shuttles to transfer electrons. After that, micro-organisms were also found to produce microbial nanowires (protein appendages) which help in electron transfer. *Pelotomaculum thermopropionicum* was found to have an electrically conductive appendage connected to *Methanothermobacter thermautotrophics*, allowing interspecies electron transfer [9]. Quorum sensing is another important mechanism of communication inside microbial biofilm. It is carried out by sensing chemicals such as fatty acyl homoserine lactones [20]. Pyocyanin is a signalling molecule generated by *P. aeruginosa* that works not only as an electron shuttle but also regulates the transcription of the quorum sensing genes [21].

This chapter discusses the influence of microbial mechanisms that benefit MFC technology, such as the formation and development of biofilm by micro-organisms, the factors influencing the biofilm development, microbial metabolism and electrogenesis, various types of electron transfer between micro-organisms and electrodes, and a description of micro-organisms suitable for electricity generation.

11.2 Microbial influence on MFCs via the production of biofilm

The essential function of the biofilm and its developmental phases on the anode surface are particularly significant in the MFC system for energy recovery. Bacteria prefer to live in a biofilm, which is a polymeric matrix (containing proteins, lipids, carbohydrates, and other components) produced by bacteria adhering to a surface.

Producing electroactive biofilms in MFCs is important for producing energy more efficiently. Biofilm formation is affected by several variables, including the electrode material, the substrates, the microbes used in the MFC, and the operational settings of the MFCs. Biofilm formation is also affected by the morphological and physiological features of the electrode surface. Micro-organisms prefer to adhere to hydrophobic surfaces rather than hydrophilic surfaces [22]. Biofilm is formed on the electrode surface because of cell-to-cell communication among microbial communities and secretion of extracellular polysaccharides (EPS) [23]. The immobilized bacteria in the biofilms on the surface of electrodes bridge the way for electron transfer and its interaction with electrodes in MFCs [24, 25]. The shorter the distance for electrons to reach the electron acceptor, the higher will be the efficiency of electron transfer [26]. The response of MFCs is significantly affected by the development of biofilm on the anode as a surface attachment, biofilm maturation, substrate diffusion, and electron transport occur in this zone [22, 27]. Therefore, the important factor of the electrochemical reaction of an MFC is a biofilm that is developed on the electrode. Several spectroscopy methods are used to study the biofilm's microbial community on the anode, for example Raman spectroscopy [28] and crystal violet assay [29], and microscopic methods such as confocal laser scanning microscopy [30], scanning electron microscopy [31], and atomic force microscopy [32]. All these spectroscopic and microscopic techniques were employed exclusively to evaluate the developed biofilm. However, the performance of MFCs is influenced by several restricting factors that can be identified by the time taken for biofilm development on the anode and the correlation with the electrochemical reaction taking place there [33].

According to earlier studies, bacteria that are unable to develop biofilms on the electrode are not capable of producing considerable current intensities in MFCs, while dense microbial biofilm formation at the anode produces steady and higher current densities. Thus dense biofilm-producing bacteria have more electricity generating potential in MFCs compared to bacteria that develop thin biofilms. Confocal imaging results showed that generally gram-positive bacteria develop thicker biofilms (~38 μm), producing a constant current density of 7–8 Am^2 [34], whereas gram-negative bacteria that construct monolayer biofilms produced significantly lower current densities [35].

11.2.1 The formation of biofilm

The formation of biofilms is caused by transferring micro-organisms to a surface, then adhering them on that surface (in MFCs, on the cathode or anode), establishing microcolonies, then developing biofilms [36]. Bacterial cells produce adhesins, which connect and encapsulate the bacteria in a biofilm made up of carbohydrates (polysaccharides), nucleic acids, and proteins [37]. The ability of electroactive biofilms to accept terminal electrons produced by metabolism and transfer them to electrode surfaces is their distinctive feature [38]. For biofilm development in *Shewanella* spp. and *Geobacter* spp., the c-type cytochromes are found to be significantly important [8, 9], while type IV pili protein, which is made up of PilA monomers, is mainly important for the formation of conductive biofilms in *Aeromonas* spp. and *Geobacter* spp. [14, 39]. Biofilm growth and consequently current generation were decreased in *Geobacter sulfurreducens* mutants in the *pilA* and *omcZ* genes, indicating a role for the protein pili and c-type cytochromes in biofilm development [40]. The oxidation–reduction active compounds, such as flavins in *Shewanella* spp., help in exocellular transport of electrons in biofilm [41]. With the help of flagellar movement, bacteria are transferred to the surface in *P. aeruginosa* biofilms. Type IV pili stimulate cellular aggregation and microcolony growth, and a maturation phase involving cell-to-cell communication leads to the production of mushroom-shaped biofilms [42–44]. Quorum sensing is an interaction and coordination process that enables bacterial populations to transmit and organize their actions. In *P. aeruginosa* and other bacteria, quorum sensing controls the expression of genes related to the biofilm and is crucial for the formation of biofilm [45, 46].

Pure culture biofilms generate lower power densities, while mixed culture biofilms provide greater power densities. An MFC of mixed culture generated approximately 20% more energy than a similar MFC using pure culture [47]. The role of non-exoelectrogens (micro-organisms that cannot generate electric current when cultivated in pure cultures) in power generation, however, remains unclear. Bacterial cells in monolayer biofilms remain close to the anodic surface and directly transfer electrons with the help of c-type cytochromes or electron shuttles to the anode. Pili have been found to help in electron transport from distant cells to the anode surface in thick multilayer biofilms [48]. Oxygen reduction on the cathode with the help of micro-organisms is a focus of research to make a cathodic biofilm as for anodic biofilms. Unlike anodic biofilms, power generation in cathodic biofilms has been shown to decrease as thickness increases [49].

11.2.2 The role of extracellular polysaccharides in biofilm regulation and formation

As a result of the development of an internal layer of inactive or dead cells in the biofilm matrix, the density of the biofilm that develops under natural conditions increases with culture age, but it is not wholly electrochemically active [50]. As a result, a method that encourages living or active micro-organisms to build more electroactive biofilms is seen as a potential way to improve MFC power performance. The major biofilm components and their function in the development of biofilm must be realized to accomplish this. Over time, our understanding of the complexity of microbial biofilms has increased [51]. Biofilms are described as a 'city of micro-organisms', with extracellular polysaccharides (EPSs) serving as the 'home of the biofilm cells', in a metaphorical depiction. The EPSs affect the porosity, charge, water content, density, hydrophobicity, absorption characteristics, and mechanical stability of cells in biofilms, dictating the immediate circumstances of life [51]. The organisms are discovered embedded in biopolymers. The density of the biofilm is proportional to the number of EPSs secreted by the organism. A denser biofilm may hold more organisms than a lighter biofilm [52]. As a result, as the age of the organisms increases, so does the density of the biofilm. As previously stated, biofilm development on the anode is critical in MFCs. The electrochemical performance of a biofilm, on the other hand, is controlled by the physical and temporal positions of the living and dead cells and not by its density within the biofilm [50]. The fast decrease in charge transmission resistance in the presence of fast increasing living cells has been shown in studies of the electroactive bacterium *G. sulfurreducens*. However, when dead cells collect in the inner layer of biofilms, the electrochemical system faces a significant diffusion resistance over time. In such situations, it is assumed that living cells on the outer surface of biofilms help in active electron transfer and, instead of the density of the biofilms, impact the system's high current generation [50]. Apart from EPSs, the biofilm contains a significant number of polysaccharides, as well as minor amounts of glycoproteins, proteins, glycolipids, a small number of nucleotides, and, in rare cases, metals [53], all of which affect the biofilm's physical and physiological outlook. The biofilm matrix, on the other hand, is mostly made up of EPSs. At least three polysaccharides with an active role have been found in biofilm development among the common bacterial EPSs. They are the polysaccharides alginate, Pel, and Psl. Extracellular polysaccharides generated by *Psl* genes are known as Psl polysaccharides. Psl polysaccharides perform a critical role in *P. aeruginosa* biofilm formation [54, 55]. The *Psl* loci include 15 co-transcribed genes, 11 of which are involved in the formation of Psl-dependent biofilms [56–58]. Preliminary attachment, permanent attachment, microcolony development, biofilm maturing, and biofilm spreading are the five sequential stages in the creation of biofilm. As a result, enhanced biofilm development by living or functionally active micro-organisms in an MFC might be attributed to the over-expression of these genes. The function of such polysaccharides during biofilm formation affects the external mobility of succeeding cells [59]. Overproduction of Psl polysaccharides has been shown to increase organism to organism contact and adhesion, facilitating the initial and most important stage in biofilm development [60, 61]. The Psl polysaccharide adheres strongly to the bacterial cell wall in a helical shape, facilitating strong intercellular connections and therefore acting as a skeleton in biofilm matrix development [61]. They are discovered connected to the perimeter of three-dimensionally constructed colonies during biofilm development, giving structural assistance and allowing biofilm distribution at the ends [60]. Extracellular polysaccharides play a critical function in the development, adhesion properties, stability, structural integrity, and longevity of biofilms, and therefore may be used as critical components in biofilm construction procedures by increasing the copy number of these EPS making genes.

11.3 External factors influencing biofilms

Although EPSs are important in biofilm development, it is also affected by external environmental variables and gene expression processes that lead to the establishment of biofilms in particular cells [62]. The development of biofilms, as well as the outer and extracellular elements of the organisms, is influenced by physical and environmental variables. To be more precise, the destiny of biofilm development is determined by exopolysaccharides (EPS) and lipopolysaccharides (LPSs) as well as quorum sensing (QS) signalling molecules [63]. External components that affect the release of EPS, QS signalling molecules, different stress molecules such as heavy metals [64], salinity [65], pH [66], nutrient starvation [53], pathogen invasion, nutrient depletion, water current in moving water bodies, and growth substrates are all identified to influence biofilms. Genetic engineering of QS signalling molecules and EPS, which are the important contributors in biofilm development, can prevail over the other environmental effects involved in biofilm formation.

11.4 The influence of quorum sensing signalling molecules on biofilm

The bacterial community interacts via cell-to-cell interactions, allowing them to coordinate their collective activity. This necessitates the discharge of autoinducers from the cells, which then ultimately produce quorum sensing. It was found that adding 100 nM quinolone type signalling molecules increased the biofilm mass of *Halanaerobium* prevalence by 95%, resulting in a 30% increase in power capacity [67]. The signalling molecules, Quinolones, that belong to the LuxR proteins family are encoded by the *hmqF* genes in *Halanaerobium* species [68]. Although the complete stimulation of biofilm formation in *P. aeruginosa* has been studied [69], nothing has been reported on the impact of autoinducers produced by other bacteria in activating QS in *Halanaerobium* sp. [67].

11.5 The influence of extracellular polysaccharides and additional physical parameters on biofilms

Separately from the QS signalling molecules, LPSs and EPSs have also been found to play a key role in biofilm development. Bacterial production of c-di-GMP produced by *diguanylate cyclases* (DGC) at elevated levels promotes the synthesis of matrix exopolysaccharides such as Pel and Alginate, which contributes to the development of biofilms. Among the three major polysaccharides, alginate, Pel, and Psl, the increased release of Psl in non-pilated organisms helps in the formation of solid surface-associated biofilms [60], whereas in gram-negative bacteria, the production of Pel, a glucose-rich extracellular matrix, helps in the formation of biofilms for non-pilated organisms [61, 70]. Furthermore, a dual-functional enzyme produced by the *algC* gene of *P. aeruginosa* is essential for the production of four polysaccharides, alginate, Pel, Psl, and LPS, all of which influence the development of biofilms [55]. Almost all the genes that generate any of the aforementioned polysaccharides contribute to the development of biofilms in an organism [55].

11.6 Bioelectrogenesis and microbial metabolism

In MFCs several micro-organisms have been tested for energy production, bioremediation, and a variety of other uses. In MFC technology different substrates have been used to grow micro-organisms, for example acetate, ethanol, glucose, lactate, sucrose, starch, and xylose, among other nutrients and wastewater such as chocolate industry wastewater, beer brewery wastewater, swine wastewater, protein-rich wastewater, and paper recycling wastewater, among others [71]. Regardless of the availability of a broad variety of carbon sources and micro-organisms, only a few microbes have been identified as producing electricity in MFCs. Exoelectrogens from a number of sources have been investigated in MFCs, including gram-negative bacteria, gram-positive bacteria, yeast, algae, cyanobacteria, and even fungus. The complex organic materials are completely oxidized in the anodic chamber into simple components by those micro-organisms, making them very efficient for power production. However, for development and energy generation, a certain exoelectrogen may oxidize specific substrates or types of substrates. Furthermore, each exoelectrogen has distinct genes, proteins, enzymes or pathways for breakdown or oxidation depending on the kind of substrate. Hence, the electricity production by MFCs is determined by the choice of bacterial consortia and preferred substrate. When an MFC was supplied with an inoculum of aerobic–anaerobic sludge along with glucose as a substrate for 3 months, the substrate to power conversion rates for the bacteria rose seven-fold [72].

Organic compounds including lipids, carbohydrates, and proteins are used as electron donors in MFCs to generate energy via redox processes at the anode. These complicated chemical compounds are subsequently converted to acetyl Co-A via glycolysis and other mechanisms, which are ultimately used in the citric acid cycle. From three nicotinamide adenine dinucleotide (NAD+), three equivalents of reduced NADH are produced in a single step of the citric acid cycle. CO_2 is created as a by-product when one flavin adenosine dinucleotide (FAD) is converted to FADH2. Both prokaryotes (bacteria) and eukaryotes (humans) have cytoplasmic metabolic pathways (the Krebs cycle and glycolysis). The electron carriers NADH and FADH2 transmit their electrons to the electron transport chain (ETC) to form adenosine triphosphate (ATP), an energy carrier molecule. The respiratory response takes place in the cell membrane of a bacterium (which includes the outer cell membrane, periplasm, and inner cell membrane), which houses the system that allows electron exchanges to take place. This electron exchange is the basis of MFCs. In yeast the ETC is found on the inner mitochondrial membrane. Cytochromes, ubiquinone, NADH dehydrogenase, and coenzyme Q are the four intermediate proteins found in the ETC (however, different species may have different intermediary proteins). The electrons are transported to the terminal electron acceptor by these proteins, while reduced protons are pushed out of the cell and given to the cathode by the PEM. Chemical facilitators were employed to catalyse the movement of electrons from the bacteria to the anode before it was discovered that bacteria can aid in electron transfer [73].

These mediators decrease when they come into contact with ETC components, then electrons are transferred to the anode after they exit the cell. Furthermore, depending on the anode capacity, the bacteria's metabolism may transition to fermentative metabolism from oxidative phosphorylation (metabolism). Bacteria adapt to oxidative metabolism as electron acceptors such as sulphate, nitrate, etc, are present, when the anode potential is low and electron acceptors receive and deposit the electrons. Bacteria prefer to use the fermentation metabolism when there are no electron acceptors available. One-third of the electrons generated during the fermentation of glucose may be utilized to generate electricity, while the extra electrons remain in the fermentation by-products, that can be anaerobically oxidized further to generate electricity in MFCs. For example *Geobacter* sp. carried out anaerobic oxidation of fermentation by-products [48, 72]. Several bacteria (*Enterococcus* sp., *Clostridium* sp.) have been injected anaerobically in MFCs to harvest fermentation products in addition to power production [74]. *Clostridium* sp. is a highly effective hydrogen generator in MFCs, just as *Geobacter* sp. is a very effective exoelectrogen [75, 76]. Certainly, as compared to pure cultures, a mixed culture biofilm in MFCs has shown higher power capabilities. This might be due to metabolic relationships between bacteria present in biofilms, however, the topic requires further exploration and experimental proof. The anode's potential is crucial in determining the bacterial metabolism. The bacteria are influenced by a negative anode potential to supply electrons via more reduced complexes [77]. Consequently, the output power and energy recovery in MFC increases due to lower energy absorption of the bacteria. A higher power potential of 45 mA m^2 at 0.6 V, compared to 15 mA m^2 at 0.2 V, was produced at negative anode potentials by a mixed culture of sulphate-reducing bacteria [78]. The potential of the cathode has also been found to enhance MFC performance. The time for start-up was lowered from 26 days to 19 days when at a constant −300 V cathodic capacity, the power potential was increased from 4.1 W m^{-3} to 6.4 W m^{-3} in a Cr (VI) reduction MFC [79].

11.7 Electron transfer by micro-organisms in MFCs

Electricity generation in MFCs requires the movement of electrons or transfer of electrons. Electrodes in MFCs are immoveable so the micro-organisms influence this transport of electrons. Electron transport primarily occurs between micro-organisms and electrodes in two directions: the first is from the micro-organisms to the electrode (at the anode), and the second is when a biocathode is used to catalyse oxygen reduction and electrons move from the electrode to the micro-organisms (at the cathode).

11.7.1 Electron transfer from micro-organisms to electrodes

Micro-organisms transfer electrons to an electrode through three different processes. First, micro-organisms contain redox-active proteins, known as cytochromes, present on their cell surfaces, which are involved in short-range electron transfer. Second, micro-organisms secrete certain soluble molecules such as flavins and pyocyanin which act as electron shuttles and aid electron transport. Third, micro-organisms have specialised micro-structures called micropili on the cell membrane which facilitate the long-range transport of electrons and are thus known as conductive pili (figure 11.2).

11.7.1.1 Cytochromes allow direct electron transfer

To better understand the processes of direct electron transfer, the bacteria *G. sulfurreducens* has been researched widely. The complete oxidation of carbon to water and carbon oxide is done anaerobically by a key metabolic enzyme present in *G. sulfurreducens*, this may involve the transfer of electrons to other electron

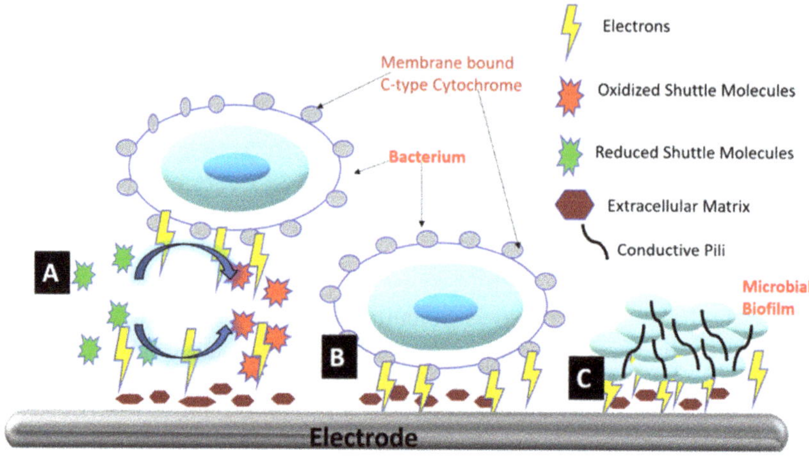

Figure 11.2. Electron transport in MFCs: (a) through micro-organism-secreted electron shuttles; (b) direct transfer through cytochromes; and (c) through microbial nanowires.

acceptors [80]. The *G. sulfurreducens* genome has also been found to have c-type cytochromes with motifs with heme groups. These motifs are involved in exposing c-type cytochromes to the cell's outer surface [40, 81]. The abundance of cytochromes in the organism is a beneficial trait that improves electron transport across the cell/electrode contact. *G. sulfurreducens* has electron transport proteins and substances in the periplasm and on the outer membrane. These additional substances or proteins include iron-sulphur proteins, quinones, and b-type cytochromes. Furthermore, c-type cytochromes have been shown to transport electrons to a variety of extracellular electron acceptors both in *in vitro* and *in vivo* conditions [40, 81, 82]. The OmcZ has a vital role in direct electron transfer, and the presence of OmcZ at the biofilm–anode interface enabling electron transfer was verified by immunogold labelling of *G. sulfurreducens* biofilms. However, between the anode and biofilm the electron transmission is hindered by the OmcZ mutant strain [40]. In 2009 Nevin *et al* examined the expression of genes in *G. sulfurreducens* biofilm cells, which are cultured on several electron acceptors. They compared the gene expression between the cells which are grown on graphite with fumarate and on simple graphite [83]. According to microarray research, C-type cytochrome encoded genes are *omcB*, *omcE*, *omcS*, *omcT*, and *omcZ*. In current harvesting cells, the cytochromes *OmcE* and *OmcZ* were found abundantly and *OmcS* was rarely present. Furthermore, cells lacking omcZ were unable to produce current or form biofilms. It demonstrates the significance of the cytochrome in electron transport. Those cells which were lacking in other genes has no effect on current generation or biofilm creation [40, 84, 85]. According to abundant evidence, in biofilms producing higher current, OmcZ appears to be the most significant cytochrome. It is a hydrophobic protein with an octaheme group and exists in two structures, OmcZ L is long and other OmcZ S is short. The latter is the principal form of OmcZ [40]. OmcZ is thought to facilitate electron transport across the biofilm, whereas OmcB is thought to mediate electron transfer across the biofilm–electrode contact in

G. sulfurreducens biofilms. In electron transport via the biofilm, the cytochromes OmcS and OmcE play a minor function [39]. The OmcF mutant strain of *G. sulfurreducens* was found to have a low current density [86]. Furthermore, the data revealed that the OmcF is involved in the electron transfer process either directly or indirectly, meaning that the OmcF is vital in the generation of electricity [86].

In *G. sulfurreducens*, the multicopper proteins OmpB and OmpC of the outer membrane are essential for the reduction of Fe (III) oxide in addition to the c-type cytochromes of the outer membrane [87, 88]. However, this would be an interesting topic of future research, how these multicopper proteins affect the production of electricity in MFCs, as it is still unknown. In *Desulfovibrio alaskensis* G20, a sulphate-reducing bacterium, electron flow components are being studied. A new model for electron transfer has been revealed, also the type I tetraheme cytochrome c 3 (TpI c 3) and transmembrane complexes (QrcA) play a key role in electron transfer across the cell membrane for sulphate reduction [89]. The direct electron transfer mechanism has also been investigated in gram-positive organisms of the genus *Thermincola potens*. During *T. potens* growth on hydrous ferric oxides, the expression of numerous multiheme c-type cytochromes (MHCs) on the cell surface is demonstrated by surface-enhanced Raman spectroscopy. The findings showed new evidence for the involvement of cytochromes associated with the cell wall and MHC in electron transport across the gram-positive bacterium's cell envelope [35]. More knowledge regarding the proteins or genes involved in direct electron transfer, as well as genetic modification, may lead to improvements in MFC output and efficiency.

11.7.1.2 Electron transfer through micro-organism secreted electron shuttles

The secretion of soluble electron shuttles allows electron transport to electron acceptors or electrodes that are soluble or insoluble. Some bacteria, such as *Shewanella oneidensis*, *P. aeruginosa*, and *Geothrix fermentans*, have been identified as mediating electron secreting soluble electron shuttles. *G. fermentans* secretes an electron shuttle that aids in Fe (III) oxide reduction [8]. To reduce Fe (III), *G. fermentans* generates two soluble redox-active electron shuttles. One is riboflavin, at a redox potential of 0.2 V, and the other is unknown, at a redox potential of 0.3 V [87]. *P. aeruginosa* produces important electron transport molecules such as phenazine-1-carboxamide and pyrocyanin. A *P. aeruginosa* mutant strain, lacking in the generation of phenazine-1-carboxamide and pyrocyanin, produced just 5% of the power output of wild-type cultures [90]. Furthermore, research found that pyocyanin, which is utilized by *P. aeruginosa* as well as other bacteria, stimulates significant electron transport [90, 91]. Exocellular electron transfer efficiency as well as power output are increased when the *phzM* (methyltransferase-encoding) gene is overexpressed as it increased pyocyanin production by 1.6-fold in *P. aeruginosa-phzM*-inoculated MFCs [92]. In conjunction with anoxic growth, *Shewanella* species produce riboflavin and flavin mononucleotides as external electron shuttles for the reduction of Fe (III) oxides [93, 94]. *S. loihica* PV-4 strain cultivated on graphite electrode, the amount of quinone derivatives and riboflavin was found to be increased in the cell-free supernatant, resulting in a higher anodic current density of 90 A cm^{-2} [95, 96]. In *S. oneidensis*, Kotloski and Gralnick (2013) found a flavin adenine dinucleotide transporter that exports flavin electron shuttles, which aids in electron transfer to intractable substrates [97]. OmcA and Decaheme c-type cytochromes MtrC are found on the external surface of *S. oneidensis* MR-1 cells; they are a portion of a multiprotein complex that aids in electron hopping across the cell membrane [89]. During biofilm development, the cytochrome OmcA is also involved in the adhesion of bacteria to the electrode surface [93]. In *S. oneidensis* MtrCAB, a mediated and direct exocellular electron transport complex was found. This MtrCAB was inserted into *Escherichia coli* utilising more flexible induction technique [89, 90]. The *E. coli* strains exhibited poor cell development, poor control of MtrCAB expression, and the data demonstrated that strains with better cell growth and fewer morphological defects provided the highest current densities, not those with more MtrC and MtrA expression [99]. *Lactococcus lactis* generates membrane-associated quinones with different colours that aids in the electron transfer towards external electron acceptors such as Cu (II) and Fe (III) [100]. The bacteria use soluble redox mediators, such as 2-amino-3-dicarboxy-1,4 naphthoquinone, to transfer electrons towards the anode [101]. In MFCs the strain of *Klebsiella pneumoniae* (L17), generates a recycling electron shuttle called 2,6-di-tert-butyl-pbenzoquinon, which transfers electrons from the cathode to the anode [102, 103].

11.7.1.3 Microbial nanowires for electron transfer

Long-range electron transmission is facilitated by the micro-organism's dense network of conductive pili, which are responsible for conductive biofilms with significant current generation. Although many microbes can produce pili, only *Shewanella* sp. [105] and *Geobacter* sp. [39, 104] can produce conductive pili. In long-range electron transport in biofilms, the function of conductive pili is well-studied in *G. sulfurreducens* and these electronic networks are found to cause a tenfold increase in power output [48]. Pili derived from *G. sulfurreducens* are type IV pili composed of PilA monomers [106]. Type IV pili are small structural proteins with a conserved N-terminal domain that forms an α-helix, a transmembrane domain, and a protein–protein interaction domain that have a molecular weight of 7–20 kDa, a length of 10–20 m, and a width of 3–5 m [88]. Furthermore, the PilA C terminus includes aromatic amino acids with a conserved sequence (Tyr, His, Trp, Met, and Phe), which contributes to the pi–pi orbitals overlapping in the pili structure, and hence causing a metal-like conductivity, which is absent in non-conductive biofilms [107]. PilR, which is a RpoN-dependent enhancer-binding protein, directly controls the function of PilA. Moreover, Juárez *et al* (2009) discovered that a strain lacking in the pilR gene exhibited decreased reduction of insoluble Fe (III) and soluble Fe (III) [108]. The hypothesis of Malvankar *et al* (2011) was ruled out, that cytochromes and pili of *G. sulfurreducens* are coupled and thus cytochromes play a role in transport of electron along with pili [39]. The conductivity by nanowires of *G. sulfurreducens* cannot be ascribed to cytochromes since the cytochrome-to-cytochrome distance was nearly 200 times greater than that required for electrification. *G. sulfurreducens* strain PA with the *pilA* gene substituted with the *pilA* gene of *P. aeruginosa* PAO1 produced equivalent pili subunits and c-type cytochrome OmcS to the control strain, but had lower current output and Fe (III) oxide reduction. Furthermore, pili conductivity is not conferred by the presence of c-type cytochrome OmcS [71, 85]. Research suggests that magnetite may help microbial extracellular electron transport; in Fe (III) oxide reduction, an OmcS-deficient strain is compensated by magnetite for extracellular electron transfers [71]. *S. oneidensis* MR-1 nanowires were shown to be conducive using the study of probe atomic force microscopy and gene deletion of *OmcA* and *MtrC* [17]. Further investigation of the electronic transport properties of *S. oneidensis* MR-1 nanowires revealed p-type, adjustable electronic behaviour with field-effect mobility [109]. For effective electron transfer and energy distribution, a multistep hopping mechanism in the methanogen *Methanothermobacter thermautotrophicus* was used [9]. Direct interspecies electron transfer (DIET) has been observed in anaerobic digester aggregates of *Methanosaeta harundinacea* and *G. metallireducens* [110]. It has been proposed that granular-activated carbon (GAC) stimulates DIET between bacteria and methanogens [71]. The function of pili and related c-type cytochrome in DIET is simulated by GAC [71]. The contribution of DIET in energy generation and its molecular mechanism are poorly known, necessitating a thorough study into the subject.

11.7.1.4 Transfer of electrons from the electrode to micro-organisms

In MFC technology, several micro-organisms have been used as biocathodes but little is known about how electrons go from the electrode to bacteria, although micro-organisms receive electrons from the cathode in a different way than they give electrons to the anode (figure 11.3). *Geobacter* species accepting electrons directly from an electrode was the first proof in this regard [111]. In the aerated cathode, *S. oneidensis* MR-1 produces a molecule riboflavin, which acts as an electron shuttle mediator for transferring electrons to Cr (VI) [112]. Pure cultures of *Acinetobacter calcoaceticus* and *S. putrefaciens* excrete a redox molecule that is comparable to pyrroloquinoline quinone (PQQ) and uses outer membrane-bound redox chemicals for extracellular electron transfer [101]. Acidophile bacteria, *Acidithiobacillus ferrooxidans*, were utilized as a biocathode to show that the outer membrane-bound cytochrome c (Cyc2) is linked to the microbial-catalyzed O_2 reduction [113]. Cyt 572, a unique membrane protein, having helical structure, isolated acidophilic micro-organisms demonstrated the capability to oxidise Fe (II) [114], although it is still unclear if the protein is involved in electron transport processes. Using cyclic voltammetry, researchers discovered an undiscovered redox-active chemical produced by *P. aeruginosa* that is contributing to electron transport from the electrode to specific azo bonds, resulting in azo dye decolourization [115]. During the dichlorination of pentachlorophenol (PCP) in MFC, a bio-cathodic microbial population dominated by Proteobacteria, Bacteroidetes, and Firmicutes transferred electrons directly, since the characterization of the medium by cyclic voltammetry revealed no redox mediator produced by the bacteria [71]. The detailed chemical process by which any microbe accepts electrons from electrodes is not completely discovered and may be considered a future feature.

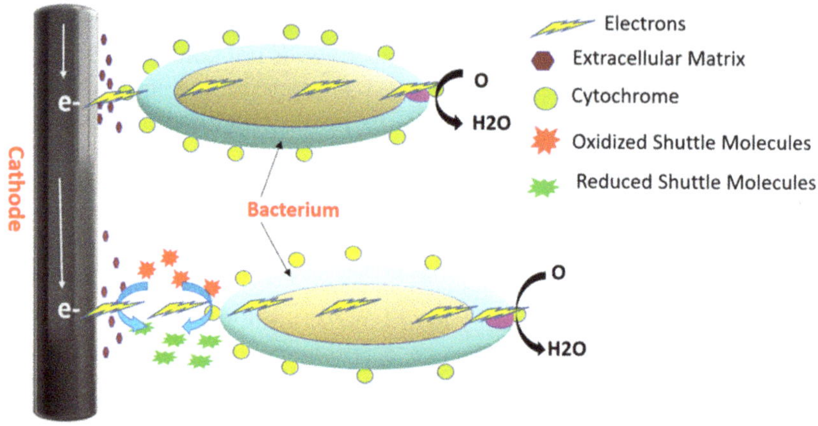

Figure 11.3. Electron transfers from the electrode to micro-organisms: direct transfer through membrane-bound c-type cytochromes and indirect transfer through electron shuttles.

11.8 Micro-organisms with higher power generating potential in MFCs

Unless the MFC design, operating circumstances, chemical solutions, and nutrients used in the process are same, the power density generated by one microbe, such as *Geobacter* sp., cannot be compared to that produced by another species, such as *Shewanella* sp. Until now, a variety of micro-organisms used in various MFCs have generated electrical energy under a variety of conditions. In MFC technology, several novel bacteria have recently been found, and this chapter only covers the most frequent micro-organisms researched (in the cathode and anode) for achieving higher power densities [7]. Tables 11.1 and 11.2 list micro-organisms with unknown and known natural electron mediators, respectively.

11.9 Future recommendations

The uses of MFC technology are still limited to within the four walls of the laboratory. To put it another way, the technology has not yet reached the point of commercialization. Only *Shewanella* spp. and *Geobacter* spp. are well-studied regarding the formation of biofilms and the transfer of electrons from exoelectrogens to electrodes, thus research on other microbes is necessary to understand and utilise their influence on MFCs. Additionally, genetic modification may enhance the exocellular electron transport rates, causing more efficient output by MFCs. It is suggested that conductive pili-producing bacteria be found, albeit such microbes can produce greater power densities. The processes of electron transmission from electrodes to microbes remains unknown. The cathode compartment will be dominated by micro-organisms that can receive electrons from the electrode. OmpB and OmpC, two outer membrane multicopper proteins, have been shown to play a significant part in Fe (III) oxide reduction, although more research into their roles in electron transport pathways is required.

Table 11.1. Micro-organisms that possess well-characterized electron transfer intermediates.

Micro-organisms	Molecules involved in electron transfers	Power density/current density	References
Thermincola ferriacetica	Anthraquinone 2, 6 disulfonate	12,000 mA m^{-2}	[34]
Pseudomonas aeruginosa	Phenazine-1-carbcxamide	4310 mW m^{-2}	[117]
	Pyocyanin		
Geobacter sulfurreducens	Type IV pili	3147 mA m^{-2}	[40]
	c-type cytochrome Z		
Shewanella oneidensis	Flavins, riboflavin	3000 mW m^{-2}	[116]
Rhodopseudomonas palustris	c-type cytochromes	2720 mW m^{-2}	[120]
Desulfovibrio desulfuricans	Tetraheme cytochrome C_3	233 mA m^{-2}	[119]
Klebsiella pneumonia	2, 6-di-tert-butyl-p-benzoquinone	199 mA m^{-2}	[102]
Geobacter metallireducens	c-type cytochromes	40 mW m^{-2}	[121]
	OmcB and OmcE		
Shewanella putrefaciens	c-type cytochromes	4.92 W m^{-3}	[118]
	MtrC and OmcA		
Desulfovibrio alaskensis	Transmembrane complexes (QrcA)		[89]
	Tetraheme cytochrome C 3		

mW m^{-2} = units of surface power density, W m^{-3} = volume power density, mA m^{-2} = current density.

Table 11.2. Power density exhibited by micro-organisms utilising unknown electron mediators.

Power density/current density	Micro-organisms	References
	Bacteria	
3390 mA m^{-2}	*Escherichia coli*	[122]
2625 mW m^{-3}	*Ochrobactrum* sp.	[123]
282 mA m^{-2}	*Saccharomyces cerevisiae*	[124]
205 mA m^{-2}	*Citrobacter* sp.	[125]
85 mW m^{-2}	*Lysinibacillus sphaericus*	[126]
	Algae	
2485 mW m^{-3}	*Chlorella vulgaris*	[127]
1926 mW m^{-2}	*Scenedesmus*	[128]
320 mW m^{-3}	*Coriolus versicolor*	[129]
14 mW m^{-2}	*Cyanobacteria* 1	[130]
10 mW m^{-3}	*Arthrospira maxima*	[131]

mW m^2 = units of surface power density, W m^3 = volume power density, mA m^2 = current density.

11.10 Conclusions

Micro-organisms serve as the powerhouses of MFCs, and those that can produce conductive biofilms are particularly important. Specific proteins produced by the bacteria, including c-type cytochromes, pili, and QS, are essential for the development of conductive biofilms. Moreover, the external polysaccharide greatly influences the biofilms, which are the key component in MFCs. In addition, a growing body of evidence indicates that the c-type cytochromes OmcB and OmcZ have a great influence on electron transport processes. *Shewanella* spp. and *Geobacter* spp. are capable of long-distance electron transport via pili. *Shewanella* spp. use flavins to transport electrons to the electrodes, whereas *Pseudomonas* spp. secrete pyocyanin. With riboflavin as an electron mediator, the exoelectrogen is demonstrated to receive electrons from electrodes. Furthermore, the transfer of electrons from the electrode to bacterial cells is mediated by two unique cytochromes, Cyt 572 and Cyt 579. Biocathodes have reduced the cost of the technology. MFC has emerged as the sole technology capable of producing renewable energy and a variety of simultaneous additional uses.

References

[1] Shewa W A, Chaganti S R and Lalman J A 2013 Electricity generation and biofilm formation in microbial fuel cells using plate anodes constructed from various grades of graphite *J. Green Eng.* **4** 13–32

[2] Mevers E, Su L, Pishchany G, Baruch M, Cornejo J, Hobert E, Dimise E, Ajo-Franklin C M and Clardy J 2019 An elusive electron shuttle from a facultative anaerobe *Elife* **8** e48054

[3] Liu X, Shi L and Gu J D 2018 Microbial electrocatalysis: redox mediators responsible for extracellular electron transfer *Biotechnol. Adv.* **36** 1815–27

[4] Costa N L, Clarke T A, Philipp L A, Gescher J, Louro R O and Paquete C M 2018 Electron transfer process in microbial electrochemical technologies: the role of cell-surface exposed conductive proteins *Bioresour. Technol.* **255** 308–17

[5] Ancona V, Caracciolo A B, Borello D, Ferrara V, Grenni P and Pietrelli A 2020 Microbial fuel cell: an energy harvesting technique for environmental remediation *Int. J. Env. Impacts* **3** 168–79

[6] Kumar S S, Kumar V, Kumar R, Malyan S K and Pugazhendhi A 2019 Microbial fuel cells as a sustainable platform technology for bioenergy, biosensing, environmental monitoring, and other low power device applications *Fuel* **255** 115682

[7] Kumar R, Singh L and Wahid Z A 2015 Role of microorganisms in microbial fuel cells for bioelectricity production *Microbial Factories* (New Delhi: Springer) pp 135–54

[8] Bond D R and Lovley D R 2003 Electricity production by *Geobacter sulfurreducens* attached to electrodes *Appl. Env. Microbiol.* **69** 1548–55

[9] Gorby Y A *et al* 2006 Electrically conductive bacterial nanowires produced by *Shewanella oneidensis* strain MR-1 and other microorganisms *Proc. Natl Acad. Sci.* **103** 11358–63

[10] Chaudhuri S K and Lovley D R 2003 Electricity generation by direct oxidation of glucose in mediatorless microbial fuel cells *Nat. Biotechnol.* **21** 1229–32

[11] Jayapriya J and Ramamurthy V 2012 Use of non-native phenazines to improve the performance of *Pseudomonas aeruginosa* MTCC 2474 catalysed fuel cells *Bioresour. Technol.* **124** 23–8

[12] Chisti Y 2007 Biodiesel from microalgae *Biotechnol. Adv.* **25** 294–306

[13] Park D H and Zeikus J 1999 Utilization of electrically reduced neutral red by *Actinobacillus succinogenes*: physiological function of neutral red in membrane-driven fumarate reduction and energy conservation *J. Bacteriol.* **181** 2403–10

[14] Pham C A, Jung S J, Phung N T, Lee J, Chang I S, Kim B H, Yi H and Chun J 2003 A novel electrochemically active and Fe (III)-reducing bacterium phylogenetically related to *Aeromonas hydrophila*, isolated from a microbial fuel cell *FEMS Microbiol. Lett.* **223** 129–34

[15] Mekuto L, Olowolafe A V, Huberts R, Dyantyi N, Pandit S and Nomngongo P 2020 Microalgae as a biocathode and feedstock in anode chamber for a selfsustainable microbial fuel cell technology: a review *S. Afr. J. Chem. Eng.* **31** 7–16

[16] Prasad D, Arun S, Murugesan M, Padmanaban S, Satyanarayanan R S, Berchmans S and Yegnaraman V 2007 Direct electron transfer with yeast cells and construction of a mediatorless microbial fuel cell *Biosens. Bioelectron.* **22** 2604–10

[17] El-Naggar M Y, Wanger G, Leung K M, Yuzvinsky T D, Southam G, Yang J, Lau W M, Nealson K H and Gorby Y A 2010 Electrical transport along bacterial nanowires from *Shewanella oneidensis* MR-1 *Proc. Natl Acad. Sci.* **107** 18127–31

[18] Dumas C, Basseguy R and Bergel A 2008 Electrochemical activity of *Geobacter sulfurreducens* biofilms on stainless steel anodes *Electrochim. Acta* **53** 5235–41

[19] Boon N, Aelterman P, Clauwaert P, De Schamphelaire L, Vanhaecke L, De Maeyer K, Höfte M, Verstraete W and Rabaey K 2008 Metabolites produced by *Pseudomonas* sp. enable a gram-positive bacterium to achieve extracellular electron transfer *Appl. Microbiol. Biotechnol.* **77** 1119–29

[20] Schaefer A L *et al* 2008 A new class of homoserine lactone quorum-sensing signals *Nature* **454** 595–9

[21] Venkataraman A, Rosenbaum M, Arends J B, Halitschke R and Angenent L T 2010 Quorum sensing regulates electric current generation of *Pseudomonas aeruginosa* PA14 in bioelectrochemical systems *Electrochem. Commun.* **12** 459–62

[22] Patil S A, Hägerhäll C and Gorton L 2012 Electron transfer mechanisms between microorganisms and electrodes in bioelectrochemical systems *Bioanal. Rev.* **4** 159–92

[23] Rollefson J B, Stephen C S, Tien M and Bond D R 2011 Identification of an extracellular polysaccharide network essential for cytochrome anchoring and biofilm formation in *Geobacter sulfurreducens J. Bacteriol.* **193** 1023–33

[24] Arulmani S R B, Ganamuthu H L, Ashokkumar V, Govindarajan G, Kandasamy S and Zhang H 2020 Biofilm formation and electrochemical metabolic activity of *Ochrobactrum* Sp JSRB-1 and *Cupriavidus* Sp JSRB-2 for energy production *Env. Technol. Innov.* **20** 101145

[25] Cecconet D, Sabba F, Devecseri M, Callegari A and Capodaglio A G 2020 *In situ* groundwater remediation with bioelectrochemical systems: a critical review and future perspectives *Env. Int.* **137** 105550

[26] Angelaalincy M J, Navanietha Krishnaraj R, Shakambari G, Ashokkumar B, Kathiresan S and Varalakshmi P 2018 Biofilm engineering approaches for improving the performance of microbial fuel cells and bioelectrochemical systems *Front. Energy Res.* **6** 63

[27] Bashir S, Mulvaney S P, Houf W, Villanueva L, Wang Z, Buck G and Liu J L 2021 Microbial fuel cells: design and evaluation of catalysts and device *Advances in Sustainable Energy* (Cham: Springer) pp 681–764

[28] Krige A, Sjöblom M, Ramser K, Christakopoulos P and Rova U 2019 On-line Raman spectroscopic study of cytochromes' redox state of biofilms in microbial fuel cells *Molecules* **24** 646

[29] Steidl R J, Lampa-Pastirk S and Reguera G 2016 Mechanistic stratification in electroactive biofilms of *Geobacter sulfurreducens* mediated by pilus nanowires *Nat. Commun.* **7** 1–11

[30] Kleine D, Chodorski J, Mitra S, Schlegel C, Huttenlochner K, Müller-Renno C, Mukherjee J, Ziegler C and Ulber R 2019 Monitoring of biofilms grown on differentially structured metallic surfaces using confocal laser scanning microscopy *Eng. Life Sci.* **19** 513–21

[31] Nandy A, Sharma M, Venkatesan S V, Taylor N, Gieg L and Thangadurai V 2019 Comparative evaluation of coated and non-coated carbon electrodes in a microbial fuel cell for treatment of municipal sludge *Energies* **12** 1034

[32] Kiran R and Patil S A 2019 Microbial electroactive biofilms *Introduction to Biofilm Engineering* (Washington, DC: American Chemical Society) pp 159–86

[33] Arulmani S R B, Jayaraj V and Jebakumar S R 2016 Long-term electricity production from soil electrogenic bacteria and high-content screening of biofilm formation on the electrodes *J. Soils Sediments* **16** 831–41

[34] Parameswaran P, Bry T, Popat S C, Lusk B G, Rittmann B E and Torres C I 2013 Kinetic, electrochemical, and microscopic characterization of the thermophilic, anode-respiring bacterium *Thermincola ferriacetica Env. Sci. Technol.* **47** 4934–40

[35] Wrighton K C, Thrash J C, Melnyk R A, Bigi J P, Byrne-Bailey K G, Remis J P, Schichnes D, Auer M, Chang C J and Coates J D 2011 Evidence for direct electron transfer by a gram-positive bacterium isolated from a microbial fuel cell *Appl. Env. Microbiol.* **77** 7633–9

[36] Arunasri K and Mohan S V 2019 Biofilms: microbial life on the electrode surface *Microbial Electrochemical Technology* (Amsterdam: Elsevier) pp 295–313

[37] Harmsen M, Yang L, Pamp S J and Tolker-Nielsen T 2010 An update on *Pseudomonas aeruginosa* biofilm formation, tolerance, and dispersal *FEMS Immunol. Med. Microbiol.* **59** 253–68

[38] Logan B E, Rossi R and Saikaly P E 2019 Electroactive microorganisms in bioelectrochemical systems *Nat. Rev. Microbiol.* **17** 307–19

[39] Malvankar N S, Tuominen M T and Lovley D R 2012 Lack of cytochrome involvement in long-range electron transport through conductive biofilms and nanowires of *Geobacter sulfurreducens Energy Env. Sci.* **5** 8651–9

[40] Inoue K, Leang C, Franks A E, Woodard T L, Nevin K P and Lovley D R 2011 Specific localization of the c-type cytochrome OmcZ at the anode surface in current-producing biofilms of *Geobacter sulfurreducens Env. Microbiol. Rep.* **3** 211–7

[41] Wang V B *et al* 2014 Comparison of flavins and a conjugated oligoelectrolyte in stimulating extracellular electron transport from *Shewanella oneidensis* MR-1 *Electrochem. Commun.* **41** 55–8

[42] Stoodley P, Sauer K, Davies D G and Costerton J W 2002 Biofilms as complex differentiated communities *Annu. Rev. Microbiol.* **56** 187–209

[43] Merritt J H, Ha D G, Cowles K N, Lu W, Morales D K, Rabinowitz J, Gitai Z and O'Toole G A 2010 Specific control of *Pseudomonas aeruginosa* surface-associated behaviors by two c-di-GMP diguanylate cyclases *MBio* **1** e00183–10

[44] Malone J G, Jaeger T, Manfredi P, Dötsch A, Blanka A, Bos R, Cornelis G R, Häussler S and Jenal U 2012 The YfiBNR signal transduction mechanism reveals novel targets for the evolution of persistent *Pseudomonas aeruginosa* in cystic fibrosis airways *PLoS Pathog.* **8** e1002760

[45] Yong Y C, Yu Y Y, Li C M, Zhong J J and Song H 2011 Bioelectricity enhancement via overexpression of quorum sensing system in *Pseudomonas aeruginosa*-inoculated microbial fuel cells *Biosens. Bioelectron.* **30** 87–92

[46] Christwardana M, Frattini D, Duarte K D, Accardo G and Kwon Y 2019 Carbon felt molecular modification and biofilm augmentation via quorum sensing approach in yeast-based microbial fuel cells *Appl. Energy* **238** 239–48

[47] Watson V J and Logan B E 2010 Power production in MFCs inoculated with *Shewanella oneidensis* MR-1 or mixed cultures *Biotechnol. Bioeng.* **105** 489–98

[48] Reguera G, McCarthy K D, Mehta T, Nicoll J S, Tuominen M T and Lovley D R 2005 Extracellular electron transfer via microbial nanowires *Nature* **435** 1098–101

[49] Behera M, Jana P S and Ghangrekar M M 2010 Performance evaluation of low cost microbial fuel cell fabricated using earthen pot with biotic and abiotic cathode *Bioresour. Technol.* **101** 1183–9

[50] Sun D, Chen J, Huang H, Liu W, Ye Y and Cheng S 2016 The effect of biofilm thickness on electrochemical activity of *Geobacter sulfurreducens Int. J. Hydrogen Energy* **41** 16523–8

[51] Kreft J U and Wimpenny J W 2001 Effect of EPS on biofilm structure and function as revealed by an individual-based model of biofilm growth *Water Sci. Technol.* **43** 135

[52] Flemming H C, Neu T R and Wozniak D J 2007 The EPS matrix: the 'house of biofilm cells' *J. Bacteriol.* **189** 7945–7

[53] Angelaalincy M, Senthilkumar N, Karpagam R, Kumar G G, Ashokkumar B and Varalakshmi P 2017 Enhanced extracellular polysaccharide production and self-sustainable electricity generation for PAMFCs by *Scenedesmus* sp. SB1 *ACS Omega* **2** 3754–65

[54] Colvin K M, Irie Y, Tart C S, Urbano R, Whitney J C, Ryder C, Howell P L, Wozniak D J and Parsek M R 2012 The Pel and Psl polysaccharides provide *Pseudomonas aeruginosa* structural redundancy within the biofilm matrix *Env. Microbiol.* **14** 1913–28

[55] Wei Q and Ma L Z 2013 Biofilm matrix and its regulation in *Pseudomonas aeruginosa Int. J. Mol. Sci.* **14** 20983–1005

[56] Friedman L and Kolter R 2004 Two genetic loci produce distinct carbohydrate-rich structural components of the Pseudomonas aeruginosa biofilm matrix *J. Bacteriol.* **186** 4457–65

[57] Jackson K D, Starkey M, Kremer S, Parsek M R and Wozniak D J 2004 Identification of psl, a locus encoding a potential exopolysaccharide that is essential for Pseudomonas aeruginosa PAO1 biofilm formation *J. Bacteriol.* **186** 4466–75

[58] Matsukawa M and Greenberg E P 2004 Putative exopolysaccharide synthesis genes influence Pseudomonas aeruginosa biofilm development *J. Bacteriol.* **186** 4449–56

[59] Zhang Y, Jiang J, Zhao Q, Gao Y, Wang K, Ding J, Yu H and Yao Y 2017 Accelerating anodic biofilms formation and electron transfer in microbial fuel cells: role of anionic biosurfactants and mechanism *Bioelectrochemistry* **117** 48–56

[60] Ma L, Jackson K D, Landry R M, Parsek M R and Wozniak D J 2006 Analysis of *Pseudomonas aeruginosa* conditional psl variants reveals roles for the psl polysaccharide in adhesion and maintaining biofilm structure postattachment *J. Bacteriol.* **188** 8213–21

[61] Ma L, Conover M, Lu H, Parsek M R, Bayles K and Wozniak D J 2009 Assembly and development of the *Pseudomonas aeruginosa* biofilm matrix *PLoS Pathog.* **5** e1000354

[62] Toyofuku M, Inaba T, Kiyokawa T, Obana N, Yawata Y and Nomura N 2016 Environmental factors that shape biofilm formation *Biosci. Biotechnol. Biochem.* **80** 7–12

[63] Nocelli N, Bogino P C, Banchio E and Giordano W 2016 Roles of extracellular polysaccharides and biofilm formation in heavy metal resistance of rhizobia *Materials* **9** 418

[64] Chen B, Li F, Liu N, Ge F, Xiao H and Yang Y 2015 Role of extracellular polymeric substances from *Chlorella vulgaris* in the removal of ammonium and orthophosphate under the stress of cadmium *Bioresour. Technol.* **190** 299–306

[65] Hong B H, Joe M M, Selvakumar G, Kim K Y, Choi J H and Sa T M 2017 Influence of salinity variations on exocellular polysaccharide production, biofilm formation and flocculation in halotolerant bacteria *J. Environ. Biol.* **38** 657

[66] Christenson L 2011 Algal biofilm production and harvesting system for wastewater treatment with biofuels by-products *MSc Thesis* Utah State University, Logan, UT

[67] Monzon O, Yang Y, Li Q and Alvarez P J 2016 Quorum sensing autoinducers enhance biofilm formation and power production in a hypersaline microbial fuel cell *Biochem. Eng. J.* **109** 222–7

[68] Agarwal A, Kahyaoglu C and Hansen D B 2012 Characterization of HmqF, a protein involved in the biosynthesis of unsaturated quinolones produced by *Burkholderia thailandensis Biochemistry* **51** 1648–57

[69] Diggle S P, Winzer K, Lazdunski A, Williams P and Cámara M 2002 Advancing the quorum in *Pseudomonas aeruginosa*: MvaT and the regulation of N-acylhomoserine lactone production and virulence gene expression *J. Bacteriol.* **184** 2576–86

[70] Vasseur P, Vallet-Gely I, Soscia C, Genin S and Filloux A 2005 The pel genes of the *Pseudomonas aeruginosa* PAK strain are involved at early and late stages of biofilm formation *Microbiology* **151** 985–97

[71] Liu X, Tremblay P L, Malvankar N S, Nevin K P, Lovley D R and Vargas M 2014 A *Geobacter sulfurreducens* strain expressing *Pseudomonas aeruginosa* type IV pili localizes OmcS on pili but is deficient in Fe (III) oxide reduction and current production *Appl. Env. Microbiol.* **80** 1219–24

[72] Rabaey K and Verstraete W 2005 Microbial fuel cells: novel biotechnology for energy generation *Trends Biotechnol.* **23** 291–8

[73] Watmough N J and Frerman F E 2010 The electron transfer flavoprotein: ubiquinone oxidoreductases *Biochim. Biophys. Acta—Bioenerg.* **1797** 1910–6

[74] Mitra R, Balachandar G, Singh V, Sinha P and Das D 2017 Improvement in energy recovery by dark fermentative biohydrogen followed by biobutanol production process using obligate anaerobes *Int. J. Hydrogen Energy* **42** 4880–92

[75] Singh L and Wahid Z A 2015 Enhancement of hydrogen production from palm oil mill effluent via cell immobilisation technique *Int. J. Energy Res.* **39** 215–22

[76] Singh L, Siddiqui M F, Ahmad A, Rahim M H A, Sakinah M and Wahid Z A 2013 Biohydrogen production from palm oil mill effluent using immobilized mixed culture *J. Ind. Eng. Chem.* **19** 659–64

[77] Zakaria B S, Lin L and Dhar B R 2019 Shift of biofilm and suspended bacterial communities with changes in anode potential in a microbial electrolysis cell treating primary sludge *Sci. Total Env.* **689** 691–9

[78] Chou T Y, Whiteley C G and Lee D J 2014 Anodic potential on dual-chambered microbial fuel cell with sulphate reducing bacteria biofilm *Int. J. Hydrogen Energy* **39** 19225–31

[79] Huang L, Chai X, Chen G and Logan B E 2011 Effect of set potential on hexavalent chromium reduction and electricity generation from biocathode microbial fuel cells *Env. Sci. Technol.* **45** 5025–31

[80] Kiely P D, Regan J M and Logan B E 2011 The electric picnic: synergistic requirements for exoelectrogenic microbial communities *Curr. Opin. Biotechnol.* **22** 378–85

[81] Leang C, Qian X, Mester T and Lovley D R 2010 Alignment of the c-type cytochrome OmcS along pili of *Geobacter sulfurreducens Appl. Env. Microbiol.* **76** 4080–4

[82] Richter K, Schicklberger M and Gescher J 2012 Dissimilatory reduction of extracellular electron acceptors in anaerobic respiration *Appl. Env. Microbiol.* **78** 913–21

[83] Nevin K P *et al* 2009 Anode biofilm transcriptomics reveals outer surface components essential for high density current production in *Geobacter sulfurreducens* fuel cells *PLoS One* **4** e5628

[84] Strycharz S M, Glaven R H, Coppi M V, Gannon S M, Perpetua L A, Liu A, Nevin K P and Lovley D R 2011 Gene expression and deletion analysis of mechanisms for electron transfer from electrodes to *Geobacter sulfurreducens Bioelectrochemistry* **80** 142–50

[85] Smith J A, Tremblay P L, Shrestha P M, Snoeyenbos-West O L, Franks A E, Nevin K P and Lovley D R 2014 Going wireless: Fe (III) oxide reduction without pili by *Geobacter sulfurreducens* strain JS-1 *Appl. Env. Microbiol.* **80** 4331–40

[86] Kim B C, Postier B L, DiDonato R J, Chaudhuri S K, Nevin K P and Lovley D R 2008 Insights into genes involved in electricity generation in *Geobacter sulfurreducens* via whole genome microarray analysis of the OmcF-deficient mutant *Bioelectrochemistry* **73** 70–5

[87] Mehta T, Childers S E, Glaven R, Lovley D R and Mester T 2006 A putative multicopper protein secreted by an atypical type II secretion system involved in the reduction of insoluble electron acceptors in *Geobacter sulfurreducens Microbiology* **152** 2257–64

[88] Holmes D E, Mester T, O'Neil R A, Perpetua L A, Larrahondo M J, Glaven R, Sharma M L, Ward J E, Nevin K P and Lovley D R 2008 Genes for two multicopper proteins required for Fe (III) oxide reduction in *Geobacter sulfurreducens* have different expression patterns both in the subsurface and on energy-harvesting electrodes *Microbiology* **154** 1422–35

[89] Keller K L, Rapp-Giles B J, Semkiw E S, Porat I, Brown S D and Wall J D 2014 New model for electron flow for sulfate reduction in *Desulfovibrio alaskensis* G20 *Appl. Env. Microbiol.* **80** 855–68

[90] Baron D, LaBelle E, Coursolle D, Gralnick J A and Bond D R 2009 Electrochemical measurement of electron transfer kinetics by *Shewanella oneidensis* MR-1 *J. Biol. Chem.* **284** 28865–73
[91] Shen H B, Yong X Y, Chen Y L, Liao Z H, Si R W, Zhou J, Wang S Y, Yong Y C, OuYang P K and Zheng T 2014 Enhanced bioelectricity generation by improving pyocyanin production and membrane permeability through sophorolipid addition in *Pseudomonas aeruginosa*-inoculated microbial fuel cells *Bioresour. Technol.* **167** 490–4
[92] Yong X Y *et al* 2014 Enhancement of bioelectricity generation by manipulation of the electron shuttles synthesis pathway in microbial fuel cells *Bioresour. Technol.* **152** 220–4
[93] Coursolle D, Baron D B, Bond D R and Gralnick J A 2010 The Mtr respiratory pathway is essential for reducing flavins and electrodes in *Shewanella oneidensis J. Bacteriol.* **192** 467–74
[94] Von Canstein H, Ogawa J, Shimizu S and Lloyd J R 2008 Secretion of flavins by *Shewanella* species and their role in extracellular electron transfer *Appl. Env. Microbiol.* **74** 615–23
[95] Jain A, Connolly J O, Woolley R, Krishnamurthy S and Marsili E 2013 Extracellular electron transfer mechanism in *Shewanella loihica* PV-4 biofilms formed at indium tin oxide and graphite electrodes *Int. J. Electrochem. Sci.* **8** 1778–93
[96] Marsili E, Baron D B, Shikhare I D, Coursolle D, Gralnick J A and Bond D R 2008 *Shewanella* secretes flavins that mediate extracellular electron transfer *Proc. Natl Acad. Sci.* **105** 3968–73
[97] Kotloski N J and Gralnick J A 2013 Flavin electron shuttles dominate extracellular electron transfer by *Shewanella oneidensis MBio* **4** e00553–12
[98] You J, Deng Y, Chen H, Ye J, Zhang S and Zhao J 2020 Enhancement of gaseous o-xylene degradation in a microbial fuel cell by adding *Shewanella oneidensis* MR-1 *Chemosphere* **252** 126571
[99] Goldbeck C P, Jensen H M, TerAvest M A, Beedle N, Appling Y, Hepler M, Cambray G, Mutalik V, Angenent L T and Ajo-Franklin C M 2013 Tuning promoter strengths for improved synthesis and function of electron conduits in *Escherichia coli ACS Synthet. Biol.* **2** 150–9
[100] Fuller S J, McMillan D G, Renz M B, Schmidt M, Burke I T and Stewart D I 2014 Extracellular electron transport-mediated Fe (III) reduction by a community of alkaliphilic bacteria that use flavins as electron shuttles *Appl. Env. Microbiol.* **80** 128–37
[101] Freguia S, Masuda M, Tsujimura S and Kano K 2009 *Lactococcus lactis* catalyses electricity generation at microbial fuel cell anodes via excretion of a soluble quinone *Bioelectrochemistry* **76** 14–8
[102] Deng L, Li F, Zhou S, Huang D and Ni J 2010 A study of electron-shuttle mechanism in *Klebsiella pneumoniae* based-microbial fuel cells *Chin. Sci. Bull.* **55** 99–104
[103] Torres C I, Marcus A K, Lee H S, Parameswaran P, Krajmalnik-Brown R and Rittmann B E 2010 A kinetic perspective on extracellular electron transfer by anode-respiring bacteria *FEMS Microbiol. Rev.* **34** 3–17
[104] Snider R M, Strycharz-Glaven S M, Tsoi S D, Erickson J S and Tender L M 2012 Long-range electron transport in *Geobacter sulfurreducens* biofilms is redox gradient-driven *Proc. Natl Acad. Sci.* **109** 15467–72
[105] Pirbadian S and El-Naggar M Y 2012 Multistep hopping and extracellular charge transfer in microbial redox chains *Phys. Chem. Chem. Phys.* **14** 13802–8

[106] Richter L V, Sandler S J and Weis R M 2012 Two isoforms of *Geobacter sulfurreducens* PilA have distinct roles in pilus biogenesis, cytochrome localization, extracellular electron transfer, and biofilm formation *J. Bacteriol.* **194** 2551–63

[107] Vargas M, Malvankar N S, Tremblay P L, Leang C, Smith J A, Patel P, Synoeyenbos-West O, Nevin K P and Lovley D R 2013 Aromatic amino acids required for pili conductivity and long-range extracellular electron transport in *Geobacter sulfurreducens MBio* **4** e00105–13

[108] Juárez K, Kim B C, Nevin K, Olvera L, Reguera G, Lovley D R and Methé B A 2009 PilR, a transcriptional regulator for pilin and other genes required for Fe (III) reduction in *Geobacter sulfurreducens J. Mol. Microbiol. Biotechnol.* **16** 146–58

[109] Leung K M, Wanger G, El-Naggar M Y, Gorby Y, Southam G, Lau W M and Yang J 2013 *Shewanella oneidensis* MR-1 bacterial nanowires exhibit p-type, tunable electronic behavior *Nano Lett.* **13** 2407–11

[110] Rotaru A E, Shrestha P M, Liu F, Markovaite B, Chen S, Nevin K P and Lovley D R 2014 Direct interspecies electron transfer between *Geobacter metallireducens* and *Methanosarcina barkeri Appl. Env. Microbiol.* **80** 4599–605

[111] Gregory K B, Bond D R and Lovley D R 2004 Graphite electrodes as electron donors for anaerobic respiration *Env. Microbiol.* **6** 596–604

[112] Xafenias N, Zhang Y and Banks C J 2013 Enhanced performance of hexavalent chromium reducing cathodes in the presence of *Shewanella oneidensis* MR-1 and lactate *Env. Sci. Technol.* **47** 4512–20

[113] Carbajosa S, Malki M, Caillard R, Lopez M F, Palomares F J, Martín-Gago J A, Rodríguez N, Amils R, Fernández V M and De Lacey A L 2010 Electrochemical growth of *Acidithiobacillus* ferrooxidans on a graphite electrode for obtaining a biocathode for direct electrocatalytic reduction of oxygen *Biosens. Bioelectron.* **26** 877–80

[114] Jeans C, Singer S W, Chan C S, VerBerkmoes N C, Shah M, Hettich R L, Banfield J F and Thelen M P 2008 Cytochrome 572 is a conspicuous membrane protein with iron oxidation activity purified directly from a natural acidophilic microbial community *ISME J.* **2** 542–50

[115] Wang Y Z, Wang A J, Zhou A J, Liu W Z, Huang L P, Xu M Y and Tao H C 2014 Electrode as sole electrons donor for enhancing decolorization of azo dye by an isolated *Pseudomonas* sp. WYZ-2 *Bioresour. Technol.* **152** 530–3

[116] Ringeisen B R, Henderson E, Wu P K, Pietron J, Ray R, Little B, Biffinger J C and Jones-Meehan J M 2006 High power density from a miniature microbial fuel cell using *Shewanella oneidensis* DSP10 Environ. Sci. Technol. **40** 2629–34

[117] Rabaey K, Boon N, Siciliano S D, Verhaege M and Verstraete W 2004 Biofuel cells select for microbial consortia that self-mediate electron transfer Appl. Environ. Microbiol. **70** 5373–82

[118] Pandit S, Khilari S, Roy S, Pradhan D and Das D 2014 Improvement of power generation using Shewanella putrefaciens mediated bioanode in a single chambered microbial fuel cell: effect of different anodic operating conditions Bioresour. Technol. **166** 451–7

[119] Kang C S, Eaktasang N, Kwon D Y and Kim H S 2014 Enhanced current production by Desulfovibrio desulfuricans biofilm in a mediator-less microbial fuel cell Bioresour. Technol. **165** 27–30

[120] Cheng S and Regan J M 2008 Electricity generation by Rhodopseudomonas palustris Environ. Sci.Technol. **42** 4146–51

[121] Min B, Cheng S and Logan B E 2005 Thermodynamic analysis of a single chamber microbial fuel cell Water Res. **39** 1675–86

[122] Qiao Y, Li C M, Bao S J, Lu Z and Hong Y 2008 Direct electrochemistry and electrocatalytic mechanism of evolved *Escherichia coli* cells in microbial fuel cells *Chem. Commun.* **11** 1290–2

[123] Li X, Zhong G Z, Qiao Y, Huang J, Hu W H, Wang X G and Li C M 2014 A high-performance xylose microbial fuel cell enabled by *Ochrobactrum sp.* 575 *cells* RSC Adv. **4** 39839–43

[124] Raghavulu S V, Goud R K, Sarma P N and Mohan S V 2011 *Saccharomyces cerevisiae* as anodic biocatalyst for power generation in biofuel cell: influence of redox condition and substrate load *Bioresour. Technol.* **102** 2751–7

[125] Xu S and Liu H 2011 New exoelectrogen *Citrobacter sp.* SX-1 isolated from a microbial fuel cell *J. Appl. Microbiol.* **111** 1108–15

[126] Nandy A, Kumar V and Kundu P P 2013 Utilization of proteinaceous materials for power generation in a mediatorless microbial fuel cell by a new electrogenic bacteria *Lysinibacillus sphaericus* VA5 *Enzyme Microb. Technol.* **53** 339–44

[127] González del Campo A, Cañizares P, Rodrigo M A, Fernández F J and Lobato J 2013 Microbial fuel cell with an algae-assisted cathode: a preliminary assessment *J. Power Sourc.* **242** 638–45

[128] Cui Y, Rashid N, Hu N, Rehman M S U and Han J I 2014 Electricity generation and microalgae cultivation in microbial fuel cell using microalgae-enriched anode and bio-cathode *Energy Conv. Manage.* **79** 674–80

[129] Wu C, Liu X W, Li W W, Sheng G P, Zang G L, Cheng Y Y, Shen N, Yang Y P and Yu H Q 2012 A white-rot fungus is used as a biocathode to improve electricity production of a microbial fuel cell *Appl. Energy* **98** 594–6

[130] Yuan Y, Chen Q, Zhou S, Zhuang L and Hu P 2011 Bioelectricity generation and microcystins removal in a blue-green algae powered microbial fuel cell *J. Hazard. Mater.* **187** 591–5

[131] Inglesby A E, Beatty D A and Fisher A C 2012 *Rhodopseudomonas palustris* purple bacteria fed *Arthrospira maxima cyanobacteria*: demonstration of application in microbial fuel cells *Rsc Adv.* **2** 4829–38

Chapter 12

Review of bioelectrochemical studies of microbial fuel cells

Mustapha Omenesa Idris

Over the years bioelectrochemical systems have become an interesting technology for the generation of bioenergy through the biological degradation of organic substances. There are a number of these systems, in practice, however, the microbial fuel cell (MFC) bioelectrochemical system has emerged as the most suitable system for multiple applications in a single continuous process, such as the production of electricity, the bioremediation of toxic substances from wastewater, the degradation of an organic pollutant utilized as a substrate for the growth of microbes, and even the incorporation of biomass waste as an electrode material, making it suitable for waste management. This chapter summarizes the bioelectrochemical roles played by each component in an MFC. The microbial activities including the electrogens, the electron transfer processes, and the chemical reactions taking place at the anode and cathode are highlighted. The configuration set-up of the system, various substrate used in MFCs, as well as the general applications in bioenergy and water treatment are also discussed.

12.1 Introduction

Microbial fuel cells (MFCs) are emerging bioelectrochemical processes that can create bioenergy while also bioremediating organic matter or wastewater in the system's anode chamber. Bioelectrochemical studies of MFCs require a detailed investigation of the bioelectrocatalytic activities of the micro-organisms and all other components of the MFC in the production of green electricity as well as the treatment of wastewater in the system. Yaqoob *et al* [1] highlighted in their review article that the MFC technique is designed to generate electrical energy through the metabolic activity of injected microbes; converting organic matter into electrical current. The goal of MFCs is to generate electricity while treating wastewater in an anodic chamber using electrogenic bacteria as a catalyst. The oxidation of organic material by removing it from the external electron-carrying circuit from the anode to the cathode generates the electricity. Under aerobic conditions (in the presence of oxygen), the electrons reach

doi:10.1088/978-0-7503-4791-4ch12

the cathode and combine with protons and oxygen to form water [2]. There are a large number of bioelectrochemical systems (BESs) including: microbial desalination cells (MDCs), microbial electrolysis cells (MECs), microbial reverse electrodialysis cells (MRCs), microbial solar cells (MSCs), enzymatic biofuel cells (EBCs), microbial electrosynthesis cells, and microbial fuel cells (MFCs). MFCs are the most advanced and intensively researched BESs [3]. The bioelectrochemical effect of an MFC has been demonstrated in various studies; some of them include bioremediation of toxic metals with energy recovery [4], the removal of nitrogen and organic matter [5], and the treatment of wastewater [6]. Because of the rapid advancements in the bioelectrochemical performance of MFCs, these systems have been regarded as one of the most renewable, sustainable, and effective BESs for generating electrical energy [7]. Hence, there is a need to comprehensively study the bioelectrochemical activities of the components and bioprocesses to provide useful information capable of enhancing research into MFC system performance and improvements.

In this chapter, we have attempted to provide a comprehensive review of the bioelectrochemical functions of each component of the MFC. The bioelectrochemistry of MFCs in terms of anodic and cathodic reactions, biofilm electrogens, biocathodes, and electron transfer processes are also discussed here. The bioenergy and bioremediation applications of MFCs are highlighted in this chapter. The structures and configurations of MFCs are also explained.

12.2 MFC layout configuration and operation

There are two main types of MFC set-ups, as shown in figure 12.1, the double chamber MFCs (DC-MFCs) and the single chamber MFCs (SC-MFCs). In general, an MFC consists of an anode electrode, a cathode electrode, the wastewater in the anode chamber where the microbes are inoculated, and an electrically conductive wire leading from the anode to the cathode electrodes. In DC-MFCs, the anodic area is usually separated from the cathodic area by a proton exchange membrane (PEM), which provides the avenue for the transportation of protons from the anode chamber to the cathode chamber [8]. The sizes and operating conditions of the cells vary in different studies, as summarized in table 12.1.

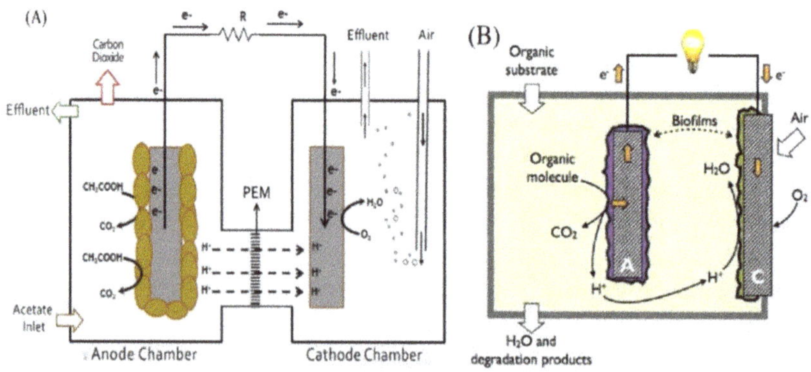

Figure 12.1. Schematic diagram showing all components of MFCs. (a) D-CMFC. (Reproduced with permission from [9]. Copyright 2020 MDPI.) (b) S-CMFC [10].

Table 12.1. Some MFC set-up and configuration conditions.

MFC type	Size of MFCs	Distance between anode and cathode	Size of electrode	Microbes or inoculum source	Reference
Double chamber	The anode and cathode chambers are, respectively, 10 cm × 9.5 cm and 10 cm × 9 cm, 1 l volume for each chamber.	7 cm	8 cm × 1.3 cm ($h \times r$)	Cd (II) supplemented wastewater	[12]
Air-cathode single chamber	The cell has a total volume of 300 ml.	NR	0.5 mg cm^{-2} thick	10% v/v fresh seed culture	[13]
Double chamber	The dimensions of the cell are 10 cm × 10 cm × 10 cm, 40 ml working volume.	NR	3 cm × 0.9 cm × 3 cm	Municipal wastewater with mixed cultures	[14]
Air-cathode single chamber	A 50 ml working volume and dimensions of 6 cm × 4 cm.	6 cm	NR	Synthetic wastewater with nourishment media	[15]
Single chamber constructed wetlands	NR	22.5 cm	Electrode diameter of 18 cm	Activated sludge from anaerobic pond	[16]
Double chamber membrane-less	The cell has a height of 60 cm with a working diameter of 15 cm.	28 cm	NR	Anaerobic sludge from septic tank	[17]
Double chamber benthic	Both chambers have dimensions of 10 cm × 9.5 cm and a total volume of 1000 ml.	7 cm	7.5 cm × 1.2 cm ($l \times w$)	Synthetic wastewater and sweetpotato waste	[18]
Single chamber	The cell volume is 850 ml with a cross sectional area of 104 cm^2.	4 cm	Electrode diameter of 280 μm	Activated sludge	[19]
Double chamber	500 ml working volume.	15 cm	0.6 cm thick	Culture media from fertilizer water	[20]

NR: not reported.

In MFCs, various types of electrode materials have been used. Some of them are materials derived from carbon derivatives, conductive polymer composites with carbon, metal and metal oxides, or metal–nanocomposite materials. The anode chamber, where microbial growth occurs, houses the wastewater and organic substrates that act as electrolytes in the system. The microbes form biofilms around the anode, which oxidize the substrate, producing electrons and protons. The most commonly used external circuits responsible for electron transfer from the anode electrode to that of the cathode are silver wires, copper wires, or aluminium. The cathode chamber is supplied with oxygen through the air diffuser, which helps as an absolute electron acceptor due to its non-toxic nature and it is preferred as an oxidizing element [11].

12.3 Bioelectrochemical roles of components

MFCs, like other bioelectrochemical systems, are new advancements that combine micro-organisms and the electrochemical approach to increase metabolism. In general, bioelectrochemical systems are processes to generate electrical energy with a specific potential produced by the transfer of electrons between the electron acceptor (cathode) and the electron donor (anode). Scientists in the field of MFCs have concentrated their research on the choice of electrode material and its optimization, the design of suitable MFC systems, and the screening of electrochemically active microbes [21]. Notably, all of these components play important bioelectrochemical roles in MFC processes. This section investigates some of the functions of these components.

12.3.1 Anode and cathode electrodes

Electrodes in MFCs have important bioelectrochemical roles in enhancing cell performance. The anode electrode, in particular, has a direct impact on performance by allowing micro-organisms to adhere to it, enhancing the electron transfer rate and fuel oxidation. Metals, conducting polymers, metal/metal oxides, and carbon-based materials are among the many materials used as anodes in MFCs. However, in order to save on costs, researchers have investigated the possibility of making a high-performance anode out of biowaste [22].

The electrons liberated by the micro-organisms are transported to the anode electrode, which assists in their transfer to the cathode through a connected external circuit (usually copper wire or aluminium). Thus, the material for the anode electrode is of paramount importance because it is responsible for increasing the electron transportation [23]. It is assumed that the resistance of the anode electrode of MFCs is exactly correlated to the ohmic loss of the system. Increasing the anode surface area can help lower the anode resistance. In addition, a highly porous anode provides more effective areas for enhancing bacterial growth, which increases the kinetic effectiveness of the electrode. Due to the biocompatibility of the anode electrode, it is able to come into direct contact with micro-organisms and their respiration process [24]. The cathode electrode, on the other hand, contributes to the operating performance of MFCs. Since the performance of MFCs is limited by the lack of catalyst activities, the substrate reduction reactions take place in the cathode chamber. There are two types of cathodes: air cathodes and aqueous air cathodes. However, air cathodes have a high tendency to improve the performance of MFCs [25].

12.3.2 Substrate (carbon energy source)

The substrate in MFCs is an organic compound that provides a food source for microbes. It serves as a carbon energy source that promotes the maturation of micro-organisms inoculated in the anode chamber. MFCs can apply a wide range of substrates for power generation, from pure compounds to complex organic mixtures in wastewater (table 12.2). The ability of bacteria to oxidize a substrate and transfer electrons from the oxidation to the anode of the electrode is directly related to the generation of electricity in an MFC. In their review article, Pant *et al* [26] provided a broad list of substrates utilized in MFC studies along with their power efficiencies. According to Zhang *et al* [27], the kind of substrate fed to the MFC is a crucial parameter for the cell's performance in terms of power generation. They found that MFCs supplemented with glucose as the substrate demonstrated rapid and higher current generation compared to when glucose was replaced with butyrate or acetate. Similarly, Ghoreyshi *et al* [28] investigated the effect of concentrations (2–20 g l^{-1}) of glucose and date syrup as a substrate in MFCs systems. They found that a concentration of 3 g l^{-1} date syrup in the MFC provided the highest power density. Treatment of substrates with mixed cultures was discovered to have a significant replacement proficiency compared to the conventional bacterial inoculum for improving power performance in MFCs [29].

12.3.3 Micro-organisms

Micro-organisms are one of the most important components of an MFC and can have a significant impact on its performance. The type of micro-organisms used in MFCs is critical. It was discovered that micro-organisms can follow separate metabolic pathways for efficient energy production that are dependent on the MFC's operating parameters. The production of bioelectricity by mixed microbial communities or pure microbes is influenced by the process architecture and the physical/chemical environment, which include the internal resistance of the process, the size of the electrode, electrode spacing, pH, and conductivity. Many bacteria are metabolically versatile organisms that can utilize almost any carbonaceous compound found in nature. They can perform aerobic or anaerobic respiration and can also ferment. Aerobic respiration necessitates the use of molecular oxygen (O_2), whereas anaerobic respiration can make use of other electron acceptors. The idea of genetically unmodified bacteria, depending on the micro-organism, is to combine organic matter's oxidation with the reduction of the terminal electron acceptor, which is usually the anode of the MFC [30]. The ability of micro-organisms to degrade complex substrates in MFC depends primarily on the conformation of the microbial cultures and its acclimation [31]. There are a few electrogenic microbes that do not require a synthetic mediator to function properly. *Shewanella* and *Geothrix* species are two examples of such microbes. These electrogenic microbes

Table 12.2. Substrates and microbes applied in some MFC studies.

Type of MFC	Micro-organism	Substrate	Size of electrode/ surface area	Purpose	Performance	Reference
Single chamber	*Escherichia coli*	Lake sediment	0.5 mg cm^{-2}/7 cm^2	Anode modification for high-performance MFC	884 mW m^{-2}	[76]
Mediator-free double chamber	*Shewanella putrefaciens*	Acetate and yeast extracts	1.8 cm × 1.8 cm/ 24 cm^2	Control modification of anode	257 mW m^{-2}	[77]
Single chamber	*Shewanella algae* (MTCC-10 608)	Dairy wastewater	0.5 mg cm^{-2}/50 cm^2	Obtaining a stable power from dairy wastewater	COD (92.21%) 50 mW m^{-2}	[13]
Double chamber	*Saccharomyces cerevisiae*	Glucose	8 cm × 4 cm × 0.4 cm	Culture of micro-organism in anaerobic chamber to generate electrons	16 mW m^{-2}	[78]
Double chamber	*Geobacter* spp.	Lactate	NR	To assess the possibility of using MFCs in biomedicals	52 mW m^{-2}	[79]
Double chamber	*Betaproteo bacterium*	Ethanol	2.5 cm × 4.5 cm/ 22.5 cm^2	Assessment of power output in various MFCs	40 mW m^{-2}	[80]
Air-cathode single chamber	*Escherichia coli*	Organic substrate	4 cm × 4 cm	Examining MFC catalyst for oxygen reduction reaction	3119 mW m^{-2}	[81]
Air-cathode single chamber	*Geobacter* sp.	Artificial wastewater	5 cm × 5 cm	Improving electricity generation in MFCs	1296 mW m^{-2}	[82]
Double chamber	*Chlorella vulgaris* AG30007 UTEX 0000265	Microalgae powder	2.5 cm *L*, 2.5 cm *D*/25 cm^2	Utilization of microalgae enriched electrode in MFCs	1930 mW m^{-2}	[83]
Single chamber	*Chryseobacterium, Bacillus, Lysinibacillus, Escherichia* strains	Rotten rice	14.9 cm × 1 cm	Utilization of rotten rice as an organic substrate to enhance bacteria growth	2.9 mW m^{-2}	[84]

can transfer electrons directly to the anode electrode [32]. In addition to electrogenic micro-organisms, a variety of microbes from methanogens, fungi, and bacteria have been used in MFCs [31].

In addition to pure plankton cultures, microbial colonies called biofilms containing a different or archetypal group of microbes have also been utilized in MFCs. Although the events that occur during the activity of a mixed culture are difficult to track and attribute, mixed cultures lend themselves to the generation of bioelectricity because of the mutualism of the organisms, the collaboration in the utilization of metabolites or substrates, and microbial electron recycling. Members of a mixed microbial consortium work together, but also compete in the derivation for carbon and energy and assist in the biodegradation of a complex substrate. Mixed cultures can adjust to a variety of environmental factors and each species in the culture can perform a certain function that has been shown to be critical for another species, resulting in greater survival and performance [33]. It was discovered that heterologous expression of the intracellular electron transfer chain (ETC) of *Shewanella* (CymA and Mtr systems) in *E. coli* can remodel the intracellular metabolism and increase the intracellular electron transportation rate by 183%. In addition, higher electrical output was obtained using methylene blue. Similarly, heterologous initiation of a synthetic flavin pathway in *Shewanella* increased the extracellular electron transfer (EET) rate [34, 35].

12.4 Bioelectrochemistry of MFCs

The bioelectrochemistry of MFCs encompasses the electrochemistry and the biophysical aspects of the system, which include the mechanism of electron/proton transport, the reactions occurring at the electrode anode and cathode, and the activities of the biofilms and the biocathode. The interaction of these different processes is responsible for the functioning of the MFC system.

12.4.1 Transfer of electron mechanisms

The mechanism of electron transfer is crucial for better understanding of bioelectrochemical studies and promoting the large-scale application of MFCs. Microbes rely on the substrate supplied to the anodic chamber, to release electrons that flow through the circuit that conducts electricity. An increase in electron transfer rate means an increase in electrons passing through the circuit, generating additional voltage and increasing the power density of the MFCs. Direct electron transfer (DET) and mediated electron transfer (MET) are two critical mechanisms for extracellular electron transfer (EET) [36]. It is known that the MET and DET mechanisms of electron transfer occur between anodic materials and micro-organisms.

In DET there is bodily contact between the microbe's cell membrane and the surface of the electrode. When microbes begin to colonize the electrode surface, thereby transforming into biofilms, they attach to the electrode surface while they transfer electrons directly, involving the redox molecule on their surfaces known as a cytochrome [37]. DET does not require the addition of an external agent or mediator for the transfer of electrons. The major drawback of this mechanism is that the bacteria must stick to the anode surface (forming a biofilm) for the electrons to be transferred. It is believed that most gram-positive bacteria participate in DET via biofilm formation through teichoic acid (TA), a compound that allows bacteria to adhere to the anode surface through direct contact [38]. Recently, numerous studies have been conducted to either isolate superior unknown electrogens or to improve the electrogenic properties of existing micro-organisms with such a capability. The ability of various molecular genetic tools to discover such strains is the driving force behind their discovery. Surface-localized factors, such as the pili of the bacterial type IV (TFP), are an important mediatorless protein which plays a vital role in some of the most studied electrogenic bacteria. A variety of exoelectrogens have been described to use the DET mechanism in electron transfer to the anode electrode, however, *Geobacter sulfurreducens* has received the bulk of attention in recent times because its genome has been sequenced [39]. Research has found that a mediatorless MFC that contains *Desulfovibrio desulfuricans* as the strain and operated with graphite felt electrodes which were previously treated electrochemically exhibited the highest current density of 233 mA m^{-2}. The cytochrome-c of the *D. desulfuricans* took part in the efficient transfer of electrons to the electrode surface. Raman microscopy, cyclic voltammetry, scanning electron microscopy, and electron impedance spectroscopy are a few of the electrochemical measurements used to study DET in exo-electrogenic biofilms [40].

In MET, the transfer of electrons to the anode occurs indirectly via mediators, with the mediators acting as electron transporters between the microbes and the anode. In this method, a redox carrier is used to function as a conduit for electron transfer to the final electron acceptor. In general, these redox mediators collect electrons from the cells of the microbes, which are then transferred to the anode in a reduced state and oxidized. These oxidized electron steps were free to begin the subsequent activity of electron transfer [41]. The mediators may be either endogenous or exogenous in nature. Endogenous mediators are redox mediators generated by the micro-organisms themselves. Exogenous mediators are compounds that can be supplied from outside to act as redox mediators. The main disadvantages of these mediators are their instability and toxicity, which limit their use in MFC systems [42]. Several factors influence electron transfer in a bioelectrochemical systems. Electrode spacing is an important factor to achieve greater performance in MFCs by ensuring increased electron transfer. It has been reported recently that a decrease in the inter-electrode spacing could improve the electrical output by reducing the ohmic losses and thus the electron transfer rate is increased. However, flat electrodes too close to each other have been shown to reduce the MFC power output due to oxygen transfer from the cathode to the anode [43]. Phenazines, flavin mononucleotides, favins, ribofavins, quinones, and pyocyanins are metabolites produced by cells and therefore serve as endogenous mediators [44]. Thionine, methylene blue, Fe (III) EDTA, various quinone derivatives, and neutral red are examples of exogenous mediators used to improve MFC performance [45]. Some characteristics of effective exogenous mediators have been identified. Exogenous mediators should be able to accept electrons from electron donors without affecting other metabolic processes. It is also established that a good exogenous mediator should be cytotoxic and able to deliver electrons within the cytoplasm for oxidation reactions. The mediator ought to be able to penetrate cell membranes smoothly and be soluble and stable in both the oxidized and reduced forms. It should not be consumed as a nutrient by the microbes in the biofilm [46].

12.4.2 Anodic and cathodic reactions

Anodic and cathodic reactions in MFCs are two half reactions occurring at the two separate chambers. At the anode electrode, oxidation occurs (i.e. transfer of electrons to the cathode) while the cathode is the terminal electron acceptor. The micro-organisms in the anode chamber of the MFCs oxidize substrates such as acetate, glucose, or other organic matter. Glucose, for example, is oxidized as follows, producing electrons, protons, and carbon dioxide [47]:

$$C_2H_{12}O_6 + 6H_2O \rightarrow 6CO_2 + 24H^+ + 24e^-. \quad (12.1)$$

The electron produced in the above reaction as a result of the activities of the electrogenic microbes is transported to the anode electrode. The anode electrode also helps transfer electrons to the cathode via an external circuit (electrically conducting wire) connecting the electrodes (anode and cathode), where they are used to reduce oxygen serving as an oxidant. The load for current collection is placed on the external circuit. Protons must normally carry the same number of positive charges from the anodic chamber through to the cathodic chamber via the PEM in order to maintain electroneutrality. Inefficient migration of protons leads to the aggregation of acidic protons in the chamber of the anode [48]. The positive charge (proton) from the above reaction travels into the cathodic chamber through the separating membrane (PEM). The PEM is an important component that plays a crucial function in preventing the diffusion of oxygen within the two chambers of the MFC [49].

The cathode reaction has a great influence on the performance of the MFC. The anode electrons that travel to the cathode electrode via the external circuit, the anodic protons through the PEM, and the oxygen present in the cathode react with the catalysts in the cathode [50]. From the anodic reaction equation (12.1), the reaction happening at the cathode is

$$24H^+ + 24e^- + 6O_2 \rightarrow 12H_2O. \quad (12.2)$$

The anodic and cathodic reactions of MFCs rely heavily on catalysts to activate the processes. In the anodic chamber, a biofilm is required to catalyze the oxidation of source carbon and electron transfer. Without a catalyst, the rate of oxygen reduction at the cathode is very slow. The efficiency of the cathodic reaction is determined by the electrode structure, the concentration of the incoming proton, the type and concentration of electron acceptors, and the capability of the catalyst [51].

12.4.3 Biofilm electrogens

The electrogenic biofilm is a critical biocatalytic component of MFCs that can have a significant impact on system performance. The growth and development of the electrogenic biofilm depend on the influence of the carbon source material and also on the anode surface area. The formation of strong biofilms with essential capacitive properties is considered critical because it can have a direct effect on the electrogenic assets such as power overshoot [52]. However, the temperature of the system, the substrate used, the anode material type, build up design, and pH are known to have an effect on the performance of biofilms. Many of these factors have been studied to understand how they affect biofilms in MFCs. For example, Venkata Mohan *et al* [53] have extensively highlighted the physicochemical factors affecting biofilm electrogens in MFCs. In their review studies, they discussed that the surface area of the electrode, the electrode material, and the electrode spacing strongly influence the growth of the electrogenic biofilm. Similarly, Michie *et al* [54] investigated the pH resistance of biofilm structure. The high level of biofilm development observed in their study was attributed to felt anodes, which facilitated higher pseudo-capacitive values on account of redox-active species, resulting in a higher level of power production while also mitigating the effects of low-pH operation. It has also been reported that the microbial community composition of biofilm is hugely influenced by the source with which it has been inoculated, the type of microbes (gram-positive or gram-negative), the nature of pre-enrichment, and whether an immaculate or blended microbial culture is used [55]. In addition, Hassan *et al* [56] reported that biofilms comprising diverse cultures are more efficient in electron transportation than single cultures because of their robust nature and a relatively slow growth rate. Biofilms composed of complex cultures are frequently found to do better than pure stains. This is because there are numerous syntrophic species in a synergistic biofilm group, with a specific role played by each organism. A consortium can be used to withstand wastewater variations and extreme environmental conditions, as a biofilm consortium is normally metabolically far stronger than a pure-culture biofilm.

12.4.4 Biocathodes

Biocathodes are suitable for the improvement of MFC sustainability without the need to introduce an artificial mediator or catalyst because of its ability to eliminate the problems associated with electron mediator replenishment, sulphur poisoning, or platinum consumption. Their greatest advantage is their promising application for cost effectiveness and potential pollution remediation [57]. Biofilms are used as catalysts in biocathodes to improve the cathode reaction. Other oxidants, such as sulphate, nitrate, carbon dioxide, Fe (III), Cr (VI), Mn (IV), tetrachloroethene, perchlorate, and trichloroethene, can be used instead of oxygen. The microbial reduction of Fe (III) and Mn (IV), which can function as terminal electron acceptors in the cathode, is well known [58]. There are two types of biocathodes, depending on the end electron acceptors used in the cathode: aerobic biocathodes and anaerobic biocathodes. The most common terminating electron acceptor for the cathode reaction in an aerobic biocathode is oxygen. This is due to its top redox potential and its surplus in the air, as well as its economic advantages. The biofilm on the surface of the cathode catalyzes the oxidation of transition metal compounds such as Fe (II) and Mn (II), allowing electrons to be released to oxygen [47]. Oxygen is absent in anaerobic biocathodes, but other compounds such as sulphate, selenate, nitrate, arsenate, fuma-rate, and CO_2 can act as terminal electron acceptors.

To eliminate the tendency of possible oxygen diffusion into the anode chamber through the PEM and preventing electron loss to oxygen instead of the electrode, is one great advantage of using an anaerobic biocathode over an aerobic biocathode [58]. However, it has been reported that MFCs that use aerobic biocathodes can generate more power density than those with anaerobic biocathodes [59].

Several studies on the use of biocathodes in MFCs have been reported. Chen *et al* [57] provide a few examples. They investigated the microbial community's activity as well as the performance of the MFCs. In a double chamber MFCs without artificial redox mediators, a biocathode was used. They discovered that using a biocathode improved MFC performance. Another study by Clauwaert *et al* [60] proves that an open air biocathode MFC effectively enhances the production of electricity in the system. The accumulation of microbial metabolites in the cathode chamber is the major drawback in biocathode configuration. This problem of accumulation can inhibit the activities of the microbes. Furthermore, bacterial metabolites that functions as electron donors can compete with the cathode, reducing MFC efficiency [61]. Either direct electron transfer (DET) or mediated electron transfer (MET) is used in biocathode electron transfer, just as it is in anode electron transfer (figure 12.2). There was insufficient information available in previous literatures on the biocathode's mechanism of transfer of electrons. Although the mechanism of biological electron transfer remains unclear, many researchers have demonstrated that micro-organisms play an important role in the biocathode electron transfer process which occurs either in a direct manner or through mediators [62]. Many different materials are used to make biocathodes, and Zhang *et al* [63] have compared some of these materials as effective biocathodes in the

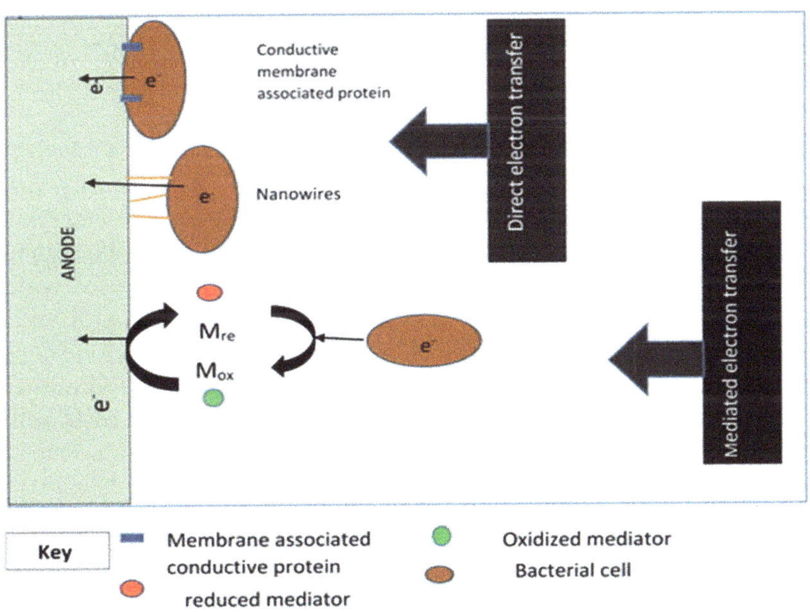

Figure 12.2. Schematic diagram of the mechanism of electron transfer in MFCs. (Reproduced with permission from [64]. Copyright 2020 Taylor and Francis.)

MFC system. They compared graphite felt, stainless-steel mesh (SSM), and carbon paper as biocathodes in MFC applications and discovered that graphite felt outperforms SSM and carbon paper in terms of catalytic activity, with the highest values for both maximum power densities and current output.

12.5 Bioelectrochemical applications

The bioelectrochemical application of MFCs is based on their ability to bioelectrochemically treat wastewater while also generating electricity from the system. There are other bioelectrochemical systems, but MFCs stand out as an emerging technology for simultaneous treatment of wastewater while producing electrical energy output. Their applications for energy generation and bioremediation of wastewater are highlighted briefly in this section.

12.5.1 Green energy generation

MFCs have emerged as one of the most promising biotechnologies for producing bioelectricity without polluting the environment. Aside from the challenges in commercial applications, researchers have recently made significant progress in improving the electrical output performance of MFCs by advancing their configuration, developing anode materials, and improving the electrogenic activities of the micro-organisms in the system [65]. For instance, Yaqoob *et al* [66] developed a cutting-edge anode to boost power generation in MFCs. They used lignocellulosic material from oil palm to create a high-powered anode electrode for MFC applications. The highest power density of 912×10^{-3} mW m^{-2} was obtained, indicating that the biomass-derived fabricated anode demonstrated a high electron transport rate and conductivity. Similarly, Treesubsuntorn *et al* [67] were able to boost MFC power generation by advancing the electrogenic activities of the inoculated microbes. They use a nitrifying bacterium (*Bacillus thuringiensis*) as a source inoculum in a wetland-MFC to increase the system's maximum power density. They observed an effective increase in maximum power density, producing approximately 20–35 mW m^{-2} max PD. Abbas *et al* [68] used an emerging MFC configuration technique to construct a sediment microbial fuel cell. Although the researchers noted that the generated energy output is not viable for commercial purposes, they did obtain significant progress in energy production with a maximum power density of approximately 400–450 mW m^{-2}. In order to examine the performance of sulphate-reducing microbes, Sharma *et al* [69] produced a biocathode of the stainless-steel mesh of an active carbon fabric electrode. In contrast to the biocathode-based active carbon fabric biocathode that does not have a stainless-steel mesh, it was noted that performance was improved with stainless-steel mesh combined with activated carbon fabric. The power density increased by 0.004 79–0.023 W m^{-2} and the current density increased by three times, 0.075–0.25 A m^{-2}. Biofilm grown on metal surfaces creates a more efficient oxygen reduction cathode in a natural marine environment. A new approach, photosynthetic MFCs, has recently attracted a great deal of attention because they produce more sustainable energy than non-photosynthetic MFCs. Photosynthesis is combined with the generation of electrical power. In photosynthetic MFCs, the microbes typically include certain specialized complexes of light collection that act as the photosynthesis devices. The light collection units can transform solar energy into chemical energy in a sustainable way and then are used for generating electricity through traditional exoelectrogens. Incorporating photosynthesis into MFC technology has opened up several new opportunities to generate sustainable bioenergy [70].

12.5.2 Bioremediation and treatment of wastewater

One major bioelectrochemical application of MFCs is their ability to bioremediate and treat wastewater; however, drawbacks such as potential self-toxicity, the high cost of MFC set-ups, and some operational technicalities are some challenges for its large-scale industrial application. However, over the years, MFC researchers have been able to apply MFC technology in the treatment of wastewater, bioremediation of toxic metals from wastewater, as well as in the degradation of organic pollutants present in industrial effluents. MFCs have been used to remedy a variety of wastewaters, including but not limited to seafood processing water, petrochemical industrial effluents, swine wastewater, vegetable and food processing waste, livestock waste, cassava mill wastewater, surgical cotton industry waste, and dairy wastewater [71]. Kook *et al* [72] studied the bioelectrochemical treatment of municipal waste effluent using MFC technology. The COD removal rate and efficiency of municipal solid waste were investigated using a DC-MFC. The maximal and average COD removal efficiencies were 94% and 87%, respectively, indicating the municipal solid waste's excellent biodegradability. In terms of COD removal rate, 1.2–1.9 kg COD m^{-3} d^{-1} was achieved. Wen *et al* [73] utilized an MFC system in treating brewery wastewater. It was built with a single air-cathode microbial fuel cell (MFC). In the anode chamber, they compared the electrical performance rate of raw brewery wastewater versus ordinary raw wastewater. The COD removal efficiency increased from 44% to 48% as substrate concentration increased, owing to increased micro-organism activity. However, the COD removal efficiency of raw brewery wastewater is clearly lower (20.7%) than that of wastewater with added PBS (47.6%) at the same concentration (COD = 1501 mg l^{-1}), indicating that PBS addition can improve the reaction rate, resulting in higher substrate removal efficiency. Catal *et al* [74] in another demonstration, used six different water sources rich in hexose, three of which were pentoses while another three were sugar derivatives, and discovered that when the substrate was used as nutrient source for a mixed bacterial culture for carbohydrate-rich wastewater treatment, over 80% COD removal efficiency was obtained in all carbon sources tested by the system. Bhowmick *et al* [75] examined the use of MFCs in the treatment of watery effluents derived from a fish market that were subjected to the anode chamber of an MFC. They compared the methanogenic activities of electrogenic bacteria in wastewater derived from the fish market and synthetic wastewater in terms of chemical oxygen demand (COD). When compared to the MFC supplied with synthetic wastewater (0.219 g COD CH_4/g volatile suspended solids (VSS)/day), it was observed that the MFC fed with the fish market water had a lower specific methanogenic activity of 0.097 g COD CH_4/g VSS/day. Heavy metal concentration is a major issue in wastewater; various wastewaters contain heavy metals that are detrimental to living organisms. Heavy metals are found in a wide variety of wastewaters. The removal of heavy metals from wastewater is extremely difficult using the convectional removal processes. MFC concepts have high potential for the removal of heavy metals from polluted wastewater. Many of them have already been demonstrated in studies.

12.6 Future prospects and conclusions

Among all known bioelectrochemical technologies, MFCs have proven to be the most appropriate system with various configurations capable of simultaneously degrading organic matter or bioremediating toxic substances in wastewater and producing electrical energy from the activities of the inoculated microbes within the system. Although their application in large-scale industry has not been fully exploited, many researchers are working to develop cost-effective MFC configurations that can be used commercially. For example, Yaqoob *et al* [85] conducted a thorough review of cost-effective MFC configurations and advancements. They focused primarily on the use of waste biomass for improved anode fabrication, as well as material modifications for improved MFC performance. Prior to this work, reports on improving various MFC components such as microbe activities, the substrate which provides a carbon source for microbe growth, the general configuration involving electrode spacing, and the introduction of an air-cathode as methods of improving MFC performance and efficiency had been published. Future research on bioelectrochemical studies of MFCs should continue to focus on maximizing their potential for industrial applications while keeping production costs and improved performance in mind. The best substrate for microbe growth, external mediators, and methods for increasing electron transport should all be investigated thoroughly. This chapter discussed the electron transfer mechanism as well as the bioelectrochemical roles that each component plays. Finally, with ongoing efforts to advance the system's bioelectrochemical processes, MFCs technology can achieve the commercialization stage with limitless potential in energy generation and control of environmental pollution.

References

[1] Yaqoob A A, Ibrahim M N M and Rodríguez-Couto S 2020 Development and modification of materials to build cost-effective anodes for microbial fuel cells (MFCs): an overview *Biochem. Eng. J.* **164** 107779

[2] Logan B E, Hamelers B, Rozendal R, Schröder U, Keller J, Freguia S, Aelterman P, Verstraete W and Rabaey K 2006 Microbial fuel cells: methodology and technology *Env. Sci. Technol.* **40** 5181–92

[3] Ivase T J P, Nyakuma B B, Oladokun O, Abu P T and Hassan M N 2020 Review of the principal mechanisms, prospects, and challenges of bioelectrochemical systems *Env. Prog. Sustain. Energy* **39** 13298

[4] Yaqoob A A, Ibrahim M N M, Yaakop A S, Umar K and Ahmad A 2021 Modified graphene oxide anode: a bioinspired waste material for bioremediation of Pb^{2+} with energy generation through microbial fuel cells *Chem. Eng. J.* **417** 128052

[5] Liu S, Feng X, Xue H, Qiu D, Huang Z and Wang N 2021 Bioenergy generation and nitrogen removal in a novel ecological-microbial fuel cell *Chemosphere* **278** 130450

[6] Nguyen D A, Pham N and Pham H T 2021 Wastewater treatment performance and microbial community of anode electrodes of membrane and membrane less MFCs under effect of sunlight *J. Water Process. Eng.* **42** 102159

[7] Ferreira Mercuri E G, Jakubiak Kumata A Y, Amaral E B and Vitule J R S 2016 Energy by microbial fuel cells: scientometric global synthesis and challenges *Renew. Sustain. Energy Rev.* **65** 832–40

[8] Yaqoob A A, Ibrahim M N M, Umar K, Parveen T, Ahmad A, Lokhat D and Setapar S H M 2021 A glimpse into the microbial fuel cells for wastewater treatment with energy generation *Desalin. Water Treat.* **214** 379–89

[9] Abu-Reesh I M 2020 Single-and multi-objective optimization of a dual-chamber microbial fuel cell operating in continuous-flow mode at steady state *Processes* **8** 839

[10] Degrenne N, Allard B, Buret F, Morel F, Adami S E and Labrousse D 2011 Comparison of 3 self-starting step-up DC:DC converter topologies for harvesting energy from low-voltage and low-power microbial fuel cells *Proc. 2011 14th Eur. Conf. Power Electron. Appl. EPE 2011*

[11] Yaqoob A A, Khatoon A, Setapar S H M, Umar K, Parveen T, Ibrahim M N M, Ahmad A and Rafatullah M 2020 Outlook on the role of microbial fuel cells in remediation of environmental pollutants with electricity generation *Catalysts* **10** 819

[12] Yaqoob A A, Serrà A, Ibrahim M N M and Yaakop A S 2021 Self-assembled oil palm biomass-derived modified graphene oxide anode: an efficient medium for energy transportation and bioremediating Cd (II) via microbial fuel cells *Arab. J. Chem* **14** 103121

[13] Choudhury P, Ray R N, Bandyopadhyay T K, Basak B, Muthuraj M and Bhunia B 2021 Process engineering for stable power recovery from dairy wastewater using microbial fuel cell *Int. J. Hydrogen Energy* **46** 3171–82

[14] Islam M A, Karim A, Mishra P, Dubowski J J, Yousuf A, Sarmin S and Khan M M R 2020 Microbial synergistic interactions enhanced power generation in co-culture driven microbial fuel cell *Sci. Total Env.* **738** 140138

[15] Khater D Z, El-Khatib K M and Hassan R Y A 2018 Effect of vitamins and cell constructions on the activity of microbial fuel cell battery *J. Genet. Eng. Biotechnol.* **16** 369–73

[16] Wang Q, Lv R, Rene E R, Qi X, Hao Q, Du Y, Zhao C, Xu F and Kong Q 2020 Characterization of microbial community and resistance gene (CzcA) shifts in up-flow constructed wetlands-microbial fuel cell treating Zn (II) contaminated wastewater *Bioresour. Technol.* **302** 122867

[17] Ghangrekar M M and Shinde V B 2007 Performance of membrane-less microbial fuel cell treating wastewater and effect of electrode distance and area on electricity production *Bioresour. Technol.* **98** 2879–85

[18] Yaqoob A A, Ibrahim M N M, Umar K, Bhawani S A, Khan A, Asiri A M, Khan M R, Azam M and Alammari A M 2021 Cellulose derived graphene/polyaniline nanocomposite anode for energy generation and bioremediation of toxic metals via benthic microbial fuel cells *Polymers* **13** 135

[19] Papillon J, Ondel O and Maire É 2021 Scale up of single-chamber microbial fuel cells with stainless steel 3D anode: effect of electrode surface areas and electrode spacing *Bioresour. Technol. Rep.* **13** 100632
[20] Khamis A, Nordin N and Mokhtar M H 2020 Performance of double chamber microbial fuel cell: effect of waste water, electrode thickness and distance *Indones. J. Electr. Eng. Comput. Sci.* **20** 619–26
[21] Zheng T, Li J, Ji Y, Zhang W, Fang Y, Xin F, Dong W, Wei P, Ma J and Jiang M 2020 Progress and prospects of bioelectrochemical systems: electron transfer and its applications in the microbial metabolism *Front. Bioeng. Biotechnol.* **8** 10
[22] Yaqoob A A, Ibrahim M N M and Umar K 2021 Biomass-derived composite anode electrode: synthesis, characterizations, and application in microbial fuel cells (MFCs) *J. Env. Chem. Eng.* **9** 106111
[23] Li S, Cheng C and Thomas A 2017 Carbon-based microbial-fuel-cell electrodes: from conductive supports to active catalysts *Adv. Mater.* **29** 1602547
[24] Kumar G G, Sarathi V G S and Nahm K S 2013 Recent advances and challenges in the anode architecture and their modifications for the applications of microbial fuel cells *Biosens. Bioelectron.* **43** 461–75
[25] Ghasemi M, Ismail M, Kamarudin S K, Saeedfar K, Daud W R W, Hassan S H A, Heng L Y, Alam J and Oh S E 2013 Carbon nanotube as an alternative cathode support and catalyst for microbial fuel cells *Appl. Energy* **102** 1050–6
[26] Pant D, Van Bogaert G, Diels L and Vanbroekhoven K 2010 A review of the substrates used in microbial fuel cells (MFCs) for sustainable energy production *Bioresour. Technol.* **101** 1533–43
[27] Zhang Y, Min B, Huang L and Angelidaki I 2011 Electricity generation and microbial community response to substrate changes in microbial fuel cell *Bioresour. Technol.* **102** 1166–73
[28] Ghoreyshi A A, Jafary T, Najafpour G D and Haghparast F 2011 Effect of type and concentration of substrate on power generation in a dual chambered microbial fuel cell *Proc. World Renewable Energy Congress (Linköping, Sweden, 8–13 May)*
[29] Sarmin S, Tarek M, Roopan S M, Cheng C K and Khan M M R 2021 Significant improvement of power generation through effective substrate-inoculum interaction mechanism in microbial fuel cell *J. Power Sources* **484** 229285
[30] Shi L, Squier T C, Zachara J M and Fredrickson J K 2007 Respiration of metal (hydr)oxides by *Shewanella* and *Geobacter*: a key role for multihaem c-type cytochromes *Mol. Microbiol.* **65** 12–20
[31] Kumar S S, Kumar V, Malyan S K, Sharma J, Mathimani T, Maskarenj M S, Ghosh P C and Pugazhendhi A 2019 Microbial fuel cells (MFCs) for bioelectrochemical treatment of different wastewater streams *Fuel* **254** 115526
[32] Choudhury P, Prasad Uday U S, Bandyopadhyay T K, Ray R N and Bhunia B 2017 Performance improvement of microbial fuel cell (MFC) using suitable electrode and bioengineered organisms: a review *Bioengineered* **8** 471–87
[33] Kumar S S, Malyan S K, Basu S and Bishnoi N R 2017 Syntrophic association and performance of Clostridium, *Desulfovibrio*, Aeromonas and Tetrathiobacter as anodic biocatalysts for bioelectricity generation in dual chamber microbial fuel cell *Env. Sci. Pollut. Res.* **24** 16019–30
[34] Sturm G, Richter K, Doetsch A, Heide H, Louro R O and Gescher J 2015 A dynamic periplasmic electron transfer network enables respiratory flexibility beyond a thermodynamic regulatory regime *ISME J.* **9** 1802–11

[35] Yang Y, Ding Y, Hu Y, Cao B, Rice S A, Kjelleberg S and Song H 2015 Enhancing bidirectional electron transfer of *Shewanella oneidensis* by a synthetic flavin pathway *ACS Synth. Biol.* **4** 815–23
[36] Song H L, Zhu Y and Li J 2019 Electron transfer mechanisms, characteristics and applications of biological cathode microbial fuel cells—a mini review *Arab. J. Chem* **12** 2236–43
[37] Khan M D, Khan N, Sultana S, Joshi R, Ahmed S, Yu E, Scott K, Ahmad A and Khan M Z 2017 Bioelectrochemical conversion of waste to energy using microbial fuel cell technology *Process Biochem.* **57** 141–58
[38] Prathiba S, Kumar P S and Vo D V N 2022 Recent advancements in microbial fuel cells: a review on its electron transfer mechanisms, microbial community, types of substrates and design for bio-electrochemical treatment *Chemosphere* **286** 131856
[39] Malvankar N S, Yalcin S E, Tuominen M T and Lovley D R 2014 Visualization of charge propagation along individual pili proteins using ambient electrostatic force microscopy *Nat. Nanotechnol.* **9** 1012–7
[40] Kang C S, Eaktasang N, Kwon D Y and Kim H S 2014 Enhanced current production by *Desulfovibrio desulfuricans* biofilm in a mediator-less microbial fuel cell *Bioresour. Technol.* **165** 27–30
[41] Ieropoulos I A, Greenman J and Melhuish C 2013 Miniature microbial fuel cells and stacks for urine utilisation *Int. J. Hydrogen Energy* **38** 492–6
[42] Yang Y, Xu M, Guo J and Sun G 2012 Bacterial extracellular electron transfer in bioelectrochemical systems *Process Biochem.* **47** 1707–14
[43] Ahn Y, Hatzell M C, Zhang F and Logan B E 2014 Different electrode configurations to optimize performance of multi-electrode microbial fuel cells for generating power or treating domestic wastewater *J. Power Sources* **249** 440–5
[44] Aiyer K S 2020 How does electron transfer occur in microbial fuel cells? *World J. Microbiol. Biotechnol.* **36** 19
[45] Ieropoulos I A, Greenman J, Melhuish C and Hart J 2005 Comparative study of three types of microbial fuel cell *Enzyme Microb. Technol.* **37** 238–45
[46] Verma J, Kumar D, Singh N, Katti S S and Shah Y T 2021 Electricigens and microbial fuel cells for bioremediation and bioenergy production: a review *Env. Chem. Lett.* **19** 2091–126
[47] Zhou M, Yang J, Wang H, Jin T, Hassett D J and Gu T 2014 Bioelectrochemistry of microbial fuel cells and their potential applications in bioenergy *Bioenergy Research: Advances and Applications* (London: Newnes)
[48] Xu J, Sheng G P, Luo H W, Li W W, Wang L F and Yu H Q 2012 Fouling of proton exchange membrane (PEM) deteriorates the performance of microbial fuel cell *Water Res.* **46** 1817–24
[49] Li W W, Sheng G P, Liu X W and Yu H Q 2011 Recent advances in the separators for microbial fuel cells *Bioresour. Technol.* **102** 244–52
[50] Pham T H, Rabaey K, Aelterman P, Clauwaert P, De Schamphelaire L, Boon N and Verstraete W 2006 Microbial fuel cells in relation to conventional anaerobic digestion technology *Eng. Life Sci.* **6** 285–92
[51] Zhou M, Wang H, Hassett D J and Gu T 2013 Recent advances in microbial fuel cells (MFCs) and microbial electrolysis cells (MECs) for wastewater treatment, bioenergy and bioproducts *J. Chem. Technol. Biotechnol.* **88** 508–18
[52] Peng X, Yu H, Yu H and Wang X 2013 Lack of anodic capacitance causes power overshoot in microbial fuel cells *Bioresour. Technol.* **138** 353–8

[53] Mohan S V, Velvizhi G, Modestra J A and Srikanth S 2014 Microbial fuel cell: critical factors regulating bio-catalyzed electrochemical process and recent advancements *Renew. Sustain. Energy Rev.* **40** 779–97

[54] Michie I S, Dinsdale R M, Guwy A J and Premier G C 2020 Electrogenic biofilm development determines charge accumulation and resistance to pH perturbation *Energies* **13** 3521

[55] Saratale G D, Saratale R G, Shahid M K, Zhen G, Kumar G, Shin H S, Choi Y G and Kim S H 2017 A comprehensive overview on electro-active biofilms, role of exo-electrogens and their microbial niches in microbial fuel cells (MFCs) *Chemosphere* **178** 534–47

[56] Hassan S H A, Kim Y S and Oh S E 2012 Power generation from cellulose using mixed and pure cultures of cellulose-degrading bacteria in a microbial fuel cell *Enzyme Microb. Technol.* **51** 269–73

[57] Chen G W, Choi S J, Lee T H, Lee G Y, Cha J H and Kim C W 2008 Application of biocathode in microbial fuel cells: cell performance and microbial community *Appl. Microbiol. Biotechnol.* **79** 379–88

[58] He Z and Angenent L T 2006 Application of bacterial biocathodes in microbial fuel cells *Electroanalysis* **18** 2009–15

[59] Srikanth S and Mohan S V 2012 Change in electrogenic activity of the microbial fuel cell (MFC) with the function of biocathode microenvironment as terminal electron accepting condition: influence on overpotentials and bio-electro kinetics *Bioresour. Technol.* **119** 241–51

[60] Clauwaert P, Van Der Ha D, Boon N, Verbeken K, Verhaege M, Rabaey K and Verstraete W 2007 Open air biocathode enables effective electricity generation with microbial fuel cells *Env. Sci. Technol.* **41** 7564–9

[61] Rismani-Yazdi H, Carver S M, Christy A D and Tuovinen O H 2008 Cathodic limitations in microbial fuel cells: an overview *J. Power Sources* **180** 683–94

[62] Rusli S F N, Abu Bakar M H, Loh K S and Mastar M S 2019 Review of high-performance biocathode using stainless steel and carbon-based materials in microbial fuel cell for electricity and water treatment *Int. J. Hydrogen Energy* **44** 30772–87

[63] Zhang Y, Sun J, Hu Y, Li S and Xu Q 2012 Bio-cathode materials evaluation in microbial fuel cells: a comparison of graphite felt, carbon paper and stainless steel mesh materials *Int. J. Hydrogen Energy* **37** 16935–42

[64] Nawaz A, Hafeez A, Abbas S Z, ul Haq I, Mukhtar H and Rafatullah M 2020 A state of the art review on electron transfer mechanisms, characteristics, applications and recent advancements in microbial fuel cells technology *Green Chem. Lett. Rev.* **13** 365–81

[65] Yaqoob A A, Ibrahim M N M and Guerrero-Barajas C 2021 Modern trend of anodes in microbial fuel cells (MFCs): an overview *Env. Technol. Innov.* **23** 101579

[66] Yaqoob A A, Mohamad Ibrahim M N and Yaakop A S 2021 Application of oil palm lignocellulosic derived material as an efficient anode to boost the toxic metal remediation trend and energy generation through microbial fuel cells *J. Clean. Prod.* **314** 128062

[67] Treesubsuntorn C, Chaiworn W, Surareungchai W and Thiravetyan P 2019 Increasing of electricity production from *Echinodosus cordifolius*-microbial fuel cell by inoculating *Bacillus thuringiensis Sci. Total Env.* **686** 538–45

[68] Abbas S Z, Rafatullah M, Ismail N and Syakir M I 2016 Removal of metals (chromium and copper) and power generation through sediment microbial fuel cell *Int. J. Env. Tech. Sci.* **2** 56–60

[69] Sharma M, Jain P, Varanasi J L, Lal B, Rodríguez J, Lema J M and Sarma P M 2013 Enhanced performance of sulfate reducing bacteria based biocathode using stainless steel mesh on activated carbon fabric electrode *Bioresour. Technol.* **150** 172–80

[70] Pillot G, Davidson S, Auria R, Combet-Blanc Y, Godfroy A and Liebgott P P 2020 Production of current by syntrophy between exoelectrogenic and fermentative hyperthermophilic microorganisms in heterotrophic biofilm from a deep-sea hydrothermal chimney *Microb. Ecol.* **79**

[71] Nancharaiah Y V, Venkata Mohan S and Lens P N L 2015 Metals removal and recovery in bioelectrochemical systems: a review *Bioresour. Technol.* **195** 102–14

[72] Koók L, Rózsenberszki T, Nemestóthy N, Bélafi Bakó K and Bakonyi P 2016 Bioelectrochemical treatment of municipal waste liquor in microbial fuel cells for energy valorization *J. Clean. Prod.* **112** 4406–12

[73] Wen Q, Wu Y, Zhao L and Sun Q 2010 Production of electricity from the treatment of continuous brewery wastewater using a microbial fuel cell *Fuel* **89** 1381–5

[74] Catal T, Li K, Bermek H and Liu H 2008 Electricity production from twelve monosaccharides using microbial fuel cells *J. Power Sources* **175** 196–200

[75] Bhowmick G D, Neethu B, Ghangrekar M M and Banerjee R 2020 Improved performance of microbial fuel cell by *in situ* methanogenesis suppression while treating fish market wastewater *Appl. Biochem. Biotechnol.* **192** 1060–75

[76] Huang L, Li X, Ren Y and Wang X 2016 *In-situ* modified carbon cloth with polyaniline/graphene as anode to enhance performance of microbial fuel cell *Int. J. Hydrogen Energy* **41** 11369–79

[77] Cui H F, Du L, Guo P B, Zhu B and Luong J H T 2015 Controlled modification of carbon nanotubes and polyaniline on macroporous graphite felt for high-performance microbial fuel cell anode *J. Power Sources* **283** 46–53

[78] Najafpour G D, Rahimnejad M, Mokhtarian N, Ramli W, Daud W and Ghoreyshi A A 2009 Low voltage power generation in a biofuel cell using anaerobic cultures *World Appl. Sci. J.* **6** 1585–88

[79] Hall H and Bettin C 2006 Applicability and feasibility of incorporating microbial fuel cell technology into implantable biomedical devices *Thesis* College of Engineering, The Ohio State University http://hdl.handle.net/1811/6443

[80] Kim J R, Jung S H, Regan J M and Logan B E 2007 Electricity generation and microbial community analysis of alcohol powered microbial fuel cells *Bioresour. Technol.* **98** 2568–77

[81] Tang H, Zeng Y, Zeng Y, Wang R, Cai S, Liao C, Cai H, Lu X and Tsiakaras P 2017 Iron-embedded nitrogen doped carbon frameworks as robust catalyst for oxygen reduction reaction in microbial fuel cells *Appl. Catal.* B **202** 550–6

[82] Yamashita T and Yokoyama H 2018 Molybdenum anode: a novel electrode for enhanced power generation in microbial fuel cells, identified via extensive screening of metal electrodes *Biotechnol. Biofuels* **11** 39

[83] Cui Y, Rashid N, Hu N, Rehman M S U and Han J I 2014 Electricity generation and microalgae cultivation in microbial fuel cell using microalgae-enriched anode and biocathode *Energy Convers. Manag.* **79** 674–80

[84] Daud N N M, Ahmad A, Yaqoob A A and Ibrahim M N M 2021 Application of rotten rice as a substrate for bacterial species to generate energy and the removal of toxic metals from wastewater through microbial fuel cells *Env. Sci. Pollut. Res.* **28** 62816–27

[85] Yaqoob A A, Ibrahim M N M, Rafatullah M, Chua Y S, Ahmad A and Umar K 2020 Recent advances in anodes for microbial fuel cells: an overview *Materials* **13** 2078

IOP Publishing

Microbial Fuel Cells
Emerging trends in electrochemical applications
Mohamad Nasir Mohamad Ibrahim, Asim Ali Yaqoob and Akil Ahmad

Chapter 13

Comprehensive cost analysis of electrochemical performance in microbial fuel cells

Al-Mustasin Abir Hossain, Nahian Masud and Mohammad Ali

Microbial fuel cells (MFCs) are regarded as one of the best potential renewable energy resources which can reduce the overall production cost of electricity generation. In this chapter, how MFCs can be beneficial in power generation and water purification is discussed. A cost–benefit analysis model is conducted here using the MFC lifecycle costs and benefits to improve the efficiency with which MFCs are used to generate power and purify wastewater. The cost per unit power and revenue per unit power for typical MFCs are calculated from the investment price and revenue from generating power during the wastewater treatment procedure. The cost–benefit ratio (CBR) calculated from the cost per unit power and revenue per unit power, is used to define the comprehensive benefits offered by MFCs. The model indicated that revenue per unit power increased faster as the cost per unit power increased for a constructed wetland MFC (CW–MFC) than for the other MFCs that were evaluated. CW–MFCs give the best comprehensive benefits of the MFCs that are assessed and then compared with microalgae MFC (MA-MFC). MA-MFCs generally provide the poorest comprehensive benefits among all the different types of MFCs. In addition, how miniature MFCs can be fabricated cheaply has been studied. The details of this chapter will provide more fruitful information.

13.1 Objective

In this chapter, cost analysis of MFC setup is discussed. This setup will be fruitful in the long term if it can be utilized in a proper manner. Socio-economic growth and excessive population growth in China are causing severe environmental problems and growing demand for energy. According to reports, China used more energy than any other country in 2018, 2019, and 2020 (previously the USA used more than any

doi:10.1088/978-0-7503-4791-4ch13

other country) [1]. Total energy consumption in China in 2016 alone was ~4.36 × 109 t standard coal equivalents, and energy consumption has augmented by 1.4% each year on average [2]. Fossil fuels such as coal and petroleum are non-renewable and will eventually run out. Water pollution is gradually becoming a serious concern in China and directly leads to water shortages. It has been identified that almost half of all water bodies in China are polluted to some extent and that almost half of the population in China has contaminated drinking water [3]. Traditional water supply and wastewater treatment processes need high energy inputs, and this can worsen water shortages. It is important to develop ways to use energy sustainably, decrease energy use, and decrease environmental pollution. MFCs, which use electrode reactions to take advantage of metabolic processes in microbes, are new systems allowing renewable energy to be produced while curing waste material. MFCs are considered to be auspicious ways of producing energy while treating wastewater [4, 5]. MFCs generate electricity through electro-active bacteria consuming organic pollutants [6].

MFCs produce electricity and purify wastewater, but presently available MFCs generate much less energy than can be generated by chemical fuel cells. High installation and operating costs (because expensive proton exchange membranes and Pt-catalytic cathodes are utilized) make currently available MFCs unappealing [7]. The gains of MFCs therefore require exploring in detail. Many previous studies of MFCs have been focused on MFC design and formation and the electrode materials, operating parameters, and bacteria used. Little effort has been applied into executing cost–benefit analyses (CBAs) for MFCs, and no complete accounting systems have been accomplished. A CBA model based on a comprehensive analysis of various MFCs is recognized here. The CBA is authenticated using MFC test cases to let the comprehensive benefits presented by different MFCs be analyzed and compared and the best operating mode recognized. In developing countries like Bangladesh, India, Pakistan and Nepal, utilizing natural resources can be an accelerator in power production. Using MFCs on a commercial scale can reduce the power supply demand in those developing countries. Since operating cost and maintenance cost of MFCs are less in comparison to other cells, overall power production cost can be minimized. Hence, in the developing countries this saving can be used in other sectors. The details of this study will provide more useful information on cost analysis of MFCs on a commercial scale.

13.2 Microbial fuel cells device

MFCs are generally composed of anode and cathode chambers separated by proton exchange membrane [8]. A schematic figure is shown in figure 13.1. Substrate in the anode chamber is oxidized under catalytic actions of microorganisms to produce electrons and protons. The electrons are transported to the anode surface before they reach the cathode through wires which join the cathode and the anode. The protons travel to the cathode through the proton exchange membrane residing between the cathode and anode chambers. Protons and electrons react with the oxidizing

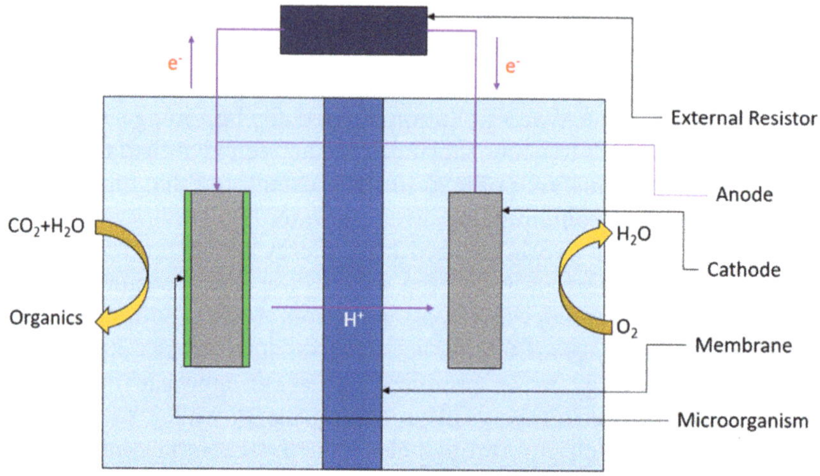

Figure-13.1 : Schematic diagram of an MFC [15, 74, 75].

materials in the cathode chamber on the cathode surface, which makes the MFC output electrical energy.

There are a few ways, using several standards, of classifying MFCs. MFCs consist of the types described in the following sections.

13.2.1 Single-chamber and double-chamber MFCs

Most MFCs presently used in workshops are double-chamber MFCs. Such an MFC has an anode chamber and a cathode chamber detached by a proton-exchange membrane [9–11]. Electrolyte solutions in the cathode and anode chambers are self-governing of each other and the cathode and anode act as electron acceptors and electron donors, correspondingly. Double-chamber MFCs are airtight, which ensures an anaerobic environment in the anode chamber and protects the anode chamber from adulteration. Double-chamber MFCs are regularly used to separate electrogenic bacteria and to test the electrogenic performances of individual bacteria species [12].

A double-chamber MFC will have a high internal resistance, as high as 900–1000 Ω [13], which will limit the electrogenic performance of the MFC. Double-chamber MFCs have complex assemblies and are exclusive and large. These shortcomings can be circumvented in single-chamber MFCs, which are established from double-chamber MFCs [14, 15]. Single-chamber MFCs do not have a cathode chamber, that means the cathode and anode are closer than in a double-chambered MFC. This design is more economical than a double-chambered MFC as no proton exchange membrane is used, and the internal resistance is inferior in a single-chambered MFC to that of a double-chambered MFC. However, it is more convenient for oxygen to reach the anode of a single-chambered MFC than the anode of a double-chambered MFC, which suggests it is hard to keep an anaerobic environment at the anode of a single-chambered MFC.

13.2.2 MFCs with different electrode materials

The electrodes determine the performance and cost of an MFC, and the main challenge is to fabricate a cost-effective and extendable MFC with appropriate electrodes [16]. The electrogenic bacteria in an MFC exist on the anode, and the catalytic reactions happen at the anode–solution boundary. A suitable anode should be conductive, chemically stable, biocompatible, and have a huge specific surface area. Carbon paper, carbon cloth, graphite sheets, and graphite plates are often used as electrodes in the laboratory [17]. Carbon-based materials are cheap, conductive, and corrosion resistant. Bacteria easily become attached to the surfaces of carbon-based materials and grow. However, carbon-based anodes generally provide bad electrogenic performances. The electrogenic performance of a carbon-based anode can be enhanced by ammoniating the anode at a high temperature, treating the anode with acid, or adding a significant number of metal ions to the carbon material. Granular activated carbon has been efficaciously used in MFCs as it has a large specific surface area and is comparatively cheaper and easier to handle [18]. Carbon nanotubes have constant pore structures, and the pores are conducive to micro-organism development. Graphene has a high catalytic activity because of its unique honeycomb lattice construction. Carbon nanotubes and graphene are extensively used in MFC anodes because of the advantages they provide, and they have been figured out to enhance the electrogenic performances of MFCs comparative to previously available anodes. Composite materials prepared by combining conductive polymers with carbon nanotubes or graphene provide higher MFC power densities than can be attained using carbon nanotubes or graphene alone.

The cathode accepts protons and electrons and governs the redox reaction rate. The cathode should have a high redox potential and be steady. A catalyst often requires to be added to the cathode to provide better electrogenic performance. Most MFCs use air cathodes, which are usually made from carbon-based materials. Activated carbon cathodes have been identified to give comparable performances to Pt-loaded cathodes in MFCs [19, 20]. Foamed nickel and stainless steel can also be utilized in MFC cathodes as they are cheap, very conductive, and physically strong [21]. The catalyst utilized in the cathode is critical to the MFC performance. Pt and Pt alloys give suitable catalytic effects. However, the price of Pt hinders the usage of Pt and Pt alloys. Transition metal oxides are globally used as cathodes as they are cheap and readily available. MnO_2 and TiO_2 are regarded as ideal cathode catalysts. Conductive polymers such as polyaniline and polypyrrole are not difficult to synthesize and have better electrical conductivities and stabilities. Polyaniline and polypyrrole used either alone or in composites with other materials (e.g., carbon nanotubes or V_2O_5) are known as good catalysts. Marked progress has been made in developing materials with performances comparable to the performance of Pt in the quick term and even longer term [6].

13.2.3 Microalgae MFCs

It has been identified that MA-MFCs merging photosynthetic microorganisms with MFCs can generate electricity efficiently using light and carbon dioxide and also purify water. Research into MA-MFCs was first performed in 1964. However, the energy usage rate is too low for early MA-MFCs to be of experimental usage [22]. Enhancements in the structures and performances of MFCs have allowed MA-MFCs to once again appeal attention. The usage of algae biocathodes in MA-MFCs has been analyzed by several researchers [23–25]. It has been shown that suitable MA-MFCs may now be formed and that many pollutants that can be detached by MFCs may be able to be removed by MA-MFCs. However, an MA-MFC needs a large surface area to allow a proper amount of light to enter the cathode chamber, and it is luxurious to harvest and process the used algal cells [7]. The electricity generation mechanism and factors affecting the electrogenic performances of MA-MFCs are not well understood and need to be researched more.

13.2.4 Mixed water MFCs

The substrate is one of the most influential biological factors affecting electricity generation in an MFC [26]. A wide range of substrates, from pure composites to wastewater containing a complex mixture of organic compounds, can be utilized in MFCs to generate power [27]. Pure components such as acetic acid, butyric acid, and glucose have low molecular masses and simple molecular constructions. Chae *et al* assessed the electrogenic performances of MFCs using acetic acid, butyric acid, glucose, and propionic acid as substrates when the MFCs were operated for a year [28]. The MFC using acetic acid had the highest coulombic efficiency, and the MFC using glucose had the highest power density. MFCs can also use compound mixtures such as domestic wastewater [29, 30], dye wastewater [31, 32], electroplating wastewater [33], swine wastewater [34], and brewery wastewater [35–37] as the substrates. These types of wastewaters have comparatively complex compositions and high organic content, and the components can be impervious to degradation and can therefore be used as electron donors and carbon sources in MFCs. Using wastewater in an MFC allows the wastewater to be purified while power is generated.

13.2.5 Constructed wetland MFCs

An MFC has an aerobic cathode and an anaerobic anode, which is steady with the natural stratification of a constructed wetland (CW), which will be aerobic at the top and anaerobic at the bottom [38]. This has led CW-MFCs to be built. Plant roots in the cathode chamber release oxygen, which acts as an electron acceptor, and a biocathode is shaped at the top of a CW. CW-MFCs are more inexpensive to build than MFCs that use a noble metal such as Pt as a catalyst. CW-MFCs are new environmentally gentle wastewater treatment systems that are viewed as economical and effective systems for harvesting bioenergy [39]. A CW-MFC can also restrict greenhouse gas emissions by acting as a carbon sink. Much research targeted at developing the electricity generation and wastewater treatment performances of CW-MFCs has been performed in recent years [40–42].

13.3 Cost model

13.3.1 Net investment cost

The investment cost when fabricating an MFC will be conquered by the procurement costs of the materials (electrodes, epoxy resin, reactors, wires, etc). MFC vessels are generally constructed of Plexiglas or polyethylene and are generally cylindrical in shape. The electrodes can be fabricated of carbon-based materials, metals, or composite materials, and can be of different shapes and sizes depending on the vessel structure. The wires (for conducting electricity) are generally constructed of Cu or Ti but are sporadically Pt. Epoxy resin is used to sequester and seal the links between the electrodes and wires.

Various materials have diverse life spans, and only some materials (e.g., Plexiglas reactors, wires, and some electrode materials) can be recycled after being used in an MFC. It has been found that Cu, Pt, Plexiglas, stainless steel, and Ti can have lifespan of 10 years or more and that these materials will have depreciated to 90% of the original price at the end of their service lives. A Nafion membrane will have a service life of ~2 years [47] and will depreciate to 50% of the original price. The depreciated price is the price obtained for a material minus the recycling cost. Some materials used in an MFC can simply be detached and cleaned before reuse, so the recycling cost will be conquered by the cost of the labor involved. A labor cost of $4.7 for recycling all materials in an MFC was used in the CBA. The net investment cost (CI) for an MFC was defined as the difference between the investment cost and the depreciated value.

13.3.2 Operating and maintenance cost

The operating and maintenance costs (CM) includes labor costs, daily operating and maintenance costs of devices, water quality monitoring costs, and routine maintenance costs. The materials required to be treated and assembled before the MFC can function, and a trial run needs to be performed to certify that the MFC has been correctly accumulated. The main costs during the trial are labor costs. Routine maintenance after operating the MFC mainly involves replacing the reaction fluid and checking that the MFC is operating normally. The routine maintenance expenditures usually include labor costs and device and instrument maintenance costs. The expenditure of the reagents used in the reaction solution can be calculated from the doses used and the operating cycle. The water quality is usually monitored twice each month to assess how effectively the MFC is treating the wastewater. The water quality monitoring costs include reaction solution costs, instrument and equipment use fees, and labor costs. Chemical oxygen demand, ammonia nitrogen, total phosphorus, and total nitrogen monitoring are each assumed to cost $24.5. Each part of the MFC is inspected and repaired once every three months at a cost of $15.8/item. The cost of replacing a material is determined by the lifespan of the material and the actual condition of the material. For example, metals have long lifespans and are replaced every year, whereas carbon-based materials such as carbon cloth and carbon paper break easily and are usually replaced after every six months. Some reactors require energy-consuming instruments such as fluorescent lamps and peristaltic pumps to be used constantly for a long time, and the electricity is usually assumed to cost \$0.13 kW h^{-1}).

13.3.3 Cost per unit power

The cost per unit power (*CU*) is regarded the cost of a unit area (1 [m^2]) of electrode material generating a unit power (1 [W]) of electricity and processing wastewater for 1 [d],

$$CU = \frac{CI}{P \times 360} + \frac{CM}{360} \tag{13.1}$$

where *CU* is in units of [\$/((W m^{-2})d)], *CI* is the net investment cost in [\$/a], *P* is defined as the power density of the device in [W m^{-2}], and *CM* is denoted as the operating and maintenance cost in [\$/a].

13.3.4 Cost analysis of a handmade MFC

A simple model of MFC was made for a practical purpose to check the performance of various electrode materials. The containers as anode and cathode chamber were made of plastic which costs 1.17\$ per unit. Three types of electrodes were used with a surface area of 8 cm^2. Graphite electrodes cost \$0.93, zinc electrodes costs \$1.05 and copper electrodes costs \$1.40. A U-tube, made of glass costs \$1. Wirings and other materials had a cost around \$1.52. So, a single handmade MFC costs approximately \$7.11. The power density was 112 mW m^{-2}. The maintenance cost was \$2 per cycle. The cost per unit power of the MFC is \$4.55/(W m^{-2})d.

13.4 Cost-benefit model

13.4.1 Revenue from electricity production

The annual power output of the MFC (for an operating year of 360 [day]) and the revenue brought in can be calculated from the power density of the MFC and the electrode area. The electricity price is assumed to be [\$0.13 (kW h^{-1})]. The revenue from electricity production (*RE*) can be calculated using the equation,

$$RE = P \times A \times 360 \times 24 \times 0.13 \tag{13.2}$$

where *RE* is the revenue from power generation in [\$/a], *P* is defined as the power density of the device in [W m^{-2}], *A* is denoted as the electrode area in [m^2], 360 × 24 is the time the MFC operates each year in [h], and 0.13 is the price of electricity in [\$/(kWh)].

13.4.2 Revenue from wastewater treatment

The wastewater treatment capacity of an MFC for one year and the revenue for treating wastewater (for an operating year of 360 [d]) can be calculated from the effective volume and operating cycle of the MFC. The revenue from treating wastewater (*RW*) was assumed to be [$0.13/$m^3$] [48]. The revenue from wastewater treatment is calculated using the equation,

$$RW = V \times \left(\frac{360}{T}\right) \times 0.13 \times 10^{-6} \tag{13.3}$$

where *RW* is defined as the annual revenue from wastewater treatment in [$/a], *V* is regarded as the effective volume of the MFC in [mL], *T* is denoted as the operating cycle of the device in [d], and 0.13 is the revenue from treating wastewater in [$/$m^3$].

13.4.3 Revenue per unit power produced

Revenue per unit power produced (*RU*) is calculated for a year because the revenue from an MFC is considered low. Revenue per unit power means the revenue for a unit area (1 [m^2]) of electrode material generating a unit of electricity (1 [W]) and processing wastewater for a year. The revenue per unit power can be calculated using the equation,

$$RU = \frac{RE + RW}{P} \tag{13.4}$$

where *RU* is in [$/((W m^{-2})a)], *RE* is the annual revenue from electricity generated by the MFC in [$/a], *RW* is the annual revenue from treating wastewater in [$/a], and *P* is the power density of the device in [W m^{-2}].

13.5 Cost–benefit ratio

The CBR [49] is defined as an indicator used in CBA to summarize the overall value for money of a project or proposal. The CBR is the ratio between the cost of the project or proposal in monetary terms and the benefit given by the project or proposal in monetary terms. The CBR takes into account the cost of executing the project and the monetary gain arising from the project. The lower the CBR the better the investment. The inverse of the CBR is the benefit–cost ratio [50].

The electrode areas, effective volumes, and operating cycles of MFCs can vary widely, so the CBR can be utilized to consistently evaluate the overall benefits of different MFCs after the costs and benefits have been estimated and will allow the costs and benefits of different MFCs to be compared equitably. The CBR can be calculated using the equation below,

$$CBR = \frac{CI}{RE + RW} \tag{13.5}$$

where *CI* is the investment cost in [$/a], and *RE* and *RW* are the revenues from generating electricity and treating wastewater, respectively, in [$/a].

13.6 Sensitivity analysis

Sensitivity analysis is a common technique to analyze the uncertainty when evaluating the economics of an investment project [44]. Sensitivity analysis includes finding which of the multiple factors most likely influence the economic benefit index of the project and analyzing the effects of variations in these factors on the economic benefit indicators for the project.

13.7 Discussion

Typical MFCs are studied. These are an anode-material-changing MFC (AC-MFC), a cathode-material-changing MFC (CC-MFC), a raw water MFC (RW-MFC), a microalgae MFC (MA-MFC), and a constructed wetland MFC (CW-MFC).

The net investment cost, operating and maintenance costs, and revenues from generating electricity and treating wastewater are projected for each MFC using the CBA model. The MFC operating parameters are shown in table 13.1. The costs calculations and CBRs for the MFCs are shown in table 13.2.

The results for each MFC are displayed in figure 13.2, with the *CU* and *RU* on the horizontal and vertical axes, respectively. It can be witnessed that the *RU* increased as the *CU* increased. The *CU* values are \$5–\$40/((W m^{-2})d), but the *RU* values varied over a broader range. The rate at which the *RU* increased as the *CU* increased is highest for the CW-MFC, next highest for the RW-MFC, then the CC-MFC, then the MA-MFC, and lowest for the AC-MFC.

The *log*(CBR) values for the MFCs are displayed as a histogram in figure 13.3. The *log*(CBR) reference value is 0, which means that the costs and benefits are equal. The CW-MFC gives the most comprehensive benefits of the MFCs that are evaluated ($p < 0.05$), the CW-MFC *log*(CBR) values being 2.7238 ± 0.1504 for experimental conditions and 2.4910 ± 0.0584 for practical conditions. The MA-MFC *log*(CBR) values are 5.0079 ± 0.1068 for experimental conditions and 4.9993 ± 0.1091 for practical conditions, both of which are meaningfully higher than the *log* (CBR) values for the other MFCs ($p < 0.05$), meaning the MA-MFC comprehensive benefits are worse than the comprehensive benefits of the other MFCs. The *log* (CBR) values for the other MFCs (4.0–4.3) are between the CW-MFC and MA-MFC *log*(CBR) values. MA-MFCs generally have high investment costs and high operating and maintenance costs because such systems require lighting equipment or photobioreactors and have low wastewater treatment capacities, meaning they have poor efficiencies. CW-MFCs are large but have high wastewater treatment capacities and generally use cheap electrodes and recycled materials. The roots of the wetland plants used in CW-MFCs can upsurge the oxygen concentration in the system, improving the water treatment effect, and the biomass generated by the plants can be harvested. These factors mean that CW-MFCs offer significant comprehensive benefits.

The MFCs labelled here are mostly used in experiments and few full-scale MFCs have been manufactured. The costs of experimental and full-scale MFCs will be different. For example, the influent of a full-scale MFC is natural sewage, so the costs of reagents used in experimental MFCs are not applied. The differences in costs of experimental and full-scale MFCs are calculated by comparing the CBRs of different MFCs under dissimilar conditions. No significant alterations are found between the CBRs for different MFCs under practical and experimental conditions.

Table-13.1: Operating parameters for typical MFCs.

	Electrode materials								
Chamber type	Anode	Cathode	Membrane	Type of substance	Anode area [cm^2]	Volume of anode chamber [mL]	Power fensity [mW m^{-2}]	Cycle (d)	Ref.
Single	Activated carbon	Pt-loaded carbon paper	Cation Exchange membrane	Wastewater from food factory	6.3	84.0	338.0	6.0	[51, 71]
Single	Carbon cloth	Pt-loaded carbon paper	Cation Exchange membrane	Wastewater from food factory	6.3	84.0	78.0	6.0	[51, 71]
Single	Graphite felt	Pt-loaded carbon cloth	—	Acetate	12.0	225.0	1256.0	3.0	[37, 71]
Single	Graphite polyaniline modified carbon cloth	Pt-loaded carbon cloth	—	Synthetic wastewater containing sodium acetate	7.0	28	831.0	2.0	[52, 71]
Single	Carbon paper	Pt-loaded carbon cloth	—	Synthetic wastewater containing sodium acetate	20.0	80	378.1	1.3	[53, 71]
Single	Carbon paper	Pt-loaded carbon paper	Proton exchange membrane	Domestic wastewater with the medium ingredients	11.3	250	38.0	10.0	[54, 71]
Single	Graphite felt	Pt-loaded carbon cloth	—	Glucose	12.0	225	1519.0	5.0	[37]
Single	Graphite felt	Carbon Paper (MnO_2/ CNTs as catalyst)	—	Synthetic wastewater containing glucose	7.0	58	210.0	6.0	[55, 71]
Single	Graphite felt	Carbon paper (Pt/C as catalyst)	—	Synthetic wastewater containing glucose	7.0	58	229.0	5.0	[55, 71]
Double	Graphite felt	Carbon Paper	—	Synthetic wastewater containing glucose	7.0	58	32.7	4.0	[55, 71]
Double	Graphite felt	Graphite felt	—	Synthetic wastewater containing glucose	7.0	58	109.5	3.0	[55, 71]
Single	Graphite felt	Carbon paper (CNTs as catalyst)	—	Synthetic wastewater containing glucose	7.0	58	8.0	5.0	[55, 71]

Double	Graphite felt	Stainless steel mesh	—	Synthetic wastewater containing glucose	7.0	58	3.1	4.0	[55, 71]
Single	Graphite felt	Pt-loaded carbon cloth	—	Brewery wastewater	12.0	225	251.0	6.0	[37, 71]
Single	Carbon paper	Pt-loaded carbon paper	Cation Exchange membrane	Wastewater from food factory	6.3	84	56.0	6.0	[51, 71]
Single	Graphite flake	Graphite rod	Proton Exchange membrane	Surplus sludge	30.0	230	44.0	2.0	[56, 71]
Double	Graphite felt	Graphite felt	Proton Exchange membrane	Synthetic wastewater containing glucose added Chlorella	30.0	256	82.6	25.0	[57, 71]
Double	Graphite felt	Pt-loaded carbon paper	Proton Exchange membrane	Synthetic wastewater containing glucose added Chlorella	49.5	500	24.4	8.0	[58, 71]
Double	Carbon felt	Pt-loaded carbon paper	Proton Exchange membrane	Synthetic wastewater containing glucose added Chlorella	49.5	500	27.5	8.0	[58, 71]
Single	Granular Activated carbon	Granular graphite	—	Synthetic wastewater containing glucose	254.4	2540	19.7	2.0	[59, 71]
Single	Activated carbon	Stainless steel mesh coupled with activated carbon	—	Synthetic wastewater containing glucose	706.5	7000	74.5	2.0	[60, 71]
Single	Stainless Steel mesh coupled with Activated carbon	Stainless steel mesh coupled with activated carbon	—	Synthetic wastewater containing sodium acetate		2544.0	9.3	1.0	[61, 71]
Single	Titanium mesh coupled with activated carbon	Titanium mesh coupled with activated carbon	—	Synthetic wastewater		4850.0	3174.1	3.0	[62, 71]

Table-13.2: Results of the Cost-Benefit Analysis of Typical Microbial Fuel Cells

Type of MFC	Cost per unit power (Cu) [$/(W m^{-2}) d]	Revenue per unit power produced (Ru) [$10^{-2}/(W m^{-2}) a]	Annual net investment		Annual revenue [$10^{-3}/a]	Cost-benefit ratio (CBR)		Ref.
			Experimental	Practical		Experimental	Practical	
AC-MFC	5.12	0.26	29.26	29.26	0.88	33 250	33 250	[51, 71]
	7.55	0.41	18.88	13.46	5.17	3652	2603	[37, 71]
	10.20	0.16	15.07	12.69	1.31	11 504	9687	[52, 71]
	15.59	3.12	74.69	39.33	1.19	62 765	33 050	[54, 71]
CC-MFC	5.44	0.27	16.67	13.46	4.15	4017	3243	[37, 71]
	7.87	0.88	14.38	9.52	0.97	14 825	9814	[55, 71]
	9.55	6.67	12.10	9.13	0.53	22 830	17 226	[55, 71]
	17.97	21.40	13.18	9.47	0.66	19 970	14 348	[55, 71]
RW-MFC	5.03	0.82	13.46	13.46	2.05	6566	6566	[37, 71]
	6.35	1.21	29.84	29.84	0.68	43 882	43 882	[51, 71]
	11.76	12.24	25.82	25.82	5.39	4790	4790	[56, 71]
MA-MFC	7.19	0.94	127.61	126.39	0.77	165 727	164 143	[57, 71]
	29.80	12.22	224.52	218.61	2.99	75 090	73 114	[58, 71]
	29.96	10.91	254.68	248.77	3.00	84 893	82 923	[58, 71]
CW-MFC	18.16	834.80	90.59	72.30	164.45	551	440	[59, 71]
	25.08	387.27	400.19	71.83	288.31	1388	249	[60, 71]
	39.11	2497.77	70.60	58.71	232.39	304	253	[61, 71]
	7.56	5.10	64.05	62.95	198.42	338	332	[62, 71]

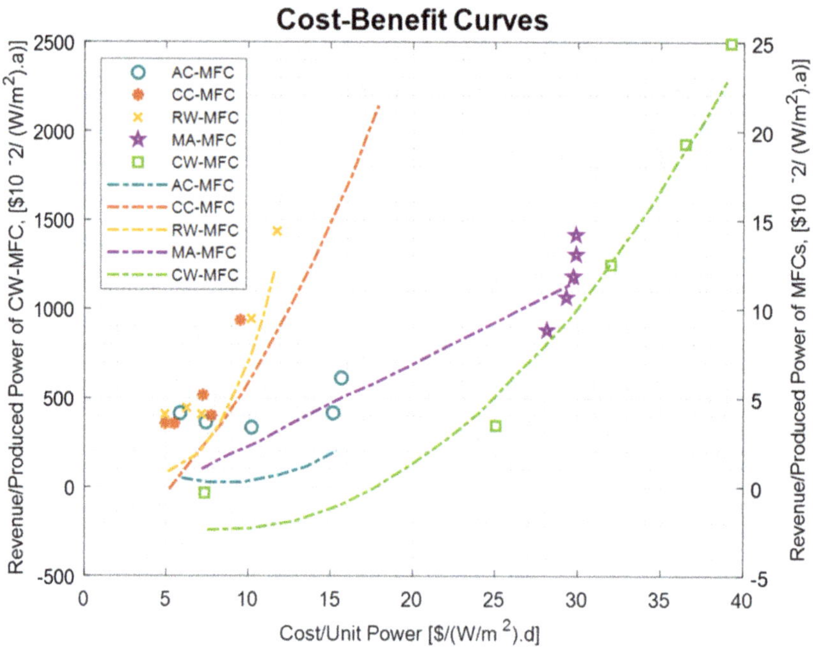

Figure-13.2 : Cost–benefit curves for typical MFCs [71].

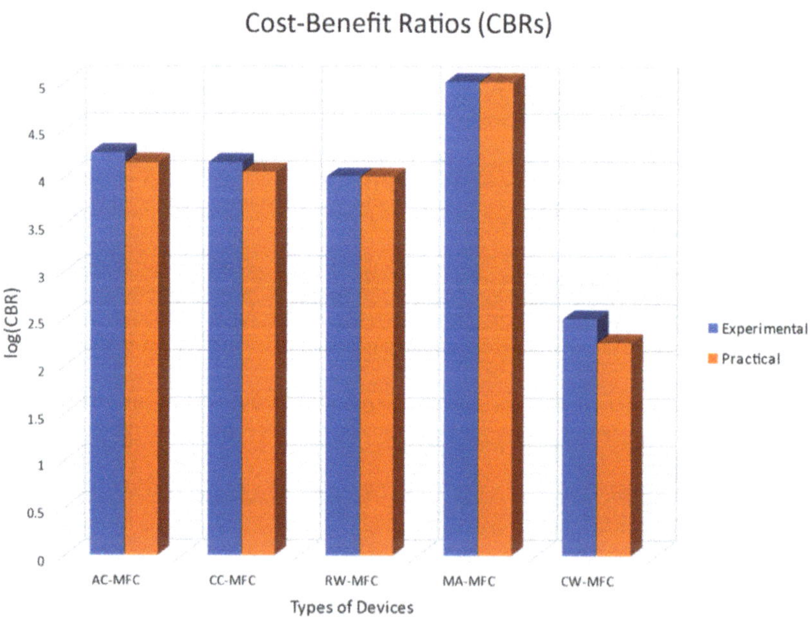

Figure-13.3 : CBRs for typical microbial fuel cells under different conditions (different letters (a, b, and c) indicate the values were significantly different at the $p = 0.05$ level) [71].

However, there are some hidden costs and benefits related to factors such as the sewage components, development of new materials, and government support. It is therefore necessary to perform sensitivity analyses. The amounts of power produced by the different MFCs, the processing costs, and the material replacement cycle are different for wastewater with different compositions, such as different chemical oxygen demands and ammonia nitrogen and nutrient concentrations. For example, MFCs treating pure acetate acid [63] and glucose [64] produce electricity more effectively in the early processing stages than MFCs treating water containing complex mixtures of pollutants. Water quality strongly affects the costs and benefits of MFCs. The MFC size also affects the benefits. Addressing all the matters described above requires data from future research.

Developing new materials may reduce the costs of electrodes. Graphene quantum dots are biocompatible. Microorganisms can adhere to and multiply on graphene quantum dots and become a good source of electrons and give a good current through an MFC [65]. Graphene quantum dots may therefore be used in improved anode materials in an MFC to increase the output efficiency. Plant fibers treated with poly (3,4-ethylenedioxythiophene) can form large directional channel conductors, which can be cheaper than conventional metal electrodes [66]. The manual maintenance costs are different in different parts of the world. Manual labor costs, for example, are different in Bangladesh and USA [67].

Governments may assist the development of 'clean energy' sources such as solar energy by providing aids [68]. Such national policies can reduce equipment costs and can progress social acceptance of clean energy production systems, making it comparatively easier to develop new facilities. Amplified demand endorses large-scale production, decreasing the unit cost.

Various types of MFCs offer different benefits, and this is reflected in the ecological services presented. Installing ecological wastewater treatment plants using MFCs to treat urine in developing countries offers social benefits that are more significant than the power generated [69]. In countries with scarce freshwater resources, it is comparatively cheaper to use microbial fuel techniques to dilute seawater than to produce potable water by reverse osmosis [70]. There can be great differences between experimental conditions and practical conditions. Other models have been studied as well [71–76] to figure out how this prescribed model can be improved. However, further studies are needed to investigate the effects of these differences.

13.8 Conclusion

In this chapter, overall cost analysis of MFCs have been discussed. Both the fabrication cost of large scale and small scale have been studied. In addition to that, the maintenance cost has been analyzed. However, depreciation cost has not been scrutinized in detail, but a CBA model and a detailed economic benefit analysis for typical MFCs are presented. As the expenditure increases, the revenue per unit power produced is identified to grow faster for CW-MFCs than the other MFCs, next fastest for RW-MFCs, then CC-MFCs, then MA-MFCs, and then AC-MFCs.

CW-MFCs are identified to offer the best comprehensive benefits of the MFCs that are evaluated. The MA-MFCs' CBR values are also discussed. These are both meaningfully higher than the CBR values for the other MFCs, which suggests MA-MFCs offer the poorest comprehensive benefits of all MFCs that are evaluated. The results specified that CW-MFCs are most suitable for practical applications.

References

[1] Tsai S B, Xue Y, Zhang J, Chen Q, Liu Y, Zhou J and Dong W 2016 Models for forecasting growth trends in renewable energy *Renew. Sust. Energ. Rev.* **77** 1169–78

[2] Liu Y, Xiao H, Lv Y and Zhang N 2017 The effect of new type urbanization on energy consumption in China: a spatial econometric analysis *J. Clean. Prod.* **163** S299–305

[3] Shao Z 2017 Current situation and countermeasures of water environment pollution *China Resour. Compr. Util.* **35** 14–15

[4] Tamakloe R Y 2015 Effect of COD and H2O2 concentration on DC– MFC, Renew *Energy* **83** 1299–304

[5] Sakdaronnarong C, Ittitanakam A, Tanubumrungsuk W, Chaithong S, Thanosawan S, Sinbuathong N and Jeraputra C 2015 Potential of lignin as a mediator in combined systems for biomethane and electricity production from ethanol stillage wastewater, Renew *Energy* **76** 242–8

[6] Winfield J, Gajda I, Greenman J and Ieropoulos I 2016 A review into the use of ceramics in microbial fuel cells *Bioresour. Technol.* **215** 296–303

[7] Lee D J, Chang J S and Lai J Y 2015 Microalgae–microbial fuel cell: A mini review *Bioresour. Technol.* **198** 891–5

[8] Elmekawy A, Hegab H M, Dominguezbenetton X and Pant D 2013 Internal resistance of microfluidic microbial fuel cell: challenges and potential opportunities *Bioresour. Technol.* **142** 672–82

[9] Cho M H, Kalathil S, Shim S H, Lee Y H and Lee J 2010 Simultaneous decolorization of mixed dye wastewater and electricity generation using a two chambered microbial fuel cell *J. Biotechnol.* **150** 239–9

[10] Oon Y S, Ong S A, Ho L N, Wong Y S, Oon Y L, Lehl H K and Thung W E 2016 Long-term operation of double chambered microbial fuel cell for bio-electro denitrification, Bioprocess *Bioprocess. Biosyst. Eng.* **39** 893–900

[11] Yu Y, Zhao J, Wang S, Zhao H, Ding X and Gao K 2017 Nitrogen removal and electricity production at a double-chamber microbial fuel cell with cathode nitrite denitrification *Environ. Technol.* **38** 3093–101

[12] Cao X X, Huang X, Zhang X, Liang P and Fan M 2009 A mini-microbial fuel cell for voltage testing of exoelectrogenic bacteria *Front. Environ. Sci. Eng. China* **3** 307–12

[13] Zhou G, Zhou Y, Zhou G, Lu L, Wan X and Shi H 2015 Assessment of a novel overflow-type electrochemical membrane bioreactor (EMBR) for wastewater treatment, energy recovery and membrane fouling mitigation *Bioresour. Technol.* **196** 648–55

[14] Wu Y, Zhao X, Jin M, Li Y, Li S, Kong F, Nan J and Wang A 2018 Copper removal and microbial community analysis in single-chamber microbial fuel cell *Bioresour. Technol.* **253** 372–7

[15] Zhang B G, Wang Z J, Zhou X, Shi C H, Guo H M and Feng C P 2015 Electrochemical decolorization of methyl orange powered by bioelectricity from single-chamber microbial fuel cells *Bioresour. Technol.* **181** 360–2

[16] Wei J, Liang P and Huang X 2011 Recent progress in electrodes for microbial fuel cells *Bioresour. Technol.* **102** 9335–44

[17] Min B and Logan B E 2004 Continuous Electricity generation from domestic wastewater and organic substrates in a flat plate microbial fuel cell *Environ. Sci. Technol.* **38** 5809–14

[18] Jiang D Q and Li B K 2009 Granular activated carbon single-chamber microbial fuel cells (GAC–SCMFCs): a design suitable for large-scale wastewater treatment processes *Biochem. Eng. J.* **47** 31–7

[19] Karra U, Muto E, Umaz R, Kölln M, Santoro C, Lei W and Li B 2014 Performance evaluation of activated carbon-based electrodes with novel power management system for long-term benthic microbial fuel cells *Int. J. Hydrogen Energ.* **39** 21847–56

[20] Santoro C, Artyushkova K, Babanova S, Atanassov P, Ieropoulos I, Grattieri M, Cristiani P, Trasatti S, Li B and Schuler A J 2014 Parameters characterization and optimization of activated carbon (AC) cathodes for microbial fuel cell application *Bioresour. Technol.* **163** 54–63

[21] Yamashita T, Ishida M, Asakawa S, Kanamori H, Sasaki H, Ogino A, Katayose Y, Hatta T and Yokoyama H 2016 Enhanced electrical power generation using flame-oxidized stainless steel anode in microbial fuel cells and the anodic community structure *Biotechnol. Biofuels* **9** 1–10

[22] Berk R S and Canfield J H 1964 Bioelectrochemical energy conversion *Appl. Microbiol.* **12** 10–12

[23] Campo A G D, Cañizares P, Rodrigo M A, Fernández F J and Lobato J 2013 *Microbial fuel cell with an algae-assisted cathode: A preliminary assessment, J. Power Sources* **242** 638–45

[24] Juang D F, Lee C H and Hsueh S C 2012 Comparison of electrogenic capabilities of microbial fuel cell with different light power on algae grown cathode *Bioresour. Technol.* **123** 23–9

[25] Venkata M S, Srikanth S, Chiranjeevi P, Arora S and Chandra R 2014 Algal biocathode for in situ terminal electron acceptor (TEA) production: Synergetic association of bacteria-microalgae metabolism for the functioning of biofuel cell *Bioresour. Technol.* **166** 566–74

[26] Liu Z D, Jing L, Zhang S P and Su Z G 2009 Study of operational performance and electrical response on mediator-less microbial fuel cells fed with carbon- and protein-rich substrates *Biochem. Eng. J.* **45** 185–91

[27] Pant D, Bogaert G V, Diels L and Vanbroekhoven K 2010 A review of the substrates used in microbial fuel cells (MFCs) for sustainable energy production *Bioresour. Technol.* **101** 1533–43

[28] Chae K J, Choi M J, Lee J W, Kim K Y and Kim I S 2009 Effect of different substrates on the performance, bacterial diversity, and bacterial viability in microbial fuel cells *Bioresour. Technol.* **100** 3518–25

[29] Feng C, Tsai C C, Ma C Y, Yu C P and Hou C H 2017 Integrating cost-effective microbial fuel cells and energy-efficient capacitive deionization for advanced domestic wastewater treatment *Chem. Eng. J.* **330** 1–10

[30] Yu J, Seon J, Park Y, Cho S and Lee T 2012 Electricity generation and microbial community in a submerged-exchangeable microbial fuel cell system for low-strength domestic wastewater treatment *Bioresour. Technol.* **117** 172–9

[31] Thung W E, Ong S A, Ho L N, Wong Y S, Ridwan F, Oon Y L, Oon Y S and Lehl H K 2015 A highly efficient single chambered up-flow membrane-less microbial fuel cell for treatment of azo dye Acid Orange 7-containing wastewater *Bioresour. Technol.* **197** 284–8

[32] Zou H and Wang Y 2017 Azo dyes wastewater treatment and simultaneous electricity generation in a novel process of electrolysis cell combined with microbial fuel cell *Bioresour. Technol.* **235** 167–75

[33] Kim C, Lee C R, Song Y E, Heo J, Choi S M, Lim D H, Cho J, Park C, Min J and Kim J R 2017 Hexavalent chromium as a cathodic electron acceptor in a bipolar membrane microbial fuel cell with the simultaneous treatment of electroplating wastewater *Chem. Eng. J.* **328** 703–7

[34] Min B, Kim J, Oh S, Regan J M and Logan B E 2005 Electricity generation from swine wastewater using microbial fuel cells *Water Res.* **39** 4961–8

[35] Zhuang L, Yuan Y, Wang Y and Zhou S 2012 Long-term evaluation of a 10-liter serpentine-type microbial fuel cell stack treating brewery wastewater *Bioresour. Technol.* **123** 406–12

[36] Wen Q, Wu Y, Cao D, Zhao L and Sun Q 2009 Electricity generation and modeling of microbial fuel cell from continuous beer brewery wastewater *Bioresour. Technol.* **100** 4171–5

[37] Yu J, Park Y, Kim B and Lee T 2015 Power densities and microbial communities of brewery wastewater-fed microbial fuel cells according to the initial substrates *Bioprocess. Biosyst. Eng.* **38** 85–92

[38] Oon Y L, Ong S A, Ho L N, Wong Y S, Dahalan F A, Oon Y S, Lehl H K, Thung W E and Nordin N 2016 Role of macrophyte and effect of supplementary aeration in up-flow constructed wetland-microbial fuel cell for simultaneous wastewater treatment and energy recovery *Bioresour. Technol.* **224** 265–75

[39] Wang J, Song X, Wang Y, Abayneh B, Li Y, Yan D and Bai J 2016 Nitrate removal and bioenergy production in constructed wetland coupled with microbial fuel cell: Establishment of electrochemically active bacteria community on anode *Bioresour. Technol.* **221** 358–65

[40] Li M, Zhou S and Xu M 2017 Graphene oxide supported magnesium oxide as an efficient cathode catalyst for power generation and wastewater treatment in single chamber microbial fuel cells *Chem. Eng. J.* **328** 106–16

[41] Song H, Zhang S, Long X, Yang X, Li H and Xiang W 2017 Optimization of bioelectricity generation in constructed wetland-coupled microbial fuel cellsystems *Water* **9** 185

[42] Wang J, Song X, Wang Y, Abayneh B, Yi D, Yan D and Bai J 2016 Microbial community structure of different electrode materials in constructed wetland incorporating microbial fuel cell *Bioresour. Technol.* **221** 697–702

[43] Dube A, David R and Ngulube P 2013 A cost-benefit analysis of document management strategies used at a financial institution in Zimbabwe: a case study *S. Afr. J. Inf. Manage.* **15** 1–10

[44] Santamaría M and Azqueta D 2015 promoting biofuels use in Spain: a cost-benefit analysis *Renew. Sust. Energ. Rev.* **50** 1415–24

[45] Djukic M, Ivanovic I J, Ivanovic O M, Lazic M and Bodroza D 2016 Cost-benefit analysis of an infrastructure project and a cost-reflective tariff: a case study for investment in wastewater treatment plant in Serbia *Renew. Sust. Energ. Rev.* **59** 1419–25

[46] Pan T C, Kao J J and Wong C P 2012 Effective solar radiation-based benefit and cost analyses for solar water heater development in Taiwan *Renew. Sust. Energ. Rev.* **16** 1874–82

[47] Sun X, Xu H, Zhu Q, Lu L and Zhao H 2015 Synthesis of Nafion®-stabilized Pt nanoparticles to improve the durability of proton exchange membrane fuel cell *J. Energ. Chem.* **24** 359–65

[48] Tan X, Shi L, Ma Z, Zhang X and Lu G 2015 Institutional analysis of sewage treatment charge based on operating cost of sewage treatment plant–empirical research of 227 samples in China *China Environ. Sci.* **35** 3833–40
[49] Coquillard P, Muzy A and Diener F 2012 Optimal phenotypic plasticity in a stochastic environment minimises the cost/benefit ratio *Ecol. Model.* **242** 28–36
[50] Coleman A and Grimes A 2010 Betterment taxes, capital gains and benefit cost ratios *Econ. Lett.* **109** 54–6
[51] Mohamed H O, Obaid M, Sayed E T, Liu Y, Lee J, Park M, Nam B and Kim H Y 2017 Electricity generation from real industrial wastewater using a single-chamber air cathode microbial fuel cell with an activated carbon anode *Bioprocess. Biosyst. Eng.* **40** 1151–61
[52] Huang L, Li X, Ren Y and Wang X 2016 In-situ modified carbon cloth with polyaniline/graphene as anode to enhance performance of microbial fuel cell *Int. J. Hydrogen Energ.* **41** 11369–79
[53] You S J, Zhang J N, Yuan Y X, Ren N Q and Wang X H 2010 Development of microbial fuel cell with anoxic/oxic design for treatment of saline seafood wastewater and biological electricity generation *J. Chem. Technol. Biotechnol.* **85** 1077–83
[54] Min B K, Cheng S A and Logan B E 2005 Electricity generation using membrane and salt bridge microbial fuel cells *Water Res.* **39** 1675–86
[55] Li W, Sun J, Hu Y, Zhang Y, Deng F and Chen J 2014 Simultaneous pH self-neutralization and bioelectricity generation in a dual bioelectrode microbial fuel cell under periodic reversion of polarity *J. Power Sources* **268** 287–93
[56] Liu Z, Li X, Jia B, Zheng Y, Fang L, Yang Q, Wang D and Zeng G 2009 Production of electricity from surplus sludge using a single chamber floating-cathode microbial fuel cell *Water Sci. Technol.* **60** 2399–404
[57] Sun J, Li W, Li Y, Hu Y and Zhang Y 2013 Redox mediator enhanced simultaneous decolorization of azo dye and bioelectricity generation in air-cathode microbial fuel cell *Bioresour. Technol.* **142** 407–14
[58] Wu X Y, Song T S, Zhu X J, Wei P and Zhou C C 2013 Construction and operation of microbial fuel cell with Chlorella vulgaris biocathode for electricity generation *Appl. Biochem. Biotechnol.* **171** 2082–92
[59] Wang L, Li X and Wang L 2017 Study on wastewater treatment by wetland type microbial fuel cell and simultaneous electricity generation *Mod. Chem. Ind.* **37** 154–7
[60] Zhou F, Xian C, Li X, Hui W and Li X 2017 Electrode and azo dye decolorization performance in microbial-fuel-cell-coupled constructed wetlands with different electrode size during long-term wastewater treatment *Bioresour. Technol.* **238** 450–60
[61] Oon Y L, Ong S A, Ho L N, Wong Y S, Dahalan F A, Oon Y S, Lehl H K and Thung W E 2016 Synergistic effect of up-flow constructed wetland and microbial fuel cell for simultaneous wastewater treatment and energy recovery *Bioresour. Technol.* **203** 190–7
[62] Xu F, Cao F Q, Kong Q, Zhou L L, Yuan Q, Zhu Y J, Wang Q, Du Y D and Wang Z D 2018 Electricity production and evolution of microbial community in the constructed wetland-microbial fuel cell *Chem. Eng. J.* **339** 479–86
[63] Xing D, Cheng S, Regan J M and Logan B E 2010 Change in microbial communities in acetate- and glucose-fed microbial fuel cells in the presence of light *Biosens. Bioelectron.* **25** 105–11

[64] Liu H and Logan B E 2004 Electricity generation using an air-cathode single chamber microbial fuel cell in the presence and absence of a proton exchange membrane *Environ. Sci. Technol.* **38** 4040–6

[65] Zheng X T, Ananthanarayanan A, Luo K Q and Chen P 2015 Glowing graphene quantum dots and carbon dots: properties, syntheses, and biological applications *Small* **11** 1620–36

[66] Stavrinidou E, Gabrielsson R, Gomez E, Crispin X, Nilsson O, Simon D T and Berggren M 2015 Electronic plants *Sci. Adv.* **1** e1501136

[67] Arias J, Artucb E, Lederman D and Rojasc D 2018 Trade, informal employment and labor adjustment costs *J. Dev. Econ* **133** 396–414

[68] Peng H T and Liu Y 2018 How government subsidies promote the growth of entrepreneurial companies in clean energy industry: An empirical study in China *J. Clean. Prod.* **188** 508–20

[69] Santoro C, Arbizzani C, Erable B and Ieropoulos I 2017 Microbial fuel cells: From fundamentals to applications *J. Power Sources* **356** 225–44

[70] Cao X X, Huang X, Liang P, Xiao K, Zhou Y J, Zhang X Y and Logan B E 2009 A new method for water desalination using microbial desalination cells *Environ. Sci. Technol.* **43** 7148–52

[71] Xu F, Yuan Q, Zhu L, Li Y, Du Y, Wang Q and Kong Q 2018 Economic benefit analysis of typical microbial fuel cells based on a cost-benefit analysis model *Desalin. Water Treatm.* **135** 59–93

[72] Hossain A 2021 Development of mathematical model to study the impact of state of charge dependent exchange current density on the generated voltage hysteresis of silicon anode-based lithium half cells *J. of Mechanical Eng. Research* **12** 37–48

[73] Hossain A, Masud N, Yasin M and Ali M 2020 Analysis of the performance of Microbial Fuel Cell as a potential energy storage device *IEICES* **6** 149–55

[74] Masud N, Hossain A, Moresalein M and Ali M 2020 Performance evaluation of microbial fuel cell with food waste solution as a potential medium *IEICES* **7** 96–102

[75] Hossain A, Masud N, Roy S and Ali M 2022 Investigation of voltage storage capacity for the variation of electrode materials in microbial fuel cells with experimentation and mathematical modelling *Int. J. Water Resourc. Environ. Eng.* **14** 97–109

[76] Hossain A and Kim S U 2020 Development of a physics-based mathematical model to of microparticle silicon-based lithium half cells *Int. Mech. Eng. Congress and Exposition*

IOP Publishing

Microbial Fuel Cells
Emerging trends in electrochemical applications
Mohamad Nasir Mohamad Ibrahim, Asim Ali Yaqoob and Akil Ahmad

Chapter 14

Application of microbial fuel cells and other bioelectrochemical systems: a comparative study

Wilgince Apollon, Alejandro Isabel Luna-Maldonado, Juan Antonio Vidales-Contreras, Humberto Rodríguez-Fuentes, Juan Florencio Gómez-Leyva and Sathish-Kumar Kamaraj

Microbial fuel cells (MFCs) are sustainable energy sources that have been designed with the main objective of generating electricity from organic materials using electrochemically active micro-organisms (EAMs). These bioelectrochemical systems (BESs) have also been used in various fields of application such as wastewater treatment, human urine treatment, the removal of heavy metals, biosensors, microalgae production, etc. Other BESs such as sediment MFCs (SMFCs) and plant-mediated MFCs (PMFCs) have also been proposed, investigated, and developed to improve not only the performance of these systems, but also their real-time applications. For example, PMFCs have been constructed to generate continuous power densities, taking advantage of the exudates released by the roots of the plants while they carry out the photosynthesis process. This chapter discusses the fundamental classification of BESs based on the energy liberated (MFCs—$\Delta G < 0$) and energy required (microbial electrolysis and microbial electrosynthesis—$\Delta G > 0$) and their real-time applications, e.g. biosensing (BOD sensors, IoT sensors, etc), bioremediation, and scalable production of biohydrogen and other valuable products (organic acids, alcohol, bioplastics, etc), including scalability. In addition, the techno-economic and sustainability assessments of BESs are also elaborated upon.

14.1 Fundamental classification of bioelectrochemical systems

14.1.1 Bioelectrochemical energy generation systems ($\Delta G < 0$)

Due to the problems faced by world energy producers, such as the depletion of fossil resources, and the negative impacts that their use has on the environment and human health [1], scientists have been impelled to implement sustainable energy production systems. Bioelectrochemical systems have since emerged as a surprising

doi:10.1088/978-0-7503-4791-4ch14

technology with huge potential, wherein research groups are interested in the microorganisms that participate in these processes, as well as the design, analysis, and engineering of these systems. BESs have shown their advantages in decreasing the problem of contamination in soil, water, etc, and also respond to the high global demand for energy. According to a report in 2007, the amount of energy consumed worldwide was estimated at 13 trillion W [2]. A vast amount of economic resources are required to produce this amount of energy. Furthermore, not all countries are capable of making large investments in energy—there are naturally some economic limitations [3] to promoting large energy production projects. Some of the alternative power generation systems developed use different types of substrates for the production of bioelectricity. The main function of such a bioelectrochemical system (BES) is to convert chemical energy into electrical energy. The sections below discuss the different types of BESs designed for low-cost and sustainable electrical power production.

14.1.1.1 Microbial fuel cells

Microbial fuel cells (MFCs) are derived from BESs that have been created for the production of bioelectricity using organic substrates through bacterial activity, such as biocatalysis in the anode chamber [4]. The idea of using bacteria in MFCs is not new and, in fact, dates back as far as 1911 when Potter reported that bacteria had the potential to produce a current [5]. Potter achieved voltages of between 0.3 and 0.5 V when using a platinum electrode placed in a liquid containing yeast and similarly *E. coli* bacteria in a glucose medium. Since then, these findings have captured the attention of many researchers and have given them the main objective of increasing efficiencies, as well as enabling large-scale applications of MFC systems. Generally, an MFC is made up of two compartments or chambers, the anode and cathode, which are separated by a membrane. The membrane is a key component for the construction of a dual chamber MFC. This means that without the presence of a membrane, the system will not work. Anion exchange membranes (AEMs), cation exchange membranes (CEMs), and proton exchange membranes (PEMs) (figure 14.1) are the most frequently used membrane types in MFCs. Other aspects to consider when planning to build an MFC are the types of anode and cathode materials [6]. Equation (14.1) illustrates the bioelectricity production process in an MFC system and equations (14.2) and (14.3) represent the reactions that occur in the anode and cathode chambers, respectively, in a dual chambered MFC:

$$C_{12}H_{22}O_{11} + 13H_2O \rightarrow 12CO_2 + 48H^+ + 48e^-, \tag{14.1}$$

$$CH_3COO^- + 2H_2O \rightarrow 2CO_2 + 7H^+ + 8e^-, \tag{14.2}$$

$$O_2 + 4e^- + 4H^+ \rightarrow 2H_2O. \tag{14.3}$$

Under anaerobic conditions, the bacteria present in the anode chamber oxidize organic matter (OM) producing protons, then release electrons into the cathode chamber via an external circuit [7]. Oxygen (O_2) is frequently used as an electron acceptor, mainly at the cathode. This is due to several aspects such as its vast

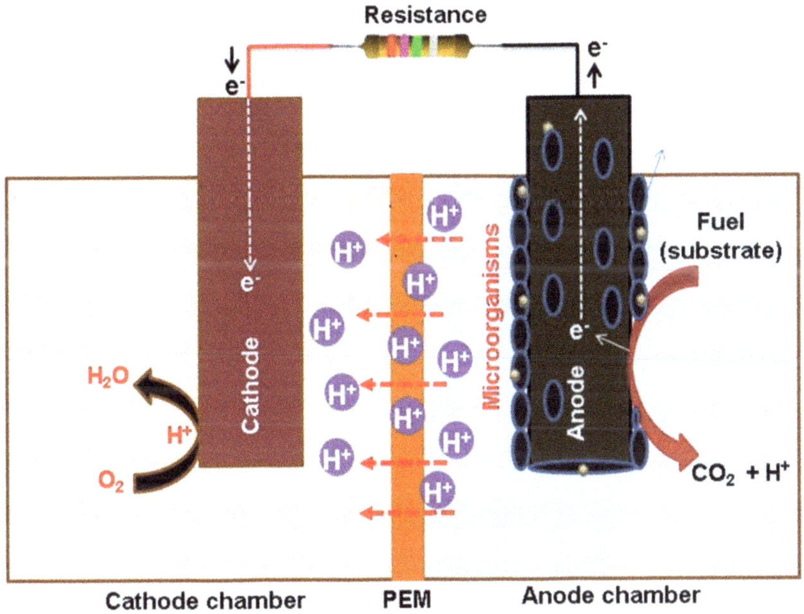

Figure 14.1. Typical dual chambered MFC using a PEM.

reduction potential, high availability, and zero cost [8]. However, the use of O_2 in the oxygen reduction reaction (ORR) is a limiting factor for the real-time and large-scale application of MFCs because of its slow reaction speed and its large overpotential [9, 10]. Conversely, platinum (Pt) has great ORR catalytic potential, however, on a laboratory scale a cathode based on platinum (Pt) represents more than 50% of the total cost of configuration of MFCs [10]. The use of Pt is not economically feasible [11] for large-scale application of MFCs.

14.1.1.2 Sediment MFCs

Like MFCs, sediment MFCs (SMFCs) are technologies that have been designed over the last decade to improve the efficiencies of BESs. The use of SMFCs has been shown to be very important due to their environmental profitability [12]. This category of BES-like MFCs produces bioelectricity in aquatic sites. SMFCs are devices that generate bioelectricity, however, their large-scale application in real time has occasioned great challenges because of the low output of power produced.

SMFCs operate by placing an anode chamber (anode electrode) embedded in an anaerobic sediment (figure 14.2) that is attached to a biocathode chamber (cathode electrode) floating in water (suspended in the water) [13, 14]. SMFCs have been used in river sediments, rice paddies, marine sediments, etc, because these aquatic sources contain a higher content of OM. Other types of SMFCs have been designed as benthic-based MFCs (BMFCs) and marine MFCs (MMFCs) [15]. One of the advantages that SMFCs have over conventional MFCs is that they operate without the need for protons to be transported by a dissolved O_2 ascent along the depth of membranes and water [16]. Compared to other types of BESs, SMFCs also use

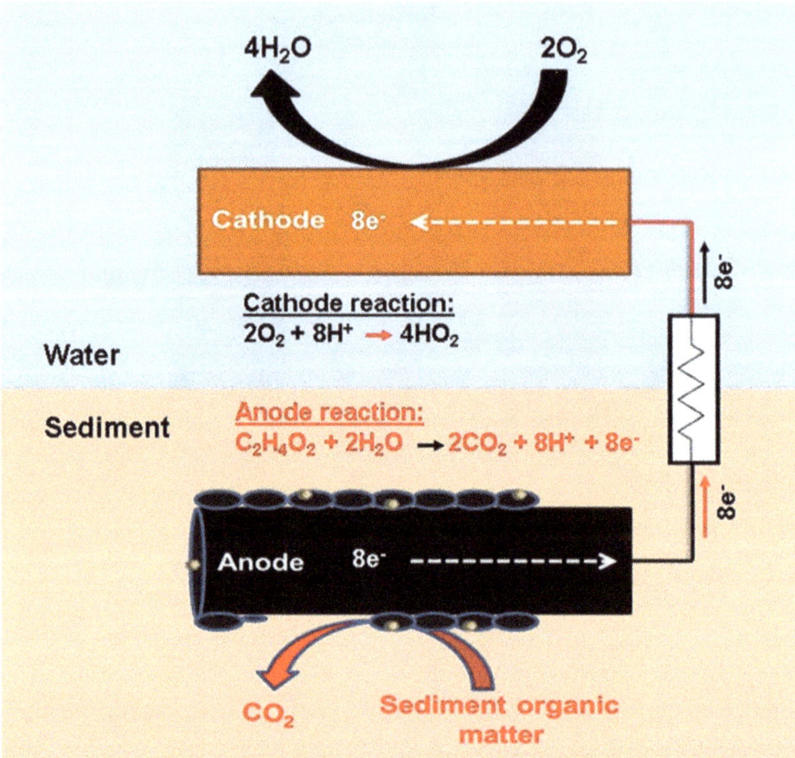

Figure 14.2. Design of a sediment MFC.

electrochemically active micro-organisms in sediments to produce bioelectricity. The main idea behind this category of BESs was to improve the power density and also to make the use of this technology, on a large scale, a reality. Wang *et al* [15] reported a maximum volumetric power density of 1 W m^{-3} using an SMFC with a floating biocathode, both for the generation of bioelectricity, as well as for organic removal. During this experiment, they observed that the biocathode used had the best performance when compared to another type of cathode (carbon paper cathode) catalyzed by Pt. This has therefore shown that SMFCs have significant ability to produce bioelectricity under laboratory conditions.

14.1.1.3 Microbial desalination cells

Microbial desalination cells (MDCs) are a novel sustainable energy technique which harness OM from seawater as the only source of energy for the desalination process. In MDCs, exoelectrogenic bacteria are capable of generating the electrical potential as the OM is degraded, which is subsequently used to desalinate the water [17]. MDCs fulfill three functions at the same time when operating in wastewater: (a) wastewater treatment, (b) bioelectricity production, and (c) desalination. A very low ion concentration of 8 g l^{-1} has been reported for domestic wastewater, when compared to the ionic concentration found in seawater, with a value of 35 g l^{-1} [18].

According to [17], these values indicate that there is supplementary energy which is accessible through the salinity gradient of each of the sources used.

Cao *et al* [19] proposed the original MDCs for the water desalination process, which were composed of various compartments. The first one was the anode chamber with a desalination medium and the second was the cathode chamber. The exoelectrogenic microbes present on the anode surface play an important function in the oxidation of the organic compounds of OM, to transfer electrons to the anode, while in the cathodic electrode compartment, O_2 plays the role of electron acceptor, for the establishment of the water molecule (H_2O). Two types of membranes were used during the desalination process: (i) an AEM was situated next to the anode and (ii) a CEM was next to the cathode (figure 14.3).

MDCs, like MFCs and SMFCs, were developed with the main objective of improving power generation. Cao *et al* [19] achieved the highest power density of up to 65 W m^{-3} in MDCs operated with the help of a ferricyanide catholyte. Later, Jacobson *et al* [20] found a volumetric power density of 31 W m^{-3} in MDC with a O_2 reduction reaction. Considering the experiment of Cao *et al* this yield is less than 52.30%. However, both cases show potential for future research taking into account the improvement of power density.

14.1.1.4 Plant-mediated MFCs

The use of living plants in bioelectrochemical systems such as plant-mediated MFCs (PMFCs) was adopted in 2008, when David Strik and co-workers first used the *Glyceria maxima* plant to produce green electricity [21]. PMFC systems have the

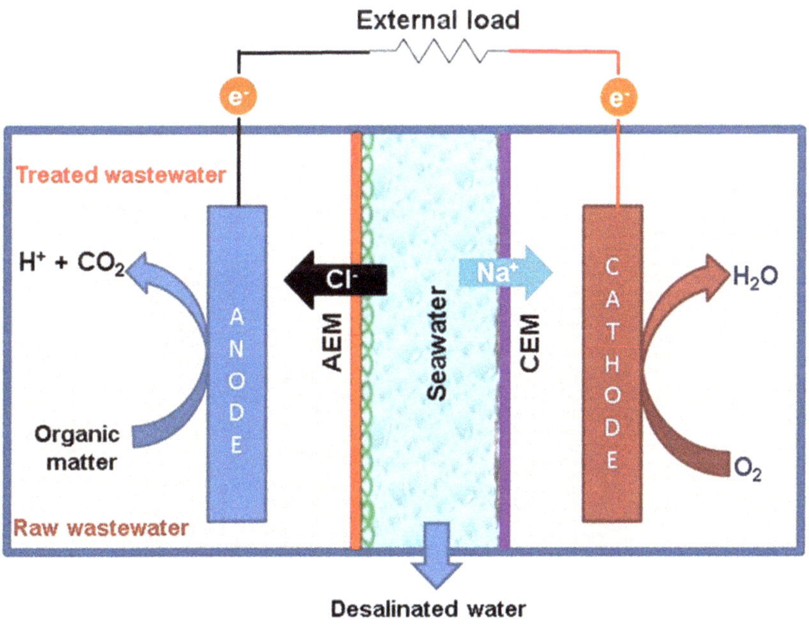

Figure 14.3. Application of MDC technology in the water desalination process.

main function of generating bioelectricity continuously. PMFCs embedded with the *G. maxima* plant indicated a maximum power density of 67 mW m^{-2}. The implementation of PMFC technology raises the possibility of producing bioelectricity while the plant grows [22]. The simultaneous production of electricity and biomass allows the production of bioelectricity that is friendly to the environment [21]. In a PMFC (figure 14.4), the microbes catalyze the anaerobic oxidation of the exudates liberated by the roots of plants, then produce bioelectricity [23]. The organic compounds are then divided into three components, electrons, protons, and CO_2, by the activities of the micro-organism in the plant rhizosphere. Micro-organisms are metabolized more efficiently when there is a higher concentration of root exudates, this implies a higher metabolic activity, and thus higher production of bioenergy. The following equation indicates the chemical reaction that occurs in a typical PMFC:

$$6CO_2 + 6H_2O \rightarrow C_6H_{12}O_6 + 6O_2 + \sim 38\ \text{ATP}. \tag{14.4}$$

Following the development of this technology, various types of PMFCs and materials have been investigated for the improvement of the generation of bioelectricity under different environmental and saturation conditions. These include tubular PMFCs embedded with *G. maxima* using graphite felt (as the anode and cathode), where the anode was placed on the cathode [24]; tubular PMFCs embedded with *Spartina anglica* [25], and flat plant MFCs embedded with *S. anglica*, with activated carbon as the anode and graphite felt as the cathode [26].

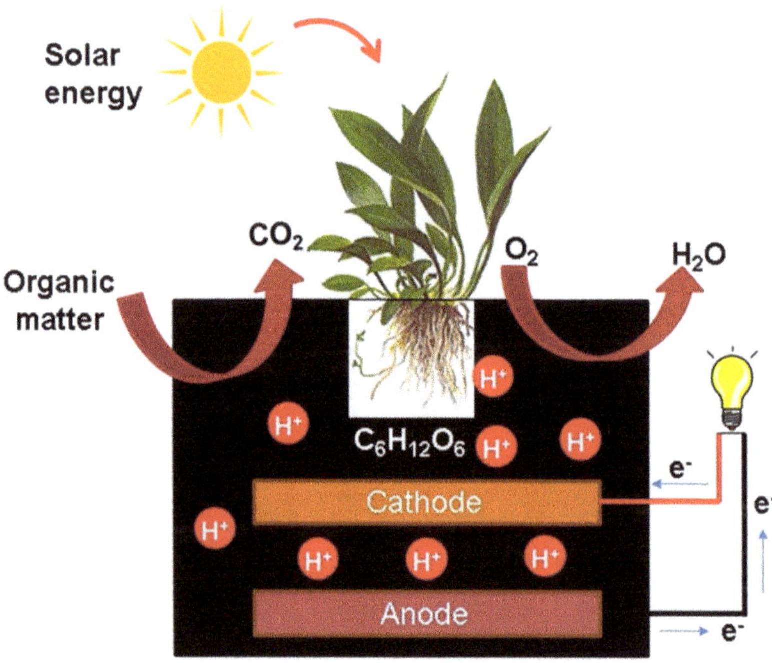

Figure 14.4. Plant microbial fuel cell.

Recently, novel PMFCs embedded with four species of semi-desert plants (*Opuntia* spp.) have been used to generate bioelectricity [27].

So far, the highest power density obtained of 679 mW m^{-2} plant growth area (PGA), was reached in a PMFC, operated in a marsh at high salinity [28]. This performance was obtained in a 10 min polarization experiment. However, a power density of 240 mW m^{-2} was obtained when the PMFC reactors operated over a longer period (i.e. for two weeks). This means that the production of bioelectricity dropped from 64.65% when applying an external load to the PMFCs. The vitality of the plant is therefore not impaired and the technology can even be combined with biomass production [29] and, as most of the system is placed below the surface, the PMFC will not alter the aesthetics of the landscape. In addition, the application of PMFCs has been shown to reduce emissions of methane from anaerobic soils, a greenhouse gas 24 times more potent than CO_2 [30]. In summary, the generation of bioenergy from these systems is a sustainable approach in which energy is produced without harming the environment. PMFC systems can be installed together with conventional agricultural crops, and do not compete with them for space.

14.1.2 Bioelectrochemical synthesis systems ($\Delta G > 0$)

Bioelectrochemical synthesis systems (BESs), like the other BES systems, are a sustainable technique, in which electro-autotrophic microbes use electrical energy as the sole source of electrons from the cathode reducing CO_2 to extracellular products through autotrophic conversion [31]. This technology has received a lot of attention since its instigation. One of the techniques derived from BESs is microbial electrosynthesis (MES), which has been used in various ways. In the section below, its different applications are discussed.

14.1.2.1 Microbial electrosynthesis of fuel and other valuable products

MES is a promising approach [32] that allows the recycling of bioelectronic CO_2 through a process of reduction in different chemical products, which include bioplastics [33] and organic acids [34]. During the process, CO_2 is reduced in the cathode chamber, which is derived from electroactive biofilm [31]. The use of MES is based exclusively on the innate ability of bacteria to transport electrons into the cathode chamber. In an MES, like all bioelectrochemical systems, two types of reaction occur: (a) oxidation (anode) and (b) reduction (cathode). The anode electrode and the cathode electrode are divided by a CEM. Unlike other biochemical processes that require high economic investment, MES systems have been designed not only for the generation of bioelectricity, but also to catalyze both the generation of alcohols, as well as the less expensive volatile fatty acids [35]. What is surprising about this innovation is the conversion of CO_2 to organic acids (i.e. the reduction of the greenhouse effect gases) [36], which are catalyzed by the activity of the microbes present in the electrode [37]. These micro-organisms absorb electrons from the electrode directly or indirectly using hydrogen (H_2) produced electrochemically as an intermediate [38, 39].

Previously, Hou *et al* [40] reported a higher production of H_2 and acetate of 0.0210 ± 0.0005 m^3 m^{-3} day^{-1} at 24 h and 29.4 ± 3.1 mg $^{1-1}$ day^{-1} at 12 h, respectively, using various types of electrochemically active bacteria (EABs) in MES systems, which were *Ochrobactrum* sp., *Pseudomonas* sp., *Pseudomonas delhiensis*, and *Ochrobactrum anthropi*. The H_2 and acetate production yields mentioned above were achieved in the presence of the EABs *Ochrobactrum anthropi* and *Pseudomonas delhiensis*, respectively. H_2 production was four times higher compared to that found elsewhere [41] using an MES system inoculated with EAB *Geobacter sulfurreducens* ($0.026 \pm 0.003 \times 10^{-4}$ m^3 m^{-3} day^{-1}). Acetate production was also higher than that found in another MES inoculated with *Sinomicrobium oceani* (8.64 mg l^{-1} day^{-1}) [42]. The above indicates the viability of this technology to help reduce environmental impacts, as well as prevent climate change by reducing greenhouse gases which currently are of concern in the entire world.

14.2 BESs in the role of bioremediation

Environmental pollution is one of the most serious concerns that the world faces today. This is due to the incessant activity of humankind and the subsequent negative effects on the environment. Many alternatives have therefore been adopted to mitigate this problem. One of these alternatives has been the use of BESs for bioremediation of contaminated sites. MFCs and microbial electrolysis cells (MECs) are types of BESs which were used in bioremediation of contaminants, such as heavy metals in wastewater and soil. Table 14.1 shows the different types of BESs used in the removal of heavy metals in wastewater.

In 2012 Tao and co-workers developed a two-chamber MFC using graphite foil as the anode and a cathode of silver (I) recovery (0.109 ± 0017 mM removal), reaching, at its highest, a power density of 109 mW m^{-2} [43]. While in an MFC with the same

Table 14.1. The application of MFCs in heavy metal removal from wastewater.

MFC configuration	Anode material	Cathode material	Metal removal	Highest removal efficiency	Highest power density	References
DC MFC	Graphite plate	Graphite plate	Ag (I)	0.109 ± 0017 mM	109 mW m^{-2}	[43]
DC MFC	Carbon brush	Carbon cloth	Ag (I)	$99.91\% \pm 0.00\%$	4.25 W m^{-2}	[49]
DC MFC	Graphite felt	Carbon paper	Hg (II)	99.54%	433.1 mW m^{-2}	[50]
SC MFC	Carbon brush	Carbon cloth/Pt	Cu (II)	98.3%	10.2 W m^{-3}	[51]
DC MFC	Carbon felt	Carbon cloth	Cr (VI)	100%	1221.91 mW m^{-2}	[45]
DC MFC	Carbon fiber felt	Carbon fiber felt	Cr (VI)	$75.4\% \pm 1.9\%$	970.2 ± 20.6 mW m^{-2}	[44]
DC MEC	Carbon felt	Nickel foam (NF)	Pb (II)	100%	702.86 mW m^{-2}	[46]

DC = dual chambered and SC = single chambered.

characteristics as previously but using carbon fiber felt both in the anode as well as in the cathode for the simultaneous reduction of chromium (VI) and vanadium (V), a 75.4% ± 1.9% a removal efficiency with a power density almost nine times higher (970.2 ± 20.6 mW m^{-2}) was obtained, when compared to the aforementioned performance [44]. Later, Li *et al* [45] reported the 100% removal of Cr (VI) with a higher maximum power density of 1221.91 mW m^{-2} when they used an MFC with the same characteristics as those used by Zhang *et al.* This indicates that the efficiency rate of removal of these heavy metals depends greatly on aspects such as the BES category and the materials used for their configuration, and also the type of membrane. For example, Li *et al* [46] reported a maximum Pb (II) removal efficiency of 100% by using a MEC, which was configured using carbon felt (CF) as the anode and nickel foam (NF) as the cathode (table 14.1). Further to previous studies, the feasibility of BES technologies such as MFCs and PMFCs in correcting soils contaminated with heavy metals has also been investigated, but, at present, studies on this highly important topic are very scarce [7]. A single chamber PMFC was designed for both the remediation of some soils contaminated with Cr (VI) as well as the generation of bioelectricity [47]. The results showed a maximum Cr (VI) removal efficiency of 99% where the PMFC indicated a voltage of 469.21 mV. In a previous study, Habibul *et al* [48] achieved the same Cr (VI) removal efficiency in a PMFC using carbon felt as the anode and cathode, which indicated a maximum current density of 55 mA m^{-2} (table 14.2).

The results above indicate that BES technologies are feasible both to reduce the problem of pollution worldwide, as well as respond to the high world-wide demand for energy consumption. MFC technologies make it possible to reduce heavy metals in both water and contaminated ground at a very low cost, compared to biological technologics. In contrast, the use of traditional techniques requires high energy consumption and a large number of chemical products. A great advantage of BESs in the bioremediation process is their profitability, and also their benefits to the environment.

Table 14.2. The application of MFCs in the removal of heavy metals from the soil.

MFC configuration	Anode material	Cathode material	Metal removal	Highest removal efficiency	Highest power production	References
SC PMFC	Graphite granules	Carbon felt	Cr (VI)	99%	55 mA m^{-2}	[48]
SC PMFC	Carbon felt	Carbon felt	Cr (VI)	99%	469.21 mV	[47]
Three-chambered MFC	Graphite mat	Graphite mesh/Pt	Zn (II)	25%	0.4 A m^{-2}	[52]
SC MFC	Granular activated carbon	Granular activated carbon	Cu (II)	N/A	65.77 mW m^{-2}	[53]
DC MFC	Graphite granules	Carbon felt	Cd (II)	31.0%	7.5 mW cm^{-2}	[54]

SC = single chambered and DC = dual chambered.

14.3 Biosensing applications

14.3.1 BOD sensors

Biochemical oxygen demand (BOD) is a measure of water quality. It concerns the amount of dissolved O_2 that is used for the oxidation of any biodegradable component as organic compounds found in water. The assessment of BOD is normally based on an old technique that determines the oxygen demand (OD) in analyzed water samples by using a highly standardized laboratory procedure [55]. Such procedures require the use of a lot of reagents and also energy consumption, making them very expensive. In 1977 a sensor supported by a biofuel cell (BFC) was designed for the first time for the evaluation of BOD [56]. Subsequently, almost three decades later, Korean researchers have made great contributions regarding the construction of BOD sensors based on free mediators supported by BFCs [57, 58], which were operated for a period of 5 years without any maintenance. Later, Kharkwal *et al* [59] designed a sensor operated with the support of a novel MFC which made it possible to continuously monitor (1.5 years) the BOD in real wastewater. The BOD results showed that the use of an MFC with β-MnO_2 catalyst performed best during the experiment. The use of a microbial biofuel cell (MBFC) as a BOD sensor is a novel approach and a sustainable technology that can reduce the high risk of contamination, as well as decrease the cost of BOD analysis in water; compared to that resulting from traditional methods.

14.3.2 IoT sensor applications

The application of the Internet of Things (IoT) opens the door to great opportunities regarding the use of intelligent systems capable of collecting data in real time. This idea was born when researchers were looking for alternatives, for example designing effective and less expensive systems to monitor water stress, humidity (environmental and soil), and also salinity. However, the use of traditional systems of wireless sensor networks (WSN) based on IoT presents a great challenge due to the difficult process of replacing batteries when they are depleted [60]. Faced with this situation, continuous energy sources such as MFCs, SMFCs, constructed wetland-MFCs (CW-MFCs) and PMFCs must be sought to power these devices. To this end, in 2019 an IoT-based PMFC system was developed by Osorio de la Rosa and co-workers with the main objective of collecting energy for self-powered IoT applications [60]. The results indicate that the PMFC used had the capacity to power the device. Previously, energy harvesting and IoT-based systems have been applied for monitoring honey bees and livestock [61], as well as pest management [62], automated irrigation [63], etc. This has shown the great potential of these systems to power remote devices in real time. Further research is, therefore, worthwhile to increase both the power, as well as the long-term operation, and large-scale application of these technologies.

14.3.3 Others

Tissue-based biosensors, DNA biosensors, enzyme-based biosensors, among others, were developed for real-time monitoring [64]. In 1967 Updike and Hicks designed the first enzyme-based sensor [65]. An enzyme biosensor was designed by Mirshra and co-workers for the quantification of three types of organophosphate pesticides in milk [66]. The biosensor consisted of a screen-printed carbon electrode, which was placed in a flow cell, where through immersion in a photocrosslinkable polymer, the enzymes were immobilized on the electrode. Previously, Akyilmaz *et al* [67] developed a tyrosinase-based biosensor as the main method of evaluating the positive impact of Cu ions on the activity of the enzyme. The study showed that the response of the biosensor depends linearly on the concentration of copper ions which range between 2.5–20.0 μM (response time was 1 min), which have an maximum recognition of 0.95 μM. On the other hand, for the determination of arginine amino acid, Rechnitz [68] constructed the first tissue-based sensor feeding through a bioelectrochemical electrode. These applications could open the door to further research for the large-scale use or application of biosensors, in order to make these technologies a reality and expand globally.

14.4 From the laboratory to field application of bioelectrochemical systems

14.4.1 Scale-up of bioelectrochemical systems

14.4.1.1 Scaled up field application of MFCs

First, MFCs have been developed at a laboratory level for the production of bioelectricity, taking advantage of bacterial activity [5]. However, the efficiencies produced have thus far been low; enriched media had to be used in order to cultivate the bacteria [69, 70]. Based on the above, researchers have made great efforts to enhance the power density of MFC systems. Cid *et al* [71] tested the efficiency of MFCs for the pretreatment of urine (human) and generating electrical energy, they achieved a maximum power density of 23 mW m^{-2} in a period of 120 days (MFC operation time). The experiment was carried out with the main goal of improving the power density of the system. The MFC designed has been shown to have adequate potential to oxidize organic matter and also transport electrons to an external circuit to produce bioelectricity. In 2008 higher power densities of 2.7 W m^{-2} [72] and 6.9 W m^{-2} [73] were reported; both results were achieved in MFCs operated with oxygen. Therefore, the power generation increased significantly depending on the conceived surface area [74]. However, moving these systems from the laboratory to field application has proved to be a great challenge. Several factors such as the cost of the electrode materials, types of materials used, type of membrane, etc, influence the limitations presented by the use of MFCs. The costs of setting up the electrodes must be reduced, and the types of cathode and anode materials is also in need of investigation in order to maximize current densities [70]. In the previous section, the different applications of MFCs were discussed, which give us an idea of the practicability of these technologies to be implemented in different field applications; for instance, to power small devices. However, more research is required for the continuous production of bioelectricity and also a suitable improvement of this technology for its application on a large scale, in real time.

14.4.1.2 Scale-up field application of SMFCs

Unlike MFCs, SMFCs have been designed in the laboratory as a method of producing practical levels of electricity by powering underwater devices. These systems were developed with the objective of testing their ability to drive, for example, communications systems, as well as oceanographic sensors that require little power (0.6 mW) for their operation in sleep mode, and in active mode require a power of 20 W [75, 76]. A performance of power densities in the range of 10^{-2} to 100 mW m^{-2} has been found in evaluating small-scale SMFCs in the laboratory [77]. Subsequently, power densities in the order of 10–50 mW m^{-2} were achieved from small-scale SMFC systems in the laboratory [76]. These values were obtained when using graphite plate anodes. The prototype SMFC systems faced great challenges with respect to the production of low power (less than 100 mW) when applied in the field [78]. For the scale-up of SMFCs by improving their efficiency (e.g. power density) the quality of the anode and cathode electrode materials, as well as the size of the electrodes (i.e. larger electrodes) should be taken into account

14.4.1.3 Scale-up field application of MDCs

Since the development of the MDC, the most problematic aspect is the scaling-up of the reactor. Several experiments have been carried out on a laboratory scale to find a solution to the resulting fundamental problems, as well as the technical difficulties of scaling. Recently, Salehmin *et al* [79], in a review article, addressed the issue of the application of MDCs in the field. In this article, they presented an overview of the main challenges faced by MDCs, the possible mitigation strategies of these systems, and also perspectives that will make this technology a great feat for worldwide application in the future.

MDC systems have been created to improve the power density produced by BESs, because other BESs have not been able to generate high power density on a large scale [80]. Based on the above, Jacobson *et al* made the first attempts to expand MDC on a laboratory scale, to evaluate both the performance as well as the generation of bioelectricity of the system, introducing an upflow microbial desalination cell (UMDC) liter scale [20]. The UMDC demonstrated potential high energy efficiency during the study. What caught the researchers' attention was the low current production, observed over the same operating time of the UMDC system. It is known that high current production favors adequate seawater desalination treatment [20], even so, the use of UMDC is ideal for the operation of a reverse osmosis water system due to its 58% desalination efficiency. In another large-scale stacked MDC system, where they used >10 l, a desalination efficiency of 96% was reported [81]. In 2015 the largest MDC system was reported [82], which was built with an AEM tube placed within a CEM tube, both of which formed a double cylinder. The tubular design has the advantage of providing a vast surface area, which is advantageous for powerful removal of salt from the intermediate compartment [79].

14.4.1.4 Scale-up field application of PMFCs

Since the implementation of the first PMFC in 2008, different configurations have been evaluated to test their implementation on a large scale. The scaling-up of a

PMFC embedded with *Physcomitrella patens* (bryophyte, a moss plant) indicated a maximum power density of 6.7 mW m^{-2} [83]. While in dual chambered MFC systems (DC MFC) embedded with *Chlorella vulgaris* (an algae species) a power density of 19151 mW m^{-3} was found [84]. Therefore, PMFCs face great challenges during scale-up. Thus the study of a large-scale PMFC should examine the issues and other limiting aspects that influence the accomplishment of these systems in the course of set-up. Osorio de la Rosa *et al* [60] built a PMFC based energy harvester (EH) with the main idea to self-power IoT applications, and also improve the performance of this category of BESs. A power density of 3.5 mW cm^{-2} was obtained for a current density of 5 mA cm^{-2} in the study. The results indicate that PMFC-EH was a good choice for powering these small devices. During the same year, Sudirjo and co-workers also developed a sensor-based PMFC system (LoRa Technology) for automatic (online) real-time data monitoring [85]. The prototype consisted of a tubular PMFC operated throughout four seasons. LoRa Technology was designed with the main objective of evaluating the performance of PMFCs. During the operation of the PMFCs, a maximum average daily density was found with a value of 9.6 mW m^{-2}, which was ideal for driving the sensor. Recently, Apollon *et al* [27] developed a prototype (12 PMFCs in series) to improve power density and power devices such as LEDs and digital clock for a week. The results indicated that the PMFCs had adequate capacity to power the devices without the need for any other reinforcement.

14.5 Techno-economic and sustainable assessment of BESs

BESs such as MFCs, MECs, MDCs, etc, have emerged in the energy world as substitutes for fossil fuels. Since their implementation, these technologies have been the targets of various studies for their global commercialization. A study has revealed that the economy and the accelerated growth of the world population are the main causes of substantial energy consumption [86]. To meet this high demand, fossil fuels are used as the main energy source due to their high energy efficiency, i.e. 80% of the total energy production worldwide is based on fossil fuels [87]. However, the excessive use of this type of non-renewable energy causes global warming due to the release of greenhouse gases [88]. Therefore, the use of BESs to produce sustainable energy is a far safer long-term strategy, which avoids the use of conventional technologies that cause damage to the environment. BESs such as microbial electrochemical systems (MES) undoubtedly produce electricity using reactive fuels such as alcohol, H_2, etc, avoiding greenhouse gas emissions [89]. MESs are an economical option because they generate electricity from wastewater (using bacterial activity). However, due to a lack of widespread knowledge, these technologies receive little attention in the bioeconomic sphere. In order to remediate this, Jourdin *et al* [90] published a study on the techno-economic evaluation of MES from CO_2 and/or organics. During the experiment, twenty-eight crucial variables were implemented. In addition, their impact on the techno-economic viability of the systems was evaluated. Jourdin *et al* took into account the production–cost ratio, where they found that the cost of the anode material was estimated at 59% (main capital and operating costs, electricity consumption rose up to 69%).

More recently, Savla *et al* [88] in a review article discussed some financial aspects of MESs and they presented an overview of the different studies carried out previously, for instance, on the economic viability of MFC, MEC, MES, and MDC systems. MESs should be investigated further both for their expansion as well as for the reduction of set-up material costs. The cost reduction will allow researchers to implement these systems efficiently in real-time fields of application. Costs for the configuration (design) of a system have been classified as (i) initial investment, (ii) operating costs, (iii) maintenance, as well as (iv) decommissioning costs and projected financial incomes [91].

14.6 Conclusions

In this chapter we presented microbial fuel cell systems and other bioelectrochemical systems that have been implemented for various applications. Since their implementation, these technologies have face great challenges that include scale-up, commercialization, real-time application, etc. Other very important aspects of these bioelectrochemical systems are the costs of the materials and the low power density produced to date. Studies have revealed that many factors can affect the performance of both microbial fuel cells, as well as other bioelectrochemical systems investigated. To boost the performance of a microbial fuel cell it is necessary to use a larger electrode. Furthermore, the generation of electricity is highly dependent on the differences in anodic and cathodic potential. MFCs from the laboratory to the field of application still face great challenges, and more research is needed to overcome all current challenges. However, the different application trials of bioelectrochemical systems such as plant microbial fuel cells have indicated the potential of these systems to power LEDs, digital clocks, the Internet of Things, and sensors.

Acknowledgments

WA extends his thanks for a PhD Scholarship from the National Council for Science and Technology (CONACYT), as well as the Universidad Autónoma de Nuevo León (UANL) through the Subdirección de Estudios de Posgrado of Faculty of Agronomy (FA-UANL) for his acceptance on the PhD program. WA also thanks, in particular, his Doctoral Thesis Committee. In addition, AILM and WA thank PAICYT (Programa de Apoyo a la Investigación Científica y Tecnológica)—(CT1519–21) from Universidad Autónoma de Nuevo León for support. KSK would like to thank the Proyectos de Desarrollo Tecnológico e Innovación-2021–TecNM-Instituto Tecnológico El Llano Aguascalientes—(IT19F245).

References

[1] Yaqoob A A, Noor N H B M, Umar K, Adnan R, Ibrahim M N M and Rashid M 2021 Graphene oxide-ZnO nanocomposite: an efficient visible light photocatalyst for degradation of rhodamine B *Appl. Nanosci.* **11** 1291–302

[2] Lewis N S 2007 Powering the planet *MRS Bull.* **32** 808–20

[3] Brown J H *et al* 2011 Energetic limits to economic growth *BioScience* **61** 19–26

[4] Logan B E, Hamelers B, Rozendal R, Schröder U, Keller J, Freguia S, Aelterman P, Verstraete W and Rabaey K 2006 Microbial fuel cells: methodology and technology *Env. Sci. Technol.* **40** 5181–92
[5] Potter M C 1911 Electrical effects accompanying the decomposition of organic compounds *Proc. R. Soc.* B **84** 260–76
[6] Apollon W, Luna-Maldonado A I, Kamaraj S-K, Vidales-Contreras J A, Rogríguez-Fuentes H, Gómez-Leyva J F and Aranda-Ruíz J 2021 Progress and recent trends in photosynthetic assisted microbial fuel cells: a review *Biomass Bioenergy* **148** 106028
[7] Fang C and Achal V 2019 The potential of microbial fuel cells for remediation of heavy metals from soil and water-review of application *Microorganisms* **7** 697
[8] Rossi R *et al* 2019 Evaluating a multi-panel air cathode through electrochemical and biotic tests *Water Res.* **148** 51–9
[9] Zhu Q, Liu B H J, Hu S, Liang S, Xiao K, Yang J and Hou H 2021 Recent advances on the development of functional materials in microbial fuel cells: from fundamentals to challenges and outlooks *Energy Env. Mater.* **5** 401–26
[10] Yuan H, Hou Y, Abu Reesh I M, Chen J and He Z 2016 Oxygen reduction reaction catalysts in microbial fuel cells for energy-efficient wastewater treatment: a review *Mater. Horiz.* **3** 382–401
[11] Rozendal R A, Hamelers H V M, Rabaey K, Keller J and Buisman C J N 2008 Towards practical implementation of bioelectrochemical wastewater treatment *Trends Biotechnol.* **26** 450–9
[12] Wen-Wei L and Han-Qing Y 2015 Stimulating sediment bioremediation with benthic microbial fuel cells *Biotechnol. Adv.* **33** 1–12
[13] Oh S E, Yoon J Y, Gurung A and Kim D J 2014 Evaluation of electricity generation from ultrasonic and heat/alkaline pretreatment of different sludge types using microbial fuel cells *Bioresour. Technol.* **165** 21–6
[14] Zabihallahpoor A, Rahimnejad M and Talebnia F 2015 Sediment microbial fuel cells as a new source of renewable and sustainable energy: present status and future prospects *RSC Adv.* **5** 94171–83
[15] Wang A, Cheng H, Ren N, Cui D, Lin N and Wu W 2012 Sediment microbial fuel cell with floating biocathode for organic removal and energy recovery *Front. Env. Sci. Eng.* **6** 569–74
[16] Babu M L and Mohan S V 2012 Influence of graphite flake addition to sediment on electrogenesis in a sediment-type fuel cell *Bioresour. Technol.* **110** 206–13
[17] Kim Y and Logan B E 2013 Microbial desalination cells for energy production and desalination *Desalination* **308** 122–30
[18] Metcalf and Eddy, Inc, Tchobanoglous G, Burton F L and Stensel H D 2003 *Wastewater Engineering, Treatment and Reuse* 4th edn (Boston, MA: McGraw-Hill) pp 1–24
[19] Cao X, Huang X, Liang P, Xiao K, Zhou Y, Zhang X and Logan B E 2009 A new method for water desalination using microbial desalination cells *Env. Sci. Technol.* **43** 7148–52
[20] Jacobson K S, Drew D M and He Z 2011 Efficient salt removal in a continuously operated upflow microbial desalination cell with an air cathode *Bioresour. Technol.* **102** 376–80
[21] Strik D P B T B, Hamelers H V M, Jan F H, Snel C J N and Buisman 2008 Green electricity production with living plants and bacteria in a fuel cell *Int. J. Energy Res.* **32** 870–6
[22] Helder M, Strik D P B T B, Hamelers H V M, Kuhn S, Blok C and Buisman C J N 2010 Concurrent bio-electricity and biomass production in three plant-microbial fuel cells using *Spartina anglica*, *Arundinella anomala* and *Arundo donax* *Bioresour. Technol.* **101** 3541–7

[23] Helder M, Strik D P B T B, Hamelers H V M, Kuijken R C P and Buisman C J N 2012 New plant growth medium for increased power output of the plant-microbial fuel cell *Bioresour. Technol.* **104** 417–23

[24] Timmers R A, Rothballer M, Strik D P B T B, Engel M, Schulz S, Schloter M, Hartmann A, Hamelers B and Buisman C 2012 Microbial community structure elucidates performance of *Glyceria maxima* plant microbial fuel cell *Appl. Microbiol. Biotechnol.* **94** 537–48

[25] Wetser K, Kim D, Buisman C J N and Strik D P B T B 2017 Electricity from wetlands: tubular plant microbial fuels with silicone gas-diffusion biocathodes *Appl. Energy* **185** 642–9

[26] Sudirjo E, Buisman C J N and Strik D P B T B 2019a Activated carbon mixed with marine sediment is suitable as bioanode material for *Spartina anglica* sediment/plant microbial fuel cell: plant growth, electricity generation, and spatial microbial community diversity *Water* **11** 1810

[27] Apollon W, Kamaraj K S-K, Silos-Espino H, Perales-Segovia C, Vlera-Montero L L, Maldonado-Ruelas V A, Vázquez-Gutiérrez M A, Ortiz-Medina R A, Flores-Benítez S and Gómez-Leyva J F 2020 Impact of *Opuntia* species plant bio-battery in a semi-arid environment: demonstration of their applications *Appl. Energy* **279** 115788

[28] Wetser K, Sudirjo E, Buisman C J N and Strik D P B T B 2015 Electricity generation by a plant microbial fuel cell with an integrated oxygen reducing biocathode *Appl. Energy* **137** 151–7

[29] De Schamphelaire L, Cabezas A, Marzorati M, Friedrich M W, Boon N and Verstraete W 2010 Microbial community analysis of anodes from sediment microbial fuel cells powered by rhizodeposits of living rice plants *Appl. Env. Microbiol.* **76** 2002–8

[30] Arends J B A, Speeckaert J, Blondeel E, De Vreize J, Boeckx P, Verstraete W, Rabaey K and Boon N 2014 Greenhouse gas emissions from rice microcosms amended with a plant microbial fuel cell *Appl. Microbiol. Biotechnol.* **98** 3205–17

[31] Jabeen G and Farooq R 2016 Bio-electrochemical synthesis of commodity chemicals by autotrophic acetogens utilizing CO_2 for environmental remediation *J. Biosci.* **41** 367–80

[32] Batlle-Vilanova P, Ganigué R, Ramió-Pujol S, Bañeras L L, Jiménez G, Hidalgo M, Balaguer M D, Colprim J and Puig S 2017 Microbial electrosynthesis of butyrate from carbon dioxide: production and extraction *Bioelectrochemistry* **117** 57–64

[33] Pepè Sciarria T, Batlle-Vilanova P, Colombo B, Scaglia B, Balaguer M D, Colprim J and Adani F 2018 Bio-electrorecycling of carbon dioxide into bioplastics *Green Chem.* **20** 4058–66

[34] Gavilanes J, Reddy C N and Min B 2019 Microbial electrosynthesis of bioalcohols through reduction of high concentrations of volatile fatty acids *Energy Fuels* **33** 4264–71

[35] Rabaey K, Butzer S, Brown S, Keller J and Rozendal R A 2010 High current generation coupled to caustic production using a lamellar bioelectrochemical system *Env. Sci. Technol.* **44** 4315–21

[36] Li H, Opgenorth P H, Wernick D G, Rogers S, Wu T-Y, Higashide W, Malati P, Huo Y-X, Cho K M and Liao J C 2012 Integrated electromicrobial conversion of CO_2 to higher alcohols *Science* **335** 1596

[37] Rabaey K and Rozendal R A 2010 Microbial electrosynthesis—revisiting the electrical route for microbial production *Nat. Rev. Microbiol.* **8** 706–16

[38] Dessì P, Rovira-Alsina L, Sánchez C, Dinesh G K, Tong W, Chatterjee P, Tedesco M, Farràs P, Hamelers H M V and Puig S 2021 Microbial electrosynthesis: towards sustainable biorefineries for production of green chemicals from CO_2 emissions *Biotechnol. Adv.* **46** 107675

[39] Nevin K P, Woodard T L, Franks A E, Summers Z M and Lovley D R 2010 Microbial electrosynthesis: feeding microbes electricity to convert carbon dioxide and water to multi-carbon extracellular organic compounds *MBio* **1** e00103–10
[40] Hou X, Huang L, Zhou P, Tian F, Tao Y and Puma G L 2019 Electrosynthesis of acetate from inorganic carbon (HCO_3^-) with simultaneous hydrogen production and Cd(II) removal in multifunctional microbial electrosynthesis systems (MES) *J. Hazard. Mater.* **371** 463–73
[41] Yates M D, Siegert M and Logan B E 2014 Hydrogen evolution catalyzed by viable and non-viable cells on biocathodes *Int. J. Hydrogen Energy* **39** 16841–51
[42] Hu J, Wang L, Zhang S, Xi X, Le Y, Fu X, Tsang Y and Gao M 2015 Interactions between autotrophic and heterotrophic strains improve CO_2 fixing efficiency of non-photosynthetic microbial communities *Appl. Biochem. Biotechnol.* **176** 1459–71
[43] Tao H C, Gao Z Y, Ding H, Xu N and Wu W M 2012 Recovery of silver from silver(I)-containing solutions in bioelectrochemical reactors *Bioresour. Technol.* **111** 92–7
[44] Zhang B, Feng C, Ni J, Zhang J and Huang W 2012 Simultaneous reduction of vanadium (V) and chromium (VI) with enhanced energy recovery based on microbial fuel cell technology *J. Power Sources* **204** 34–9
[45] Li M *et al* 2018 Simultaneous Cr(VI) reduction and bioelectricity generation in a dual chamber microbial fuel cell *Chem. Eng. J.* **334** 1621–9
[46] Li M, Zhou S and Xu Y 2019 Performance of Pb(II) reduction on different cathodes of microbial electrolysis cell driven by Cr(VI)-reduced microbial fuel cell *J. Power Sources* **418** 1–10
[47] Guan C-Y, Tseng Y-H, Tsang D C W, Hu A and Yu C-P 2019 Wetland plant microbial fuel cells for remediation of hexavalent chromium contaminated soils and electricity production *J. Hazard. Mater.* **365** 137–45
[48] Habibul N, Hu Y, Wang Y-K, Chen W, Yu H-Q and Sheng G-P 2016a Bioelectrochemical chromium(VI) removal in plant-microbial fuel cells *Env. Sci. Technol.* **50** 3882–9
[49] Choi C and Cui Y 2012 Recovery of silver from wastewater coupled with power generation using a microbial fuel cell *Bioresour. Technol.* **107** 522–5
[50] Wang Z, Lim B and Choi C 2011 Removal of Hg^{2+} as an electron acceptor coupled with power generation using a microbial fuel cell. *Bioresour. Technol.* **102** 6304–7
[51] Wu Y, Zhao X, Jin M Y, Li S, Li F, Kong J, Nan and Wang A 2018 Copper removal and microbial community analysis in single-chamber microbial fuel cell *Bioresour. Technol.* **253** 372–7
[52] Chen Z, Zhu B K, Jia W W-F, Liang J-H and Sun G-X 2015 Can electrokinetic removal of metals from contaminated paddy soils be powered by microbial fuel cells? *Env. Technol. Innov.* **3** 63–7
[53] Wang H, Song H, Yu R, Cao X, Fang Z and Li X 2016 New process for copper migration by bioelectricity generation in soil microbial fuel cells. *Env. Sci. Pollut. Res.* **23** 13147–54
[54] Habibul N, Hu Y and Sheng G-P 2016b Microbial fuel cell driving electrokinetic remediation of toxic metal contaminated soils *J. Hazard. Mater.* **318** 9–14
[55] Reshetilove A, Arlyapove V, Alferov V and Reshetilov T 2013 BOD biosensor: application of novel technologies and prospects for the development *State of the Art in Biosensors: Environmental and Medical Applications* (London: IntechOpen) pp 57–77
[56] Karube T, Matsunaga S, Mitsuda and Suzuki S 1977 Microbial electrode BOD sensors *Biotechnol. Bioeng.* **19** 1535–47

[57] Moon H, Chang I S, Kang K H, Jang J K and Kim B H 2004 Improving the dynamic response of a mediator-less microbial fuel cell as a biochemical oxygen demand (BOD) sensor *Biotechnol. Lett.* **26** 1717–38

[58] Chang I S, Jang J K, Gil G C, Kima M, Kima H J and Cho B W B H K 2004 Continuous determination of biochemical oxygen demand using microbial fuel cell type biosensor *Biosens. Bioelectron.* **19** 607–13

[59] Kharkwal S, Tan Y, Lu M and Ng H 2017 Development and long-term stability of a novel microbial fuel cell BOD sensor with MnO_2 catalyst *Int. J. Mol. Sci.* **18** 276

[60] Osorio de la Rosa E, Vázquez-Castillo J, Carmona-Campos M, Barbosa-Pool G R, Becerra-Núñez G, Castillo-Atoche A and Ortegón-Aguilar J 2019 Plant microbial fuel cells-based energy harvester system for self-powered IoT applications *Sensors* **19** 1378

[61] Huircán J J, Munoz C, Young H, Von Dossow L, Bustos J, Vivallo G and Toneatti M 2010 ZigBee-based wireless sensor network localization for cattle monitoring in grazing fields *Comput. Electron. Agric.* **74** 258–64

[62] Sreekantha D K and Kavya. M 2017 Agricultural crop monitoring using IoT—a study *11th Int. Conf. on Intelligent Systems and Control (ISCO)* pp 134–9

[63] Gutiérrez J, Villa-Medina J F, Nieto-Garibay A and Porta-Gándara M A 2014 Automated irrigation system using a wireless sensor network and GPRS module *IEEE Trans. Instrum. Meas.* **63** 166–76

[64] Mehrotra P 2016 Biosensors and their applications—a review *J. Oral Biol. Craniofac. Res.* **6** 153–9

[65] Updike S and Hicks G 1967 The enzyme electrode *Nature* **214** 986–8

[66] Mishra R, Dominguez R, Bhand S, Munoz R and Marty J 2012 A novel automated flow-based biosensor for the determination of organophosphate pesticides in milk *Biosens. Bioelectron.* **32** 56–61

[67] Akyilmaz E, Yorganci E and Asav E 2010 Do copper ions activate tyrosinase enzyme? A biosensor model for the solution *Bioelectrochemistry* **78** 155–60

[68] Rechnitz G A 1978 Biochemical electrodes uses tissues slice *Chem. Eng. News* **56** 16–21

[69] Allen M J 1972 Cellular electrophysiology *Methods Microbial* ed J R Norris and D W Ribbon (New York: Academic) pp 247–83

[70] Logan B E 2010 Scaling up microbial fuel cells and other bioelectrochemical systems *Appl. Microbiol. Biotechnol.* **85** 1665–71

[71] Cid C A, Stinchcombe A, Ieropoulos I and Hoffmann M R 2018 Urine microbial fuel cells in a semi-controlled environment for onsite urine pre-treatment and electricity production *J. Power Sources* **400** 441–8

[72] Xing D, Zuo Y, Cheng S, Regan J M and Logan B E 2008 Electricity generation by *Rhodopseudomonas palustris* DX-1 *Env. Sci. Technol.* **42** 4146–51

[73] Fan Y, Sharbrough E and Liu H 2008 Quantification of the internal resistance distribution of microbial fuel cells *Env. Sci. Technol.* **42** 8101–7

[74] Logan B E 2009 Exoelectrogenic bacteria that power microbial fuel cells *Nat. Rev. Microbiol.* **7** 375–81

[75] Nielsen M E, Reimers C E, White H K, Sharma S and Girguis P R 2008 Sustainable energy from deep ocean cold seeps *Energy Env. Sci.* **1** 584–93

[76] Hsu L, Chadwick B, Kagan J, Thacher R, Wotawa-Bergen A and Richter K 2013 Scale up considerations for sediment microbial fuel cells *RSC Adv.* **3** 15947

[77] He Z, Shao H and Angenent L T 2007 Increased power production from a sediment microbial fuel cell with a rotating cathode *Biosens. Bioelectron.* **22** 3252–5

[78] Gong Y M, Radachowsky S E, Wolf M, Nielsen M E, Girguis P R and Reimers C E 2011 Benthic microbial fuel cell as direct power source for an acoustic modem and seawater oxygen/temperature sensor system *Env. Sci. Technol.* **45** 5047–53

[79] Salehmin M N I, Su L S, Satar J and Daud W R W 2020 Pushing microbial desalination cells towards field application: prevailing challenges, potential mitigation strategies, and future prospects *Sci. Total Env.* **759** 143485

[80] Clauwaert P, Aelterman P, De Schamphelaire L, Carballa M, Rabaey K and Verstraete W 2008 Minimizing losses in bio-electrochemical systems: the road to applications. *Appl. Microbiol. Biotechnol.* **79** 901–13

[81] Dong Y, Liu J, Sui M Y, Qu J J, Ambuchi H, Wang and Feng Y 2017 A combined microbial desalination cell and electrodialysis system for copper-containing wastewater treatment and highsalinity-water desalination *J. Hazard. Mater.* **321** 307–15

[82] Zhang F and He Z 2015 Scaling up microbial desalination cell system with a post-aerobic process for simultaneous wastewater treatment and seawater desalination *Desalination* **360** 28–34

[83] Bombelli P, Dennis R J, Felder F, Cooper M B, Iyer D M R, Royles J, Harrison S T L, Smith A G, Harrison C J and Howe C J 2016 Electrical output of bryophyte microbial fuel cell systems is sufficient to power a radio or an environmental sensor *R. Soc. Open Sci.* **3** 160249

[84] Hou Q, Pei H, Hu W, Jiang L and Yu Z 2016 Mutual facilitations of food waste treatment, microbial fuel cell bioelectricity generation and *Chlorella vulgaris* lipid production *Bioresour. Technol.* **203** 50–5

[85] Sudirjo E, de Jager P, Buisman C J N and Strik D P B T B 2019b Performance and long-distance data acquisition via LoRa technology of a tubular plant microbial fuel cell located in a paddy field in west Kalimantan, Indonesia *Sensor* **19** 46–7

[86] Fouquet R 2016 Historical energy transitions: speed, prices and system transformation *Energy Res. Soc. Sci.* **22** 7–12

[87] Davidson D J 2019 Exnovating for a renewable energy transition *Nat. Energy* **4** 254–6

[88] Savla N Suman, Pandit S, Verma J P, Awasathi A K, Sana S S and Prasad R 2021 Techno-economical evaluation and life cycle assessment of microbial electrochemical systems: a review *Curr. Res. Green Sust. Chem.* **4** 100111

[89] Ellis M W, Von Spakovsky M R and Nelson. D J 2001 Fuel cell systems: efficient, flexible energy conversion for the 21st century *Proc. IEEE* **89** 1808–18

[90] Jourdin L, Sousa J, van Stralen N and Strik D P B T B 2020 Techno-economic assessment of microbial electrosynthesis from CO_2 and/or organics: an interdisciplinary roadmap towards future research and application *Appl. Energy* **279** 115775

[91] Zhou Z, Chi Y, Dong J, Tang Y and Ni M 2019 Model development of sustainability assessment from a life cycle perspective: a case study on waste management systems in China *J. Clean. Prod.* **210** 1005–14

Chapter 15

Commercialization of bio-electrochemical cell (microbial fuel cells): challenges and future perspectives

Idzni Nashuha Mohd Rashidy, Nur Syamimi Ab Jalil, Mohammad Tajuddin Sabri, Nor Asniza Ishak and Muaz Mohd Zaini Makhtar

The bio-electrochemical system (BES) technology such as microbial fuel cell (MFC) technology in the laboratory stage had shown a high current density generation which are close to those necessary for practical applications. On the other hand, full-scale implementation of BES in wastewater treatment is challenging due to unique microbiological, technological, and economic challenges that have never been addressed in any other wastewater treatment system. The main challenge is to take these innovations out of the lab and develop commercially viable bioenergy production systems. Some of the most recent successes and possible problems in terms of actual usage of these MFCs for renewable energy production and other reasons are discussed below. What are the broad and issues that it faces, both in terms of research and in terms of commercialization? The chapter will emphasize a prevalent worry about microbial activity, which frequently causes misunderstanding among researchers in the MFC and BES fields. Thus, in this chapter, we present both supportive and opposing viewpoints on the issue.

15.1 Introduction—microbial activity in BES

Microorganisms function as biological, live catalysts in BESs, and their electron transfer rates are one of the most difficult problems to solve. The natural microbial metabolic rates, which have upper limitations, regulate electron production rates. During the power generation process, a wide range of microbial species have been discovered in contact with the anode. Members of the Geobacteraceae family, in general, dominate such biofilms and are frequently linked to greater levels of power production from a range of environmental inoculum. However, in the case of complex organic waste streams, a prefermentation step is necessary to break down

doi:10.1088/978-0-7503-4791-4ch15

the organics before they can be used in the anode chamber, which reduces the process efficiency. Little is currently known regarding the positive and negative impacts of mixed cultures on BES power production. Because the growth and metabolism of microorganisms must continue, environmental parameters such as pH, temperature, salinity, organic feed, and shear rate must be maintained so that the complex electroactive biofilms do not get disrupted. Alternative energy transfer routes, such as methanogenesis in the anode compartment, might cause organics to be diverted away from the intended product. Electron shuttles have been effectively provided as redox mediators to stimulate microbial interactions with an electrode, however their long-term viability has been questioned.

Increases in electron transfer rates to an electrode have been achieved by modifying microbial electron transfer capacities by overproduction of microbial nanowires. Furthermore, using an amphiphilic conjugated oligoelectrolyte to modify the Shewanella lipid membrane permitted current consumption for succinate synthesis via fumarate reduction. It is necessary to get a better understanding of the microorganisms connected with the electrodes, as well as their synergistic relationships and the creation of genetic engineering methods. Electron transport to and from an electrode are two distinct processes, each with its own set of difficulties. *Shewanella oneidensis* uses both electron shuttles and the mtr cytochrome route when using an anode as an electron acceptor.

15.2 Revolution on BES configuration

BES such a MFC systems confront several physical and electrochemical challenges in terms of engineering. MFC power generation has been anticipated to reach 17 to 19 W m^2 (Logan 2006), although average maximum power densities are 2 to 3 W m^{-2} of projected electrode area, which is significantly less than a conventional hydrogen fuel cell (1 W cm^2). (Hays *et al* 2011) found that the greatest volumetric power densities produced in an MFC utilising residential wastewater were only 12 W m^{-3}, far less than the potential energy contained in the wastewater. Ohmic voltage losses due to both ionic and electronic resistance to charge transport; activation overpotentials due to bacteria-to-electrode charge transfer energy barriers; concentration overpotentials due to mass transport resistance; and coulombic losses are all examples of internal losses in MFCs.

For BES, a variety of designs have been proposed. A single chamber membraneless system can occasionally replace the traditional two-chamber design. These characteristics minimise material costs depending on the BES's requirements, but they must be tested for specific applications. Graphite rods, paper, foil, granules, and fibre brush; carbon felt and foam; activated carbon cloth; catalysed and non-catalyzed; electrodes modified with conductive polymers, nanoparticles, and pre-treatment; metals such as platinum, aluminium, nickel, and stainless steel have all been tested and used in various setups (Zhou *et al* 2011). High electrical conductivity, non-corrosive characteristics suited for long-term usage, and machining and production options are all available. A number of new composite materials, such as textile coated carbon nanotubes, macroporous graphene/polyaniline, and hierarchically porous chitosan/graphene, have been created to improve biocompatibility. Specific pore diameters in the pore-coating of electrodes have been used to generate three-dimensional support structures that encourage microbial adhesion and biofilm growth. Cathodes have also been treated with Pt to enhance interactions with oxygen, protons, and microorganisms.

15.3 A typical debatable issue in BES technology

Due to BES technology encompasses a wide range of topics such as microbiology, chemistry, electrical engineering, and technology, various researchers have differing perspectives on some topics. As a result, this chapter focused on seven (7) common BES concerns, with both side arguments backed up by research data from throughout the world.

15.3.1 Electrogenic bacteria survival

(A) EB cannot survive in waste with heterogenous substrates

EB cannot easily survive in the application of MFC that convert to waste with heterogenous qualities. Pure culture bacteria were not preferred as it is not practical for industrial effluent treatment and involved a considerable cost (Makhtar *et al* 2018). Some researchers also claimed that EB can only provide complete oxidation of simple substrates (Huarachi-Olivera *et al* 2018) as waste with heterogeneous qualities, such as waste with a cell and colloid structure that distinguishes it from general solid waste and makes dewatering and anaerobic digestion difficult (W. Qiao *et al* 2011).

In addition to that, a lot of paper also stated that the application of MFC to waste is more test and usually done with synthetic wastewater. This is supported by a study by Kumar *et al* (2019), that initial research with respect to MFC technology was carried out with synthetic wastewater; however, recently they have been demonstrated to treat industrial effluents with high efficiencies (Kumar *et al* 2019). One of the factors could be the acidity of wastewater would increase the pH that was associated with the removal of sCOD (Leiva *et al* 2016).

(B) EB can survive in heterogenous waste substrates

EB has an enzyme that can break down waste to produce power (Muaz *et al* 2019). Organic substrates containing carbohydrates, lipids, and proteins are employed as electron donors in MFCs to generate energy through redox processes at the anode, this is the case (Kumar *et al* 2019). Proteins or other substrates in the medium can be digested and used as intermediates in glycolytic or other metabolic pathways, leading to the production of acetyl co-A, a citric acid cycle intermediary (Nandy *et al* 2013) (figure 15.1). The citric acid cycle brings the linked oxidation–reduction process to a close, resulting in reduced forms of electron carriers that carry electrons from the cytoplasm to the cell membrane, passing through multiple membrane complexes before reaching a terminal electron acceptor.

Furthermore, the anode compartment produces electrons and protons as a result of microorganisms oxidizing organic compounds (Mohd Zaini Makhtar and

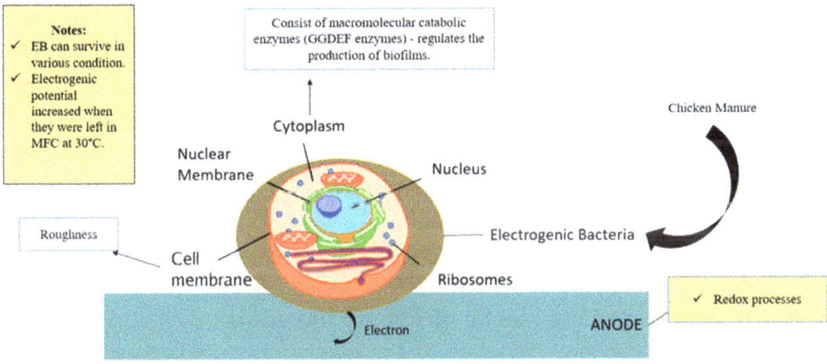

Figure 15.1. Figure of electron donors in MFC to produce energy through redox processes at the anode.

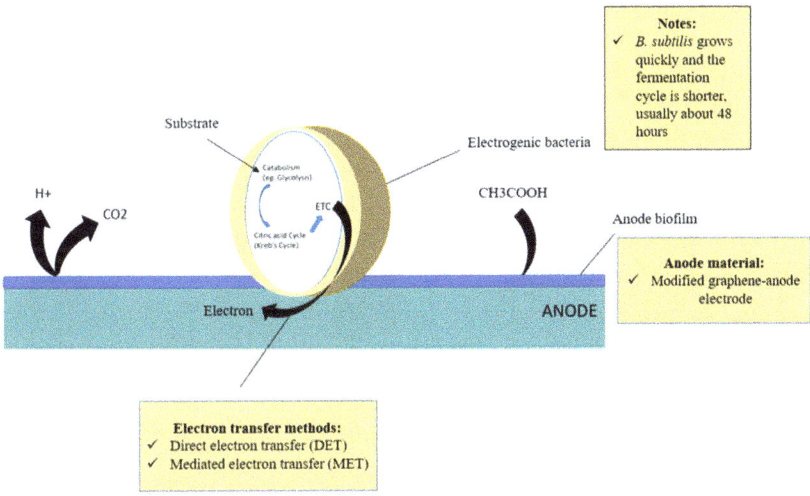

Figure 15.2. Figure of oxidation and electron transport chain at positive anode, material of anode and electron transfer methods.

Vadivelu 2019). The difference in redox potential between the two different electrodes determines the MFC voltage. Redox potential is a property of electrogenic chemical species: the higher the value, the greater the electron affinity and the greater the tendency to be reduced (figure 15.2).

Additionally, EB has also been shown to be effective in the treatment of real wastewater. Based on Zhuang *et al* (2012), the ability of simultaneous actual wastewater treatment and bioelectricity generation under continuous flow mode was examined using a tubular air-cathode microbial fuel cell (MFC) stack with great scalability and cheap material cost (Zhuang *et al* 2012). Five non-Pt MFCs coupled in series and parallel circuit modes treating swine wastewater can enhance the voltage and current at the two organic loading rates (OLRs) studied (1.2 and 4.9 kg COD/m^3 d) (Zhuang *et al* 2012). Due to the substrate cross-conduction effect, the parallel stack maintained high power output while the series link lost energy and it shows that the parallel stack removed 83.8% COD and 90.8% COD at 1.2 kg COD/m^3 d with continuous electricity production, and 77.1% COD and 80.7% COD at 4.9 kg COD/m^3 d with continuous electricity production (Zhuang *et al* 2012).

15.3.2 Electrogenic bacteria (EB) kinetic growth

(A) EB growth is slow and the kinetic of EB are unknown

EB growth is slow as the slow kinetics of the electrochemical oxygen reduction reaction (ORR) is a crucial bottleneck in the development of microbial fuel cells (MFCs) (Mohd Zaini Makhtar and Tajarudin 2020). According to the journal Erable *et al* (2012), employing oxygen as the final electron acceptor would be the most convenient method for developing MFCs with broad use, however the kinetics of the oxygen reduction reaction (ORR) is slow (Erable *et al* 2012) Furthermore, according to their research Rabbani *et al* (2021), delayed microbe growth and adhesion at the start of the MFC operation will result in a low power density (Rabbani *et al* 2021).

Similarly, the kinetics of EB growth are unknown, and the growth and kinetics of exoelectrogens in MFCs are poorly understood (Torres 2014). This is because measuring exoelectrogen numbers, growth rates, and growth yields in MFCs is inherently difficult due to a variety of factors, including a lack of suitable primers and probes, a lack of knowledge about which organisms are electrogenic or what makes them electrogenic, and the fact that sampling biomass on the electrodes destroys the bio-film required for successful operation (Heidrich *et al* 2016).

(B) EB grow well and the kinetic of EB are identified

Microbial fuel cells (MFC) are bioelectrical devices that use exoelectrogenic bacteria's normal metabolism to generate electricity (Ren *et al* 2014). According to Naureen *et al* (2016), certain strains were capable of swiftly creating electricity while others enriched slowly and showed a rise in electricity output after 7 and 21 days (Naureen *et al* 2016). As a result, *Bacillus subtilis* grows quickly and the fermentation cycle is shorter, usually about 48 h (Su *et al* 2020). In addition to that, the kinetic model can determine the EB growth. The kinetic parameters may be obtained by graphing microbial growth rates ($dX = dt$) against cell concentration (X), and the induction and transient phases of growth were followed by an exponential phase in which $dX = dt$ increased linearly with cell concentration (Mohd Zaini Makhtar and Vadivelu 2019) (figure 15.2).

Over and above, the kinetics of electron transfer affects the performance and power capability of MFCs. Direct electron transfer (DET) and mediated electron transfer (MET) are two types of electron transfer methods (Mohd Zaini Makhtar and Vadivelu 2019). Over and above, DET can occur directly through membrane-bound cytochromes or indirectly through electrically conductive nanowires (pili).

Redox mediators or oxidation of reduced secondary metabolites can facilitate electron transport in the MET (Feng *et al* 2014).

There are a few maintenances at the anode to increase the growth of EB. For example, a modified graphene-anode electrode improved the oxidation process of the substrate by supporting the healthy development of bacteria on the anode electrode's surface, resulting in a significant increase in electron flow (Yaqoob *et al* 2020). The use of waste-derived graphene for the anode electrode lowers the cost of MFC operation while also improving its electrochemical performance (Yaqoob *et al* 2020). Furthermore, the addition of metal oxides to graphene, such as TiO_2, ZnO, AgO, CuO, and other metal oxides, has a significant impact on cost reduction and corrosion avoidance (Yaqoob *et al* 2020). A study by Najafabadi *et al* (2016), showed that by using graphene-anode, and carbon cloth cathode with the surface area of anode 4 cm^2, will produce a maximum power of 2850 mW m^{-2} (Najafabadi *et al* 2016).

15.3.3 The electrode-bacterium interaction

(A) The interaction is not well described

The interaction and/or coexistence of bacteria and solid electrodes in electron transfer mechanisms is poorly understood, especially in complex situations where a variety of microbial species (electroactive and non-electroactive) can be found on the electrodes (Mohd Zaini Makhtar *et al* 2021). Several forms of organic waste have been employed as fuel for microbial anodes. According to Santoro *et al* (2017), but electroactive bacteria kinetics remain weak, and the interaction between electrode and bacteria is still unknown (Santoro *et al* 2017). Furthermore, the lack of interaction between the electrode and the bacterium could be attributed to the use of traditional carbon-based MFC electrodes (Delord *et al* 2017). These carbons have a limited bacterial adhesion activity, resulting in a small amount of biofilm growth and weak contacts between bacteria and the electrode surface, resulting in a low power output efficiency (S. Li *et al* 2017).

(B) The interaction is understandable

The exoelectrogenic bacteria can transfer electrons to the anode in the anodic chamber as exoelectrogenic bacteria that uses four basic methods to deliver electrons to the anode in the anodic chamber. To begin, direct contact of exoelectrogens cells with the anode for electron transfer using C-Type Cytochromes (CTCs), followed by the use of soluble electron shuttles like flavins, solid conductive components like nanowires or pili, and finally the production of electro-active biofilms (Guang *et al* 2020) (figure 15.3). On top of that, it is said that outer-membrane cytochromes (OMCs) and their function in both *Geobacter* and *Shewanella* spp., conductive pili or cell appendages (referred to as nanowires) play a role in promoting direct electron transfer (DET) to minerals (Patil *et al* 2014).

To increase the interaction of the electrode and the bacterium, in terms of anode-bacteria interactions, the surface morphology of anode materials should be enhanced. Electrochemical oxidation, chemical, and thermal treatment can be employed to expand the surface area of an anode, improving microbe adhesion

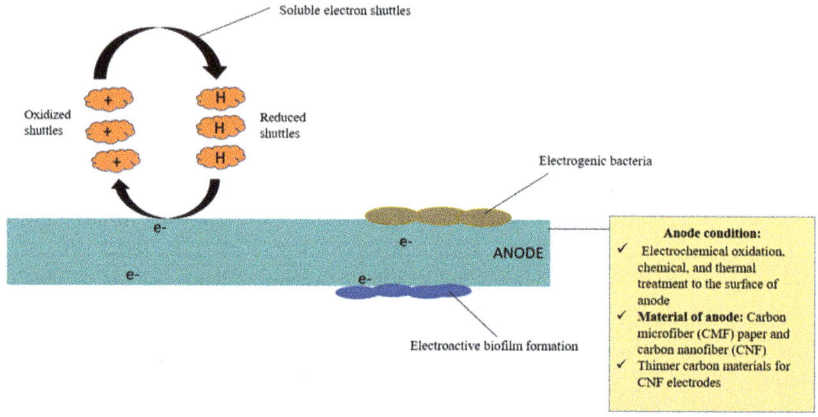

Figure 15.3. Main recognized mechanisms of electron transfer from exoelectrogens and the condition at the anode.

and cell output power (Nosek *et al* 2020). Carbon microfiber (CMF) paper and carbon nanofiber (CNF) mats have both been explored as anode materials in MFC in recent years. Thinner carbon materials for CNF electrodes led in a bigger specific surface area and improved morphology of the electrode surface, which enhanced bacterial adherence and the creation of a thick and stable biofilm, resulting in increased energy output (Nosek *et al* 2020).

Over and above, the interaction between the electrode and the bacterium can be determined by using cyclic voltammetry. According to López Zavala *et al* (2019), cyclic voltammetry is an electrochemical technique that is often used to explain the electron transport interactions in the interface electrode-biomass/biofilm-liquor in the anode chamber of MFC (López Zavala *et al* 2019). These methods can distinguish between direct and indirect electron transfer processes in anode-associated biofilms (Mohd Zaini Makhtar *et al* 2021).

15.3.4 Electron transportation from EB

(A) Limiting factor for BES performance

The transport of electrons from the microorganism to the anode is too sluggish and not effective. As concern, the quantity of electron production is determined by microorganisms in the anode chamber. Additionally, oxygen has a higher redox potential than many other electron acceptors, which is why it is widely regarded as a good cathode electron acceptor. However, poor oxygen contact with the electrode and a slow rate of oxygen reduction on the normal carbon electrode are drawbacks that limit the use of oxygen in MFCs (Ucar *et al* 2017). Although the cathodic reaction can be improved by using catalytic-coated electrodes, catalysts are often expensive and rare metals are scarce (Zhou *et al* 2011). Furthermore, according to the journal Velvizhi and Venkata Mohan (2012), the main limitations are slow biocatalyst metabolic rates and inefficient electron transfer from the biocatalyst to the anode surface and then to the cathode due to high internal resistance in the system, which includes internal losses at the anode, cathode, proton exchange membrane, and electrolyte (Velvizhi and Venkata Mohan 2012).

(B) The electron transportation not solely the issue

The transport of electrons from the microorganism to the anode is not solely the core effect for MFC's performance. Electrons must reach the cell's outer membrane, and physical contact between the outer membrane and the anode is essential for direct electron transfer. On the anode surface, electricigens create biofilms (electrically conductive nanowires pili and flagella) (Cao *et al* 2019). Without the use of diffusional electron mediators, electrons are transferred directly through the outer membrane cytochrome and nanowires or trans-membrane electron transport proteins (Mohd Zaini Makhtar and Tajarudin 2020). The nanowires connect to membrane-bound cytochromes, allowing electricigens to employ an electron acceptor that is not in direct touch with the cell (Cao *et al* 2019). Electron transport proteins move electrons from the cytoplasm to the outer membrane and then to the anode, which is a crucial part of direct electron transfer (figure 15.4).

Not to mention, due to their excellent electrical conductivity, metal electrodes such as silver, SS, aluminum, nickel, molybdenum, titanium, gold, and copper provide strong performance in MFCs (Makhtar *et al* 2018). Stainless steel is a widely available industrial material with great mechanical and corrosion resistance, high conductivity, and low cost, making it an excellent electrode material for MFC anodes (Nosek *et al* 2020). Moreover, the macroporosity of SS anodes allows for the attachment of carbon nanoparticles, which increases anode biocompatibility and favors internal bacteria colonization, which could improve electrode responses and it is said that, an increase in the anode area from 2.5 to 20.1 cm^2 raised the maximum power density from 0.3 to 0.67 mW m^{-2} when employing copper wires (Nosek *et al* 2020).

Over and above, because it contains electrogenic bacteria, electron transfer from microorganisms will be more efficient, as EB will aid in the decomposition of carbon. According to the publication Shah *et al* (2019), electrogenic bacteria will aid in the decomposition or oxidation of carbon-rich organic materials by gaining electrons through cellular respiration and transferring them to the anode electrode (Shah *et al* 2019). Additionally, from the anode, electrons are transferred to the cathode through a resistor or charge that generates energy (Mohd Zaini Makhtar and Tajarudin 2020).

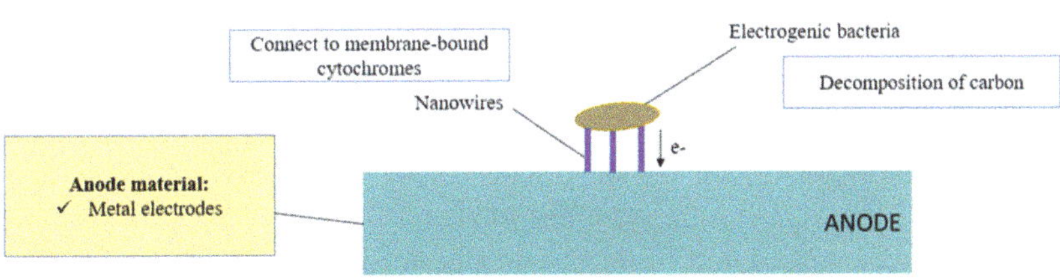

Figure 15.4. Electrons from electricigens flow to the anode directly through nanowires and material at the electrodes.

15.3.5 Environment condition for EB to grow

(A) Unknown of specific condition for EB

There is no specific condition for EB to growth. This is because EB did survive for growth, but it did not completely oxide the substrate properly. According to Sacco *et al* (2017), one of the EBs, *S. oneidensis*, may survive in the presence of oxygen, but one drawback is that it does not completely oxidize the organic substrate commonly employed in MFCs (lactate), leaving electrons unutilized and waste products (Sacco *et al* 2017). Furthermore, according to Sacco *et al* (2017), the range of carbon substrates that these bacteria can utilize is limited (Sacco *et al* 2017). However, when microorganisms living in an MFC system can totally oxidize organic substrates to CO_2, higher columbic efficiencies have been found. Furthermore, bacterial growth in MFCs is dependent on a temperature and pH range that is not specified. It said that weak EB growth observations were conducted at 30 °C and 55 °C, as well as pH 4 and pH 11 (Naureen *et al* 2016).

(B) EB's best condition has been established

EB has a specific condition for growth. Firstly, EB of *B. subtilis* has a higher oxidation rate as compared to other microbes (Mohd Zaini Makhtar and Tajarudin 2020). A study found that the rate of oxidation of substrates by microorganisms in terms of COD and BOD was higher in MFC inoculated with *B. subtilis* than in MFC injected with mixed cultures (Ismail and Jaeel 2013). Over and above, several methods for establishing and proving electrochemical communication between electroactive bacteria and the anode have been devised. Chemical anode treatment, as well as mediator and chemical/functional group modification of anode surfaces, are examples (Katuri *et al* 2020). In general, electrodes modified with charged, hydrophilic functional groups will improve biofilm attachment, reduce start-up times, and improve MFC or MEC performance (Guo *et al* 2013) (figure 15.5).

Not only that, not to mention that every bacterium has their specific range of temperature/ pH and condition to growth. For example, a study found that most EB isolates were able to Anode surface modification with mediators and chemical/ functional groups (Naureen *et al* 2016). Furthermore, the specific growth conditions for EB in an ML-MFC are pH 6.0, electrode distance of 3 cm, moisture content of 30% (v/w), and temperature of 35 °C (Mohd Zaini Makhtar and Vadivelu 2019).

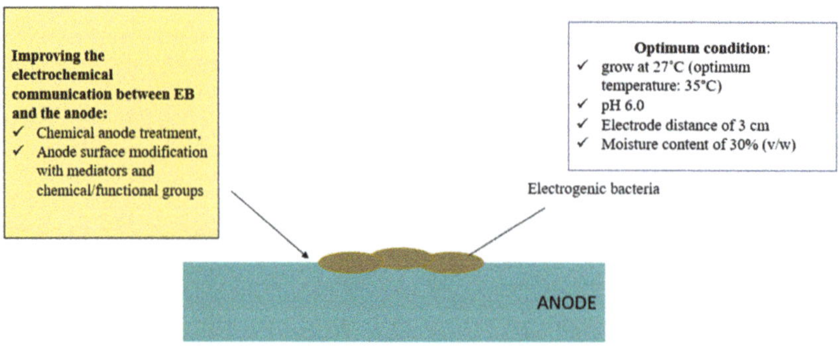

Figure 15.5. Optimum condition of EB and the electrochemical communication between EB and anode.

15.3.6 EB biofilm: growth progress and biofouling

(A) Slow development of biofilm at anode surface

Slow development of biofilm growth should be no problem as it will rely on the bacteria. This is supported by a study Y J Qiao *et al* (2017), which found that the cell density on the anode increased slowly, especially during the first 36 h, and that there are only scattered biofilms on the anode even after 60 h, and that no biofilm can be found on the anode with continuous Sodium houttuyfonate addition, even after 180 h (Y J Qiao *et al* 2017). In this case, the current generation of this MFC should be relied on the bacteria cells. In addition to that, if biofilm fouling at anode will produce it will also lead to current increasing. According to Eaktasang *et al* (2013), the formation of the primary biofilm resulted in a gradual increase in current production, and when more biofilm was formed, current production was more rapid and the maximum current production levels were higher than those obtained in the first cycle, implying that current performance tends to inflate due to biofilm accumulation (Eaktasang *et al* 2013).

(B) Fasten the biofilm, the better MFC performance

Anode biofilm is an important part of microbial fuel cell (MFC) for electricity generation. The bacteria responsible for generating electrical current in the MFC grow in the biofilm attached to the surface of the anode, and these microbial cells release electrons to the anode (Mohd Zaini Makhtar *et al* 2021). This indicates that slowly or no biofilm present will decrease the electron transfer. According to Wu *et al* (2020), with such a low initial OD value, the rate of cells attaching to the electrode was slow, resulting in slow biofilm growth and low current generation (Wu *et al* 2020). In addition to that, increase in biofilm fouling at the anode also will affect the electron transfer as more biofilm fouling will decrease the electron transfer (figure 15.6).

There are a few factors that affect the biofilm formation. Firstly, to begin with, the characteristics of the substratum may have a significant impact on the rate and extent of microorganism attachment as the rougher and more hydrophobic materials

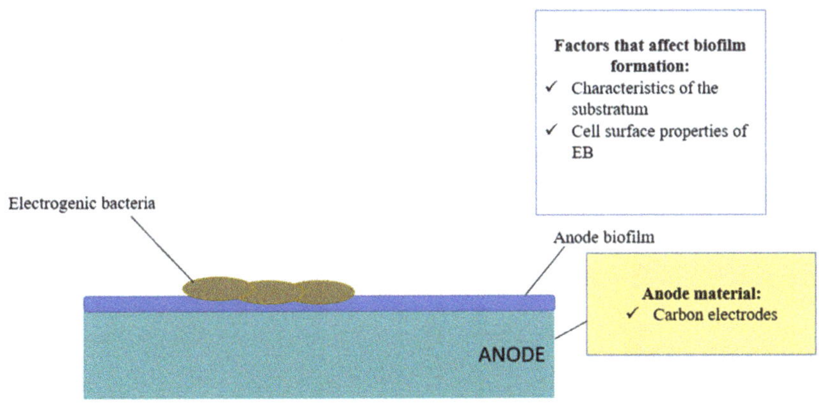

Figure 15.6. Factors that affect biofilm formation and the material at the anode electrode and the material used at the anode.

will develop biofilms more quickly in general (though there are exceptions) (Donlan, n.d.). Second, the substratum and cell surface properties are crucial. Flagella, pili, fimbriae, and glycocalyx, for example, may influence the rate of microbial attachment because the microbial cell must overcome the repulsive forces common to all materials once drawn to the surface, and these appendages allow the cell to stay attached until more permanent attachment mechanisms are in place.

In addition to that, carbon-based electrodes are widely employed in MFCs because they are chemically stable, resistant to environmental conditions in MFCs, have a high electrical conductivity, and a large specific surface area, all of which promote biofilm formation (Nosek *et al* 2020). Carbon/graphite felt (CF/GF), CC, carbon paper (CP), graphite paper (GP), and activated carbon are all examples of carbon-based electrodes (AC) (Nosek *et al* 2020). High porosity, electrical conductivity, and a particular surface area describe the most used CF and CC; nevertheless, CF has greater chemical and mechanical stability and is less expensive. On top of that, high porosity, electrical conductivity, and a particular surface area describe the most commonly used CF and CC; nevertheless, CF has greater chemical and mechanical stability and is less expensive (Nosek *et al* 2020).

15.3.7 EB competition in mixed cultured MFC

(A) High competition of 'food'

The mixed cultured MFC has resulted in more competitiveness and a decrease in current. This is because, the mixed culture will compete food, life and many more for them to survive. According to Jamialahmadi *et al* (2018), total bacterial growth may increase in natural environments with multicultural bacterial communities, whereas individual species growth may be limited (Jamialahmadi *et al* 2018). Furthermore, as the organic content is reduced, the COD is reduced, resulting in a reduction in the electron yield. This is supported by the findings of Prabowo *et al* (2016), who found that multicultural rumen microorganisms in MFC might reduce COD by up to 67.9% (Prabowo *et al* 2016).

(B) Mixed culture EB boost the MFC performance

Employing multicultural microorganisms generates the highest current values in MFC since they are more efficient than microorganisms accumulated in monocultures (Production and Cells 2019). This is because in MFCs with multicultural organisms, hundreds of species strains interact in a symbiotic way to offer free electrons (Vajda *et al* 2014) (figure 15.7), and thus increasing the current. In addition

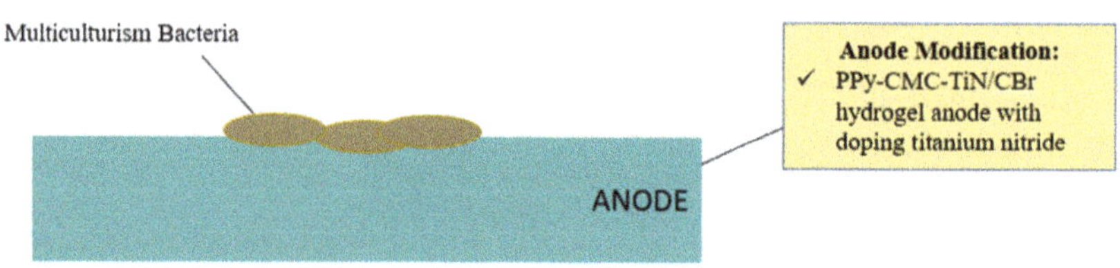

Figure 15.7. Multiculturism bacteria attached to the anode modification of PPy-CMC-TiN/CBr hydrogel anode with doping titanium nitride.

to that, most mixed cultures have dominant species that are easily adaptable to the environment of the MDC system's tri-functional process. Based on the journal Guang *et al* (2020), it is said that though the majority of electrons are produced by the oxidation of organic matter in the anode chamber by the dominant species, other exoelectrogenic species also produce electrons through their respiratory activities and comparison to MDC systems with pure cultures, which may utilise few pathways in electron transfer, this boosts power generating capacity while also improving wastewater treatment and desalination (Guang *et al* 2020).

It is also said that anode modification by using PPy-CMC-TiN/CBr hydrogen anode, showed a better power output in maximum power density (Nosek *et al* 2020). Based on a study of Kataky *et al* (2020), shows that, when compared to the MFC equipped with the other two anodes, the CMC-TiN/CB hydrogel anode produced more power in terms of maximum power density (14.11 W m^3) and average peak current density (7.80 A m^2) (Kataky *et al* 2020). This is because, due to strong adsorption and chemical stability, titanium nitride was conducive to the interface connection between the conductive polymer and substrate, as well as between the microorganisms and the conductive polymer, following doping (Wang *et al* 2020). It aided the electrostatic effect between the electrons created by the bacteria and the anode surface, thereby enhancing ion diffusion and electron transfer rates inside and outside the electrode, and therefore improving the MFCs' electrical performance (Wang *et al* 2020).

15.4 Reliability of MFC from polarization curve perspective

The working voltage of the MFC is plotted as a function of current or current density on a polarization curve. The polarization curve, despite being one of the most frequent ways of testing a fuel cell, lacks specificity (Spiegel 2008). This is demonstrated through a study that establishes a foundation for characterizing MFC polarization curves, and the simulation results reveal the limiting physical, chemical, and electrochemical events, as well as their effects on cell performance (Feng *et al* 2014). Furthermore, 'power overshoot' is a common problem that arises from an unexpectedly high reduction in voltage with rising current density while assessing polarization curves.

There are two types of curves (polarization curve and power curve) are usually derived from polarization data and are most used to report MFC data (Mohd Zaini Makhtar and Tajarudin 2020). The polarization curve shows the operating voltage of the MFC as a function of current or current density. Although these curves may not accurately represent MFC performance for technical design, they do provide useful information for characterization of the chemistry and dynamics of MFC operation (Simeon *et al* 2020). Even more, power density and polarization curve directly reflect MFC performance. They are the most often used methods for calculating MFC internal losses and MAP (Simeon *et al* 2020). Furthermore, the most striking property of the polarisation curve is that there is a straight linear relationship between the voltage produced and the current density over the usable range of current (Virdis *et al* 2011).

15.5 BES (MFC and MEC) at real site setup

To powering distant devices in marine applications, MFCs have been demonstrated. But there have been lack reports of aboveground pilot-scale research with MFCs or MECs that have been published. Based on online posts, conference presentations, and conversations with academics, it appears that at least three pilot-scale testing of MFCs or MECs have been conducted. The Advanced Water Management Center at the University of Queensland conducted the first large-scale test of MFCs at Foster's brewery in Yatala, Queensland, under the supervision of Jurg Keller and Korneel Rabaey (Australia) (www.microbiafuelcell.org). Carbon fibre brush anodes were used in tubular reactors, with flow up through the tubes and out over the reactor's exterior, which was coated with graphite fibre brush cathodes. This was a similar concept to one that had been tested in the lab using ferricyanide catholyte (Logan *et al* 2006). Other than the fact that solution conductivity was low, limiting current generation, and that the high biochemical oxygen demand in the wastewater leaving the anode chamber resulted in excessive biofilm buildup on the cathodes as the wastewater was exposed to air, there isn't much information about the MFC's performance at the site.

There was also researchers from the University of Connecticut and their collaborators (Fuss & O'Neill and Hydroqual Inc.) are also working on an MFC and BES project in a real-world location in the United States (Baikun Li, personnel communication). The researchers suggested a reactor architecture in which the anode is granular graphite and the cathode is Pt-catalyzed carbon fabric (Jiang and Li 2009). The systems treat wastewater by removing up to 80% of the chemical oxygen demand seen in the 300–600 mg l^{-1} range.

15.6 Conclusion

MFCs and BES are a promising technology for the future, with applications in energy and the environment. However, due to many constraints and hurdles, such as material costs and scalability to plant size, MFC is currently struggling to be deployed on a wide scale. The researchers' current focus is on developing low-cost electrode and membrane replacements. There are some acceptable future ideas, such as low-power production for industrial applications, and the use of a power storage device to enhance the power output during discharge. Furthermore, in rural areas, it is hoped that a practical personal MFC would be available for everyday usage to power low-energy equipment. Finally, MFC and BES studies are needed to realise the objective of powering wastewater treatment plant facilities using waste-generated electricity, resulting in energy savings.

References

Borole A P, Reguera G, Ringeisen B, Wang Z W, Feng Y and Kim B H 2011 Electroactive biofilms: current status and future research needs *Energy Environ. Sci.* **4** 4813–34

Cao Y, Hui M, Wei L, Rubing Z, Jing G, Mo X and Huizhou L 2019 Electricigens in the anode of microbial fuel cells: pure cultures versus mixed communities *Microb. Cell Fact.* **18** 1–14

Delord B, Neri W, Bertaux K, Derre A, Ly I, Mano N and Poulin P 2017 Carbon nanotube fiber mats for microbial fuel cell electrodes *Bioresour. Technol.* **243** 1227–31

Eaktasang N, Kang C S, Ryu S J, Suma Y and Kim H S 2013 Enhanced current production by electroactive biofilm of sulfate-reducing bacteria in the microbial fuel cell *Environ. Eng. Res.* **18** 277–81

Erable B, Féron D and Bergel A 2012 Microbial catalysis of the oxygen reduction reaction for microbial fuel cells: a review *ChemSusChem* **5** 975–87

Feng C, Jiangwei L, Dan Q, Lixiang C, Feng Z, Shaohua C, Hongbo H and Chang Ping Y 2014 Characterization of exoelectrogenic bacteria enterobacter strains isolated from a microbial fuel cell exposed to copper shock load. *PLoS One* **9** e113379

Fornero J J, Miriam R and Angenent T 2010 Electric power generation from municipal, food, and animal wastewaters using microbial fuel cells *Electroanalysis* **22** 832–43

Franks A E and Nevin K P 2010 Microbial fuel cells, a current review *Energies* **3** 899–919

Generating electricity from bacteria 2019 https://elifesciences.org/digests/43959/generating-electricity-from-bacteria

Guang L, Desmond Ato K, Huang J, David E-M and Nicholas M 2020 Performance of exoelectrogenic bacteria used in microbial desalination cell technology *Int. J. Environ. Res. Public Health* **17** 10–2

Guo K, Stefano F, Paul G D, Xin C, Bogdan C D, Jurg K, Gooding J J and Korneel R 2013 Effects of surface charge and hydrophobicity on Anodic Biofilm formation, community composition, and current generation in bioelectrochemical systems *Environ. Sci. Technol.* **47** 7563–70

Hays S, Zhang F and Logan B E 2011 Performance of two different types of anodes in membrane electrode assembly microbial fuel cells for power generation from domestic wastewater *J. Power Sourc.* **196** 8293–300

Heidrich E S, Curtis T P, Woodcock S and Dolfing J 2016 Quantification of effective exoelectrogens by most probable number (MPN) in a microbial fuel cell *Bioresour. Technol.* **218** 27–30

Huarachi-Olivera R, Dueñas-Gonza A, Yapo-Pari U, Vega P, Romero-Ugarte M, Tapia J, Molina L, Lazarte-Rivera A, Pacheco-Salazar D G and Esparza M 2018 Bioelectrogenesis with microbial fuel cells (MFCs) using the microalga *Chlorella vulgaris* and bacterial communities *Electron. J. Biotechnol.* **31** 34–43

Ismail Z Z and Ali J J 2013 Sustainable power generation in continuous flow microbial fuel cell treating actual wastewater: influence of biocatalyst type on electricity production *Sci. World J.* **2013** 1–7

Jamialahmadi N, Safari E and Baghdadi M 2018 Interaction of graphene oxide nano-sheets and landfill leachate bacterial culture *Environ. Technol.* **39** 2457–66

Jiang D and Li B 2009 Novel electrode materials to enhance the bacterial adhesion and increase the power generation in microbial fuel cells *(MFCs) Water Sci. Technol.* **59** 557–63

Kataky K P, Amaresh Dalal G B and Chin T W 2020 Numerical aimulation of three-dimensional microbial fuel cell *IOP Conf. Ser.: Earth Environ. Sci.* **463** 012062

Katuri K P, Kamireddy S, Kavanagh P, Muhammad A, Conghaile P, Kumar A, Saikaly P E and Leech D 2020 Electroactive biofilms on surface functionalized anodes: the anode respiring behavior of a novel electroactive bacterium, *Desulfuromonas acetexigens Water Res.* **185** 116284

Kumar S S, Kumar V, Malyan S K, Sharma J, Mathimani T, Maskarenj M S, Ghosh P C and Pugazhendhi A 2019 Microbial fuel cells (MFCs) for bioelectrochemical treatment of different wastewater streams *Fuel* **254** 115526

Leiva E, Leiva-Aravena E and Vargas I 2016 Acid water neutralization using microbial fuel cells: An alternative for acid mine drainage treatment *Water (Switzerland)* **8** 1–9

Li M, Zhou M, Tian X, Tan C, McDaniel C T, Hassett D J and Gu T 2018 Microbial fuel cell (MFC) power performance improvement through enhanced microbial electrogenicity *Biotechnol. Adv.* **36** 1316–27

Li S, Cheng C and Thomas A 2017 Carbon-based microbial-fuel-cell electrodes: from conductive supports to active catalysts *Adv. Mater.* **29** 1–30

López Zavala M Á, González Peña O I, Ruelas H C, Mena C D and Guizani M 2019 Use of cyclic voltammetry to describe the electrochemical behavior of a dual-chamber microbial fuel cell *Energies* **12** 3532

Logan B E, Hamelers B, Rozendal R, Schröder U, Keller J, Freguia S, Aelterman P, Verstraete W and Rabaey K 2006 Microbial fuel cells: methodology and technology *Environ. Sci. Technol.* **40** 5181–92

López Z, Miguel Á, Omar Israel G P, Héctor Cabral R, Cristina Delgado M and Mokhtar G 2019 Use of cyclic voltammetry to describe the electrochemical behavior of a dual-chamber microbial fuel cell *Energies* **12** 3532

Mohd Zaini Makhtar M, Don M M and Tajarudin H A 2018 Microbial fuel cell (MFC) development from anaerobic digestion system *Green Energy and Technology* vol 0 9–31 (Berlin: Springer)

Mohd Zaini Makhtar M and Tajarudin H A 2020 Electricity generation using membrane-less microbial fuel cell powered by sludge supplemented with lignocellulosic waste *Int. J. Energy Res.* **44** 3260–65

Mohd Zaini Makhtar M, Tajarudin H A, Samsudin M D M, Vadivelu V M, Shoparwe N F and Zainuddin N I 2021 Membrane-less microbial fuel cell: Monte Carlo simulation and sensitivity analysis for COD removal in dewatered sludge *AIP Adv.* **11** 65016

Mohd Zaini Makhtar M and Vadivelu V M 2019 Membraneless microbial fuel cell: characterization of electrogenic bacteria and kinetic growth model. *J. Environ. Eng.* **145** 04019015

Mohd Zaini Makhtar M, Abdul R and Vadivelu V M 2019 Recovery of energy and simultaneous treatment of dewatered sludge using membrane-less microbial fuel cell *Environ. Prog. Sustain. Energy* **38** 208–19

Najafabadi A T, Norvin N and Elod G 2016 Electrochemically exfoliated graphene anodes with enhanced biocurrent production in single-chamber air-breathing microbial fuel cells *Biosens. Bioelectron.* **81** 103–10

Naureen Z, Rashid Al Matani Z A, Nasser Al Jabri M, Al Housni S K, Abdullah Gilani S, Mabood F, Farooq S, Hussain J and Al Harrasi A 2016 Generation of electricity by electrogenic bacteria in a microbial fuel cell powered by waste water *Adv. Biosci. Biotechnol.* **07** 329–35

Noori M T, Ghangrekar M M, Mukherjee C K and Min B 2019 Biofouling effects on the performance of microbial fuel cells and recent advances in biotechnological and chemical strategies for mitigation *Biotechnol. Adv.* **37** 107420

Nosek D, Piotr J and Agnieszka C-K 2020 Anode modification as an alternative approach to improve electricity generation in microbial fuel cells *Energies* **13** 1–22

Patil S A, Hägerhäll C and Gorton L 2014 Electron transfer mechanisms between microorganisms and electrodes in bioelectrochemical systems *Advances in Chemical Bioanalysis* (Cham: Springer) pp 71–129

Prabowo A K, Tiarasukma A P, Christwardana M and Ariyanti D 2016 Microbial fuel cells for simultaneous electricity generation and organic degradation from slaughterhouse wastewater *Int. J. Renew. Energy Dev.* **5** 107–12

Qiao W, Peng C, Wang W and Zhang Z Z 2011 Biogas production from supernatant of hydrothermally treated municipal sludge by upflow anaerobic sludge blanket reactor *Bioresour. Technol.* **102** 9904–11

Qiao Y J, Qiao Y, Zou L, Wu X S and Liu J H 2017 Biofilm promoted current generation of *Pseudomonas aeruginosa* microbial fuel cell via improving the interfacial redox reaction of phenazines *Bioelectrochemistry* **117** 34–9

Ren L, Yongtae Ahn and Logan B E 2014 A two-stage microbial fuel cell and anaerobic fluidized bed membrane bioreactor (MFC-AFMBR) system for effective domestic wastewater treatment *Environ. Sci. Technol.* **48** 4199–206

Ren J, Li N, Du M, Zhang Y, Hao C and Hu R 2021 Study on the effect of synergy effect between the mixed cultures on the power generation of microbial fuel cells *Bioengineered* **12** 844–54

Rabbani A, Zainith S, Deb V K, Das P, Bharti P, Rawat D S, Kumar N and Saxena G 2021 Microbial technologies for environmental remediation potential issues, challenges, and future prospects Microbe Mediated Remediation of Environmental Contaminants (Cambridge: Woodhead)

Shah S, Venkatramanan V and Ram P 2019 Microbial fuel cell: sustainable green technology for bioelectricity generation and wastewater treatment *Sustainable Green Technologies for Environmental Management* (Singapore: Springer) pp 199–218

Shoparwe, Noor F, Wee Seng K, Suhairi S, Muaz Zaini M, Maryana Mohamad N and Tunmise Ayode O 2022 Study on the performance of microbial fuel cells through electrochemical impedance spectroscopy method by Geobacter sulfurreducens as a biocatalys *AIP Conf. Proc.* **2454** 030015

Simeon M I, Felix U A, Aliyu M, Raji O A and Freitag R 2020 Polarization and power density trends of a soil-based microbial fuel cell treated with human urine *Int. J. Energy Res.* **44** 5968–76

Spiegel C S 2008 *PEM Fuel Cell Modeling and Simulation Using MATLAB* (New York: Elsevier)

Sacco N J, Bonetto M C and Cortón E 2017 Isolation and characterization of a novel electrogenic bacterium, *Dietzia* sp. RNV-4 *PLoS One* **12** 1–20

Santoro C, Arbizzani C, Erable B and Ieropoulos I 2017 Microbial fuel cells: from fundamentals to applications. A review *J. Power Sources* **356** 225–44

Shah S, Venkatramanan V and Prasad R 2019 Microbial fuel cell: sustainable green technology for bioelectricity generation and wastewater treatment *Sustainable Green Technologies for Environmental Management* ed S Shah, V Venkatramanan and R Prasad (Singapore: Springer)

Su Y, Liu C, Fang H and Zhang D 2020 *Bacillus subtilis*: a universal cell factory for industry, agriculture, biomaterials and medicine *Microb. Cell Fact.* **19** 1–12

Torres C I 2014 On the importance of identifying, characterizing, and predicting fundamental phenomena towards microbial electrochemistry applications *Curr. Opin. Biotechnol.* **27** 107–14

Ucar D, Zhang Y and Angelidaki I 2017 An overview of electron acceptors in microbial fuel cells *Front. Microbiol.* **8** 1–14

Vajda B, Nándor Nemestóthy P B and Bélafi-Bakó K 2014 Xylose substrate as the only nutrient in the operation of microbial fuel cells *Environ. Protect. Eng.* **40** 131–41

Velvizhi G and Venkata Mohan S 2012 Electrogenic activity and electron losses under increasing organic load of recalcitrant pharmaceutical wastewater *Int. J. Hydrogen Energy* **37** 5969–78

Virdis B, Freguia S, Rozendal R A, Rabaey K, Yuan Z and Keller J 2011 Microbial fuel cells *Treatise on Water Science* ed P Wilderer (Oxford: Elsevier) pp 641–65

Wang Y, Qing Wen Y C and Li W 2020 Conductive polypyrrole-carboxymethyl cellulose-titanium nitride/carbon brush hydrogels as bioanodes for enhanced energy output in microbial fuel cells *Energy* **204** 117942

Watson V J and Bruce E L 2011 Analysis of polarization methods for elimination of power overshoot in microbial fuel cells *Electrochem. Commun.* **13** 54–6

Wlodarczyk P P and Wlodarczyk B 2019 Preparation and analysis of Ni–Co catalyst use for electricity production and COD reduction in microbial fuel cells *Catalysts* **9** 1042

Wu Y, Luo X, Luo X, Luo X, Qin B, Li F, Häggblom M M, Liu T and Liu T 2020 Enhanced current production by exogenous electron mediators via synergy of promoting biofilm formation and the electron shuttling process *Environ. Sci. Technol.* **54** 7217–25

Yaqoob A A, Mohamad Ibrahim M and Rodríguez-Couto S 2020 Development and modification of materials to build cost-effective anodes for microbial fuel cells (MFCs): an overview *Biochem. Eng. J.* **164** 107779

Zhou M, Chi M, Luo J, He H and Jin T 2011 An overview of electrode materials in microbial fuel cells *J. Power Sources* **196** 4427–35

Zhuang L, Yu Z, Shungui Z, Yong Y, Haoran Y and Yong C 2012 Scalable microbial fuel cell (MFC) stack for continuous real wastewater treatment *Bioresour. Technol.* **106** 82–8

www.ingramcontent.com/pod-product-compliance
Ingram Content Group UK Ltd.
Pitfield, Milton Keynes, MK11 3LW, UK
UKHW051904100926
8556IPUK00005B/755

9 780750 347891